Quantum Probability
and Related Topics

Proceedings of the 28th Conference

QP–PQ: Quantum Probability and White Noise Analysis*

Managing Editor: W. Freudenberg
Advisory Board Members: L. Accardi, T. Hida, R. Hudson and K. R. Parthasarathy

QP–PQ: Quantum Probability and White Noise Analysis

Vol. 23: Quantum Probability and Related Topics
eds. J. C. García, R. Quezada and S. B. Sontz

Vol. 22: Infinite Dimensional Stochastic Analysis
eds. A. N. Sengupta and P. Sundar

Vol. 21: Quantum Bio-Informatics
From Quantum Information to Bio-Informatics
eds. L. Accardi, W. Freudenberg and M. Ohya

Vol. 20: Quantum Probability and Infinite Dimensional Analysis
eds. L. Accardi, W. Freudenberg and M. Schürmann

Vol. 19: Quantum Information and Computing
eds. L. Accardi, M. Ohya and N. Watanabe

Vol. 18: Quantum Probability and Infinite-Dimensional Analysis
From Foundations to Applications
eds. M. Schürmann and U. Franz

Vol. 17: Fundamental Aspects of Quantum Physics
eds. L. Accardi and S. Tasaki

Vol. 16: Non-Commutativity, Infinite-Dimensionality, and Probability at the Crossroads
eds. N. Obata, T. Matsui and A. Hora

Vol. 15: Quantum Probability and Infinite-Dimensional Analysis
ed. W. Freudenberg

Vol. 14: Quantum Interacting Particle Systems
eds. L. Accardi and F. Fagnola

Vol. 13: Foundations of Probability and Physics
ed. A. Khrennikov

QP–PQ

Vol. 11: Quantum Probability Communications
eds. S. Attal and J. M. Lindsay

Vol. 10: Quantum Probability Communications
eds. R. L. Hudson and J. M. Lindsay

Vol. 9: Quantum Probability and Related Topics
ed. L. Accardi

Vol. 8: Quantum Probability and Related Topics
ed. L. Accardi

*For the complete list of titles in this series, please visit the website at
http://www.worldscibooks.com/series/qqpwna_series.shtml

QP-PQ
Quantum Probability and White Noise Analysis
Volume XXIII

Quantum Probability and Related Topics

Proceedings of the 28th Conference

CIMAT-Guanajuato, Mexico   2–8 September 2007

Editors

J. C. García
Universidad Autonoma Metropolitana-Iztapalapa, Mexico

R. Quezada
Universidad Autonoma Metropolitana-Iztapalapa, Mexico

S. B. Sontz
CIMAT-Guanajuato, Mexico

NEW JERSEY • LONDON • SINGAPORE • BEIJING • SHANGHAI • HONG KONG • TAIPEI • CHENNAI

Published by

World Scientific Publishing Co. Pte. Ltd.
5 Toh Tuck Link, Singapore 596224
USA office: 27 Warren Street, Suite 401-402, Hackensack, NJ 07601
UK office: 57 Shelton Street, Covent Garden, London WC2H 9HE

Library of Congress Cataloging-in-Publication Data
QP Conference on Quantum Probability and Related Topics (28th : 2007 :
 Guanajuato, Mexico)
 Quantum probability and related topics : proceedings of the 28th
conference CIMAT-Guanajuato, Mexico, 2–8 September 2007 / edited by J.C. García,
R. Quezada & S.B. Sontz.
 p. cm. -- (QP-PQ: quantum probability and white noise analysis ; v. 23)
 Includes bibliographical references and index.
 ISBN-13: 978-981-283-526-0 (hardcover : alk. paper)
 ISBN-10: 981-283-526-1 (hardcover : alk. paper)
 1. Probabilities--Congresses. 2. Quantum theory--Congresses. I. García, J. C.
II. Quezada, R. III. Sontz, S.B. IV. Title.
QC174.17.P68Q26 2007
519.2--dc22
                                              2008036176

British Library Cataloguing-in-Publication Data
A catalogue record for this book is available from the British Library.

Copyright © 2008 by World Scientific Publishing Co. Pte. Ltd.

All rights reserved. This book, or parts thereof, may not be reproduced in any form or by any means, electronic or mechanical, including photocopying, recording or any information storage and retrieval system now known or to be invented, without written permission from the Publisher.

For photocopying of material in this volume, please pay a copying fee through the Copyright Clearance Center, Inc., 222 Rosewood Drive, Danvers, MA 01923, USA. In this case permission to photocopy is not required from the publisher.

Printed in Singapore.

PREFACE

The 28th Conference on Quantum Probability and Related Topics was held September 2-8, 2007, in the city of Guanajuato, Mexico. This was the first time this important international event took place in the Americas. So all of us in Mexico who were involved in the local organization are gratified to have been able to welcome our academic guests, many of them visiting Mexico for their first time, to the truly marvelous colonial era city of Guanajuato, which has been recognized by UNESCO since 1988 as a World Heritage Site. We were also fortunate to host this conference in the Centro de Investigación en Matemáticas (CIMAT), a national research facility devoted to the advancement of mathematics. The participants were able to enjoy CIMAT's user friendly architecture and ambience that are so conducive to those ever important after talk chats over a cup of coffee. And with a spectacular panoramic view down on the town and its surroundings from CIMAT's perch on the side of a hill thrown in to boot!

The success of a meeting of this size (with 91 participants) is the result of the combined efforts of many people. In particular, special thanks go to the members of the international organizing committee as well as to the staff and support personnel of the local institutions.

After such a successful meeting we hope to see continued valuable academic contacts between the participants from so many countries (Australia, Austria, Belgium, Canada, Chile, France, Germany, Hungary, India, Italy, Japan, Poland, Russia, Slovakia, Spain, Tunisia, United Kingdom and the United States) and those from Mexico.

The conference itself was divided into eleven special sessions, each one with its own organizers. However, we were fortunate in not having to schedule parallel sessions, so the participants were never forced to choose between two talks at the same time. The topics of these special sessions were:

(1) Interacting Fock Space and Orthogonal Polynomials
(2) Log-Sobolev Inequalities
(3) Infinitely Divisible Processes
(4) White Noise

(5) Free Probability
(6) Quantum Statistics and Stochastic Control
(7) Classical and Quantum Models in Biology
(8) Dilations
(9) Applications of Quantum Probability in Physics
(10) Quantum Markov Semigroups
(11) Quantum Information

The twenty-two articles in these Proceedings of the 28th Conference on Quantum Probability and Related Topics were accepted for publication only upon successfully gaining approval after a careful reading by anonymous peer reviewers. These articles reflect the scope of the topics presented in the talks of the eleven special sessions in a variety of areas: Quantum Markov Semigroups, Quantum Measurements, White Noise Analysis, Random Walks, von Neumann Algebras, Orthogonal Polynomials, Free Convolution Semigroups, Quantum Controls, Quantum Information, Stochastic Analysis, Schrödinger Equations, Deformed Quantum Mechanics, Protein Folding and Binding, and Quantum Models of Brain Activity.

Finally, we would like to thank these Mexican institutions for their generous financial support:

- Consejo de Ciencia y Tecnología del Estado de Guanajuato, CONCYTEG, Convenio 07-02-K121-085, Anexo 2 (State of Guanajuato)
- Centro de Investigación en Matemáticas, CIMAT (Guanajuato)
- Universidad Autónoma Metropolitana-Iztapalapa, UAM-I (Mexico City)
- Departamento de Matemáticas Aplicadas y Sistemas de la Universidad Autónoma Metropolitana-Cuajimalpa (Mexico City)
- Consejo Nacional de Ciencia y Tecnología, CONACYT, Grants 49510-F and 49187-F (Mexico)
- Subsecretaría de Educación Superior, SEP, PIFI 3.3, UAM-I-CA-52 and UAM-I-CA-56 (Mexico)
- Facultad de Ciencias, UNAM, (Mexico City)
- Sociedad Matemática Mexicana (Mexico)

We hope the readers of these Proceedings find these articles to be interesting and beneficial to them in their own scientific research activities, whether in mathematics or in applied areas.

The editors:
Julio César García
Roberto Quezada
Stephen Bruce Sontz

August, 2008
Mexico City and Guanajuato

CONTENTS

Preface ... v

Linear Independence of the Renormalized Higher Powers of
White Noise ... 1
  L. Accardi & A. Boukas

Brownian Dynamics Simulation for Protein Folding and Binding ... 11
  T. Ando

Quantum Fokker-Planck Models: The Lindblad and
Wigner Approaches ... 23
  A. Arnold, F. Fagnola & L. Neumann

Hilbert Space of Analytic Functions Associated with a
Rotation Invariant Measure ... 49
  N. Asai

Quantum Continuous Measurements: The Spectrum of the Output ... 63
  A. Barchielli & M. Gregoratti

Characterization Theorems in Gegenbauer White Noise Theory ... 77
  A. Barhoumi, A. Riahi & H. Ouerdiane

A Problem of Powers and the Product of Spatial Product Systems ... 93
  B. V. R. Bhat, V. Liebscher & M. Skeide

Free Martingale Polynomials for Stationary Jacobi Processes ... 107
  N. Demni

Accardi Complementarity for $-1/2 < \mu < 0$ and Related Results   120
L. A. Echavarría-Cepeda, C. J. Pita-Ruiz-Velasco & S. B. Sontz

Quantum Models of Brain Activities I. Recognition of Signals   135
K.-H. Fichtner & L. Fichtner

Kallman Decompositions of Endomorphisms of
von Neumann Algebras   145
R. Floricel

Schrödinger Equation, L^p-Duality and the Geometry of
Wigner-Yanase-Dyson Information   157
P. Gibilisco, D. Imparato & T. Isola

Multiplicative Renormalization Method for Orthogonal Polynomials   165
H.-H. Kuo

Quantum L_p and Orlicz Spaces   176
L. E. Labuschagne & W. A. Majewski

Vicious Random Walkers and Truncated Haar Unitaries   190
J. I. Novak

Incoherent Quantum Control   197
A. Pechen & H. Rabitz

Compromising Non-Demolition and Information Gaining for
Qubit State Estimation   212
L. Ruppert, A. Magyar & K. M. Hangos

Positivity of Free Convolution Semigroups   225
M. Schürmann & S. Voß

Quasi-Free Stochastic Integrals and Martingale Representation   236
W. J. Spring

A Method of Recovering the Moments of a Probability Measure from Commutators 242
   A. I. Stan & J. J. Whitaker

Description of Decoherence by Means of
Translation-Covariant Master Equations and Lévy Processes 254
   B. Vacchini

Minimum Back-Action Measurement via Feedback 266
   M. Yanagisawa

Author Index 277

LINEAR INDEPENDENCE OF THE RENORMALIZED HIGHER POWERS OF WHITE NOISE

L. ACCARDI

Centro Vito Volterra, Università di Roma Tor Vergata
via Columbia 2, 00133 Roma, Italy, E-mail: accardi@volterra.mat.uniroma2.it
http://volterra.mat.uniroma2.it

A. BOUKAS

Department of Mathematics and Natural Sciences, American College of Greece
Aghia Paraskevi, Athens 15342, Greece, E-mail: andreasboukas@acgmail.gr

The connection between the Lie algebra of the Renormalized Higher Powers of White Noise (RHPWN) and the centerless Virasoro (or Witt)-Zamolodchikov-w_∞ Lie algebras of conformal field theory, as well as the associated Fock space construction, have recently been established in Ref.[1-6] In this note we prove the linear independence of the RHPWN Lie algebra generators.

1. Introduction: Renormalized Higher Powers of White Noise

The quantum white noise functionals $a_t^\dagger$ and a_t satisfy the Boson commutation relations

$$[a_t, a_s^\dagger] = \delta(t-s), \; ; \; [a_t^\dagger, a_s^\dagger] = [a_t, a_s] = 0 \qquad (1)$$

where $t, s \in \mathbb{R}$ and δ is the Dirac delta function, as well as the duality relation

$$(a_s)^* = a_s^\dagger \qquad (2)$$

Here (and in what follows) $[x, y] := xy - yx$ is the usual operator commutator. For all $t, s \in \mathbb{R}$ and integers $n, k, N, K \geq 0$ we have (Ref.[6])

$$[a_t^{\dagger^n} a_t^k, a_s^{\dagger^N} a_s^K] = \qquad (3)$$

$$\epsilon_{k,0}\,\epsilon_{N,0} \sum_{L\geq 1} \binom{k}{L} N^{(L)} a_t^{\dagger n} a_s^{\dagger N-L} a_t^{k-L} a_s^K \delta^L(t-s)$$

$$-\epsilon_{K,0}\,\epsilon_{n,0} \sum_{L\geq 1} \binom{K}{L} n^{(L)} a_s^{\dagger N} a_t^{\dagger n-L} a_s^{K-L} a_t^k \delta^L(t-s)$$

where for $n,k \in \{0,1,2,...\}$ we have used the notation $\epsilon_{n,k} := 1 - \delta_{n,k}$, where $\delta_{n,k}$ is Kronecker's delta and $x^{(y)} = x(x-1)\cdots(x-y+1)$ with $x^{(0)} = 1$. In order to consider the smeared fields defined by the higher powers of a_t and $a_t^\dagger$, for a test function f and $n,k \in \{0,1,2,...\}$ we define the sesquilinear form

$$B_k^n(f) := \int_{\mathbb{R}} f(t)\, a_t^{\dagger n} a_t^k \, dt \qquad (4)$$

with involution

$$(B_k^n(f))^* = B_n^k(\bar{f}) \qquad (5)$$

In[1] and[2] we introduced the convolution type renormalization of the higher powers of the Dirac delta function

$$\delta^l(t-s) = \delta(s)\,\delta(t-s) \ ; \quad l = 2,3,.... \qquad (6)$$

By multiplying both sides of (3) by test functions $f(t)\,g(s)$ such that $f(0) = g(0) = 0$ and then formally integrating the resulting identity (i.e. taking $\int\int \ldots ds\,dt$ of both sides), using (6), we obtained the RHPWN Lie algebra commutation relations

$$[B_k^n(f), B_K^N(g)]_{RHPWN} := (kN - Kn)\, B_{k+K-1}^{n+N-1}(fg) \qquad (7)$$

As shown in[1] and,[2] for $n,k \in \mathbb{Z}$ with $n \geq 2$, the white noise operators

$$\hat{B}_k^n(f) := \int_{\mathbb{R}} f(t)\, e^{\frac{k}{2}(a_t - a_t^\dagger)} \left(\frac{a_t + a_t^\dagger}{2}\right)^{n-1} e^{\frac{k}{2}(a_t - a_t^\dagger)} \, dt \qquad (8)$$

with involution

$$\left(\hat{B}_k^n(f)\right)^* = \hat{B}_{-k}^n(\bar{f}) \qquad (9)$$

satisfy the commutation relations of the second quantized Virasoro-Zamolodchikov-w_∞ Lie algebra (Ref.[7]), namely

$$[\hat{B}_k^n(f), \hat{B}_K^N(g)]_{w_\infty} = (k(N-1) - K(n-1))\, \hat{B}_{k+K}^{n+N-2}(fg) \qquad (10)$$

In particular,

$$\hat{B}_k^2(f) := \int_\mathbb{R} f(t)\, e^{\frac{k}{2}(a_t - a_t^\dagger)} \left(\frac{a_t + a_t^\dagger}{2}\right) e^{\frac{k}{2}(a_t - a_t^\dagger)}\, dt \qquad (11)$$

is the white noise form of the centerless Virasoro algebra generators.

We may analytically continue the parameter k in the definition of $\hat{B}_k^n(f)$ to an arbitrary complex number $k \in \mathbb{C}$ and to $n \geq 1$ and we can show (Ref.[3]) that the RHPWN and w_∞ Lie algebras are connected through

$$\hat{B}_k^n(f) = \frac{1}{2^{n-1}} \sum_{m=0}^{n-1} \binom{n-1}{m} \sum_{p=0}^{\infty} \sum_{q=0}^{\infty} (-1)^p \frac{k^{p+q}}{p!\, q!}\, B_{n-1-m+q}^{m+p}(f) \qquad (12)$$

and

$$B_k^n(f) = \sum_{\rho=0}^{k} \sum_{\sigma=0}^{n} \binom{k}{\rho}\binom{n}{\sigma} \frac{(-1)^\rho}{2^{\rho+\sigma}} \frac{\partial^{\rho+\sigma}}{\partial z^{\rho+\sigma}}\bigg|_{z=0} \hat{B}_z^{k+n+1-(\rho+\sigma)}(f) \qquad (13)$$

For $n \geq 1$ we define the n-th order RHPWN $*$–Lie algebras $\mathcal{L}_n$ as follows: (i) $\mathcal{L}_1$ is the $*$–Lie algebra generated by B_0^1 and B_1^0 i.e., $\mathcal{L}_1$ is the linear span of $\{B_0^1, B_1^0, B_0^0\}$ (ii) $\mathcal{L}_2$ is the $*$–Lie algebra generated by B_0^2 and B_2^0 i.e., $\mathcal{L}_2$ is the linear span of $\{B_0^2, B_2^0, B_1^1\}$ (iii) For $n \in \{3, 4, ...\}$, $\mathcal{L}_n$ is the $*$–Lie algebra generated by B_0^n and B_n^0 through repeated commutations and linear combinations. It consists of linear combinations of creation/annihilation operators of the form B_y^x where $x - y = kn$, $k \in \mathbb{Z} - \{0\}$, and of number operators B_x^x with $x \geq n - 1$. Through white noise and norm compatibility considerations, the action of the RHPWN operators on Φ was defined in[4] as

$$B_k^n(f)\,\Phi := \begin{cases} 0 & \text{if } n < k \text{ or } n \cdot k < 0 \\ B_0^{n-k}(f)\,\Phi & \text{if } n > k \geq 0 \\ \frac{1}{n+1}\int_{\mathbb{R}} f(t)\,dt\,\Phi & \text{if } n = k \end{cases} \qquad (14)$$

In what follows, for all integers n, k we will use the notation $B_k^n := B_k^n(\chi_I)$ where I is some fixed subset of $\mathbb{R}$ of finite measure $\mu := \mu(I) > 0$. Moreover, for all $t \in [0, +\infty)$ and for all integers n, k we will use the notation $B_k^n(t) := B_k^n(\chi_{[0,t]})$.

To avoid ghosts (i.e., vectors of negative norm) appearing in the cases $n \geq 3$ in the Fock kernels $\langle (B_0^n)^k\,\Phi, (B_0^n)^k\,\Phi \rangle$ where $k \geq 0$, in[4] we defined

$$B_{n-1}^{n-1}(B_0^n)^k\,\Phi := \left(\frac{\mu}{n} + k\,n\,(n-1)\right)(B_0^n)^k\,\Phi \qquad (15)$$

and were able to show that for all $k, n \geq 1$

$$\langle (B_0^n)^k\,\Phi, (B_0^n)^m\,\Phi \rangle = \delta_{m,k}\,k!\,n^k\,\prod_{i=0}^{k-1}\left(\mu + \frac{n^2\,(n-1)}{2}i\right) \qquad (16)$$

Therefore, the $\mathcal{F}_n$ inner product $\langle \psi_n(f), \psi_n(g) \rangle_n$ of the exponential vectors

$$\psi_n(\phi) := \prod_i e^{a_i\,B_0^n(\chi_{I_i})}\,\Phi \qquad (17)$$

where $\phi := \sum_i a_i\,\chi_{I_i}$ is a test function, for $n = 1$ is

$$\langle \psi_1(f), \psi_1(g) \rangle_1 := e^{\int_{\mathbb{R}} \bar{f}(t)\,g(t)\,dt} \qquad (18)$$

while for $n \geq 2$ it is

$$\langle \psi_n(f), \psi_n(g) \rangle_n := e^{-\frac{2}{n^2\,(n-1)}\int_{\mathbb{R}} \ln\left(1 - \frac{n^3\,(n-1)}{2}\bar{f}(t)\,g(t)\right)\,dt} \qquad (19)$$

where $|f(t)| < \frac{1}{n}\sqrt{\frac{2}{n\,(n-1)}}$ and $|g(t)| < \frac{1}{n}\sqrt{\frac{2}{n\,(n-1)}}$.

The n-th order truncated RHPWN (or TRHPWN) Fock space $\mathcal{F}_n$ is the Hilbert space completion of the linear span of the exponential vectors $\psi_n(f)$

under the inner product $\langle \cdot, \cdot \rangle_n$. The full TRHPWN Fock space $\mathcal{F}$ is the direct sum of the $\mathcal{F}_n$'s.

The Fock representation of the TRHPWN generators B_0^n and B_n^0 obtained in[4] is

$$B_n^0(f)\psi_n(g) = n\int_{\mathbb{R}} f(t)g(t)\,dt\ \psi_n(g) + \frac{n^3(n-1)}{2}\frac{\partial}{\partial \epsilon}\Big|_{\epsilon=0}\psi_n(g+\epsilon f g^2) \quad (20)$$

$$B_0^n(f)\psi_n(g) = \frac{\partial}{\partial \epsilon}\Big|_{\epsilon=0}\psi_n(g+\epsilon f) \quad (21)$$

where $f := \sum_i a_i \chi_{I_i}$ and $g := \sum_i b_i \chi_{I_i}$ with $I_i \cap I_j = \emptyset$ for $i \neq j$ and $f(0) = g(0) = 0$.

As shown in,[4] for all $s \in [0, \infty)$

$$\langle e^{s(B_0^1(t)+B_1^0(t))}\Phi, \Phi\rangle_1 = e^{\frac{s^2}{2}t} \quad (22)$$

i.e., $\{x_1(t) := B_0^1(t) + B_1^0(t)\}_{t\geq 0}$ is Brownian motion, while for $n \geq 2$

$$\langle e^{s(B_0^n(t)+B_n^0(t))}\Phi, \Phi\rangle_n = \left(\sec\left(\sqrt{\frac{n^3(n-1)}{2}}s\right)\right)^{\frac{2nt}{n^3(n-1)}} \quad (23)$$

i.e., for each $n \geq 2$, $\{x_n(t) := B_0^n(t) + B_n^0(t)\}_{t\geq 0}$ is a continuous binomial/Beta process.

2. Linear independence of the RHPWN generators

Lemma 2.1.
  For all integers $m \geq 0$

$$\sum_{n=0}^{m} c_n B_0^n(f_n) = 0 \implies c_n = 0\ \forall n \in \{0, 1, ..., m\} \quad (24)$$

where we assume that the test functions f_n are such that for all $n \in \{0, 1, ..., m\}$

$$\int_{\mathbb{R}} f_n(t)\,a_t^{\dagger n}\,dt \neq 0 \quad (25)$$

Proof.
For $m = 0$,

$$c_0 B_0^0(f_0) = 0 \implies c_0 \int_{\mathbb{R}} f_0(t)\, dt = 0 \implies c_0 = 0 \tag{26}$$

and so (24) holds. Suppose that it holds for $m = M$. We will show that it is true for $m = M + 1$ also. So suppose that

$$\sum_{n=0}^{M+1} c_n B_0^n(f_n) = 0 \tag{27}$$

Then

$$\sum_{n=0}^{M+1} c_n [B_1^0(g), B_0^n(f_n)] = 0 \tag{28}$$

where g is any test function such that

$$\int_{\mathbb{R}} g(t)\, f_n(t)\, a_t^{\dagger n}\, dt \neq 0 \tag{29}$$

for all n, i.e.,

$$\sum_{n=0}^{M+1} n\, c_n\, B_0^{n-1}(g\, f_n) = 0 \tag{30}$$

which is equivalent to

$$\sum_{n=1}^{M+1} n\, c_n\, B_0^{n-1}(g\, f_n) = 0 \tag{31}$$

or, letting $N := n - 1$, to

$$\sum_{N=0}^{M} (N+1)\, c_{N+1}\, B_0^N(g\, f_{N+1}) = 0 \tag{32}$$

which, by the induction hypothesis, implies that

$$(N+1)\, c_{N+1} = 0 \implies c_{N+1} = 0 \implies c_n = 0 \tag{33}$$

for all $n \in \{1, 2, ..., M+1\}$. But then (27) reduces to $c_0 B_0^0(f_0) = 0$ which, as we have already seen, implies that $c_0 = 0$ as well. □

Lemma 2.2. *For all integers $m \geq 0$*

$$\sum_{k=0}^{m} c_k B_k^0(f_k) = 0 \implies c_k = 0 \ \forall k \in \{0, 1, ..., m\} \tag{34}$$

where we assume that the arbitrary test functions f_k are such that for all $k \in \{0, 1, ..., m\}$

$$\int_{\mathbb{R}} f_k(t) \, a_t^k \, dt \neq 0 \tag{35}$$

Proof. Taking the adjoint of equation (34) we obtain

$$\sum_{k=0}^{m} \bar{c}_k B_0^k(\bar{f}_k) = 0 \tag{36}$$

which by Lemma 2.1 implies that $\bar{c}_k = 0$, and so $c_k = 0$, for all $k \in \{0, 1, ..., m\}$. □

Theorem 2.1. *The generators $B_k^n(f)$ of the RHPWN Lie algebra are linearly independent, i.e., for all integers $m \geq 0$*

$$\sum_{n=0}^{m} \sum_{k=0}^{m} c_{n,k} B_k^n(f_{n,k}) = 0 \implies c_{n,k} = 0 \ \forall n, k \in \{0, 1, ..., m\} \tag{37}$$

where we assume that the arbitrary test functions $f_{n,k}$ are such that

$$\int_{\mathbb{R}} f_{n,k}(t) \, a_t^{\dagger^n} a_t^k \, dt \neq 0 \tag{38}$$

Note: By filling in with zero coefficients if necessary, every finite linear combination of the RHPWN generators can be put in the form

$$\sum_{n=0}^{m} \sum_{k=0}^{m} c_{n,k} B_k^n(f_{n,k}) \tag{39}$$

Proof. We will proceed by induction on m. For $m = 0$, equation (37) becomes

$$c_{0,0} B_0^0(f_{0,0}) = 0 \implies c_{0,0} = 0 \tag{40}$$

which is true by (38). Suppose that equation (37) holds for $m = M$. We will show that it is true for $m = M + 1$ also. So suppose that

$$\sum_{n=0}^{M+1} \sum_{k=0}^{M+1} c_{n,k} B_k^n(f_{n,k}) = 0 \tag{41}$$

Taking the commutator of (41) first with $B_0^1(g)$ and then with $B_1^0(g)$, where g is any test function such that

$$\int_{\mathbb{R}} g(t) f_{n,k}(t) a_t^{\dagger n} a_t^k \, dt \not\equiv 0 \tag{42}$$

for all n, k, we obtain

$$\sum_{n=0}^{M+1} \sum_{k=0}^{M+1} k\, n\, c_{n,k} B_{k-1}^{n-1}(g^2 f_{n,k}) = 0 \tag{43}$$

which is equivalent to

$$\sum_{n=1}^{M+1} \sum_{k=1}^{M+1} k\, n\, c_{n,k} B_{k-1}^{n-1}(g^2 f_{n,k}) = 0 \tag{44}$$

which, letting $N := n - 1$ and $K := k - 1$, is equivalent to

$$\sum_{N=0}^{M} \sum_{K=0}^{m} (K+1)(N+1) c_{N+1,K+1} B_K^N(g^2 f_{N+1,K+1}) = 0 \tag{45}$$

which, by the induction hypothesis, implies that

$$(K+1)(N+1) c_{N+1,K+1} = 0 \implies c_{N+1,K+1} = 0 \implies c_{n,k} = 0 \tag{46}$$

for all $n, k \in \{1, 2, ..., M+1\}$. If $n = 0$ and/or $k = 0$ then equation (41) reduces to

$$c_{0,0} B_0^0(f_{0,0}) + \sum_{n=1}^{m+1} c_{n,0} B_0^n(f_{n,0}) + \sum_{k=1}^{m+1} c_{0,k} B_k^0(f_{0,k}) = 0 \qquad (47)$$

Taking the commutator of (47) with $B_0^1(g)$, where g is as above, we obtain

$$\sum_{k=1}^{m+1} k\, c_{0,k} B_{k-1}^0(g\, f_{0,k}) = 0 \qquad (48)$$

which by Lemma 2.2 implies that $k\, c_{0,k} = 0$ for all $k \in \{1, 2, ..., M+1\}$ and so $c_{0,k} = 0$ for all $k \in \{1, 2, ..., M+1\}$. Similarly, taking the commutator of (47) with $B_1^0(g)$ we obtain

$$\sum_{n=1}^{m+1} n\, c_{n,0} B_0^{n-1}(g\, f_{n,0}) = 0 \qquad (49)$$

which by Lemma 2.1 implies that $n\, c_{n,0} = 0$ for all $n \in \{1, 2, ..., M+1\}$ and so $c_{n,0} = 0$ for all $n \in \{1, 2, ..., M+1\}$. So, (41) reduces to

$$c_{0,0} B_0^0(f_{0,0}) = 0 \qquad (50)$$

which by (38) implies that $c_{0,0} = 0$. Therefore $c_{n,k} = 0$ for all $n, k \in \{0, 1, 2, ..., M+1\}$. □

References

1. L. Accardi and A. Boukas, *Renormalized higher powers of white noise (RHPWN) and conformal field theory*, Infinite Dimensional Anal. Quantum Probab. Related Topics **9** (3) (2006), 353-360.
2. L. Accardi and A. Boukas, *The emergence of the Virasoro and w_∞ Lie algebras through the renormalized higher powers of quantum white noise*, International Journal of Mathematics and Computer Science **1** (3) (2006), 315–342.
3. L. Accardi and A. Boukas, *Renormalized Higher Powers of White Noise and the Virasoro–Zamolodchikov–w_∞ Algebra*, Reports on Mathematical Physics, **61** (1) (2008), 1-11, http://arxiv.org/hep-th/0610302.
4. L. Accardi and A. Boukas, *Fock representation of the renormalized higher powers of white noise and the Virasoro–Zamolodchikov–w_∞ *-Lie algebra*, to appear in Journal of Physics A: Mathematical and Theoretical, arXiv:0706.3397v2 [math-ph].
5. L. Accardi and A. Boukas, *Lie algebras associated with the renormalized higher powers of white noise*, Communications on Stochastic Analysis **1** (1) (2007), 57-69.

6. L. Accardi, A. Boukas and U. Franz, *Renormalized powers of quantum white noise*, Infinite Dimensional Analysis, Quantum Probability, and Related Topics, **9** (1) (2006), 129–147.
7. S. V. Ketov, *Conformal field theory*, World Scientific, 1995.

BROWNIAN DYNAMICS SIMULATION FOR PROTEIN FOLDING AND BINDING

TADASHI ANDO*

Department of Biological Science and Technology, Tokyo University of Science 2641 Yamazaki, Noda, Chiba, 278-8510, Japan
*E-mail: tando@r4s.noda.tus.ac.jp

A protein with a certain amino acid sequence folds into its unique three-dimensional shape spontaneously and binds to its ligand or partner proteins to play its own role in a living system. This relationship between a protein sequence and its three-dimensional structure constitutes the second part of the genetic code that links DNA sequence information of a gene with the function of its product, which remains an enigma in biology. Computer simulation is a powerful tool that can analyze and calculate energies of various conformations of proteins at atomic resolution. However, when the protein folding and binding are pursued using this method, we face to two barriers: efficient sampling of protein conformations and improvement of force field. To break these barriers, I have developed an atomistic Brownian dynamics (BD) simulation method that uses physics-based energy terms and force field. The BD method does not treat water molecules explicitly and can adopt a long time step, resulting reduction of the computation time greatly. Some peptides folded into their native structures from extended conformations without statistical information obtained from databases of proteins using the BD method. I have also developed an umbrella sampling method combined with the BD to calculate absolute binding affinity of protein-protein interaction. By using the BD/umbrella sampling method, binding free energy of WW-domain/Pro-rich peptide could be estimated quantitatively. In these respects, the BD method would be effective for analysis of protein folding and binding.

1. Introduction

Proteins play pivotal roles in living organisms. They are synthesized as unbranched long-chain polymers of just twenty kinds of amino acids in a cell. Unlike most of polymers, each chain can fold spontaneously into a well-ordered three-dimensional structure depending on its amino acid sequence encoded by the gene (Figure 1). Furthermore, they can bind to other molecules to expresses their biological functions. Remarkable precision and fidelity of protein folding and binding form the basis of all living systems.

Now thanks to tremendous progress of various high-throughput projects, such as genome sequencing and structural genomics, in recent years, a huge number of genes, protein structures, and relating data are accumulating. Under these circumstances, understanding the mechanism of the protein folding and binding and predicting these dynamical reactions are a crucial step to shift the today's genomic biology to new phase where many biological phenomena can be predicted in a computer based on the collected data.

Molecular dynamics (MD) simulation is an essential tool that can analyze protein folding and binding. However, when these problems are approached using this method, we face to two difficult problems. The first is the quantitative uncertainty of the free energy function describing both protein's intramolecular interactions and intermolecular interactions with solvent for arbitrary conformations. The second is the insufficiency of simulation time that should be necessary for following the whole process of folding and binding: ten nanosecond simulation of a protein-solvent system necessitates about a week of computation time even with up-to-date computers, though protein folding and binding take place from sub-microseconds to seconds.

In order to overcome the difficulty of these two obstacles, various approaches have been developed, such as replica-exchange,[1] generalized-Born model,[2] and so on. In our group, an atomistic Brownian dynamics (BD) simulation with multiple time step algorithm and a new implicit solvent model to describe the protein folding process at atomic resolution have been developed.[3–6] In this report, I will describe our BD approach to long time folding simulation firstly. Then, I would like to introduce an example of binding affinity calculation using umbrella sampling with BD method briefly.

2. Methods and Models

2.1. *Brownian dynamics simulation algorithm*

By treating the effects of solvent as a dissipative random force, the Langevin equation can be expressed as

$$m_i \frac{d^2\mathbf{r}_i}{dt^2} = -\zeta_i \frac{d\mathbf{r}_i}{dt} + \mathbf{F}_i + \mathbf{R}_i. \tag{1}$$

Here, m_i and $\mathbf{r}_i$ represent the mass and position of atom i, respectively. ζ_i is a frictional coefficient and is determined by the Stokes' law, that is, $\zeta_i = 6\pi a_i^{\text{Stokes}} \eta$ in which a_i^{Stokes} is a Stokes radius of atom i and η

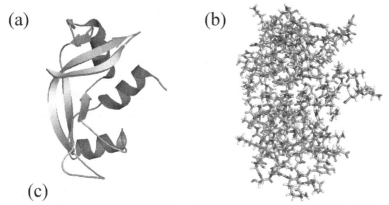

(c)
KETAAAKFERQHMDSSTSAASSSNYCNQMMKSRNLTKDRCKP
VNTFVHESLADVQAVCSQKNVACKNGQTNCYQSYSTMSITDC
RETGSSKYPNCAYKTTQANKHIIVACEGNPYVPVHFDASV

Fig. 1. Three-dimensional structure of ribonuclease A (PDB ID: 1fs3). (a) Schematic view of the protein. α-Helices and β-sheets are drawn in red and yellow, respectively. (b) The structure represented by sticks. Carbon atoms are in green; hydrogen atoms are in white; oxygen atoms are in red; nitrogen atoms are in blue; and sulfur atoms are in orange. (c) The amino acid sequence of the protein. Each amino acid is written with one letter representation: A = alanine, C = cysteine, D = aspartic acid, E = glutamic acid, F = phenylalanine, G = glycine, H = histidine, I = isoleucine, K = lysine, L = leucine, M = methionine, N = asparagine, P = proline, Q = glutamine, R = arginine, S = serine, T = threonine, V = valine, W = tryptophan and Y = tyrosine. Figures of the protein are generated with PyMOL.[17]

is the viscosity of water. $\mathbf{F}_i$ is the systematic force on atom i. $\mathbf{R}_i$ is a random force on atom i having a zero mean $\langle \mathbf{R}_i(t) \rangle = 0$ and a variance $\langle \mathbf{R}_i(t)\mathbf{R}_i(0) \rangle = 2\zeta_i k_B T \delta_{ij} \delta(t) \mathbf{I}$ where k_B is the Boltzmann's constant, T is absolute temperature, δ_{ij} is the Kronecker delta, $\delta(t)$ is the Dirac delta, and $\mathbf{I}$ is 3×3 unit tensor; this derives from the effects of solvent.

For the overdamped limit (the solvent damping is large and the inertial memory is lost in a very short time), we set the left side of Eq. 1 to zero,

$$\zeta_i \frac{d\mathbf{r}_i}{dt} = \mathbf{F}_i + \mathbf{R}_i. \qquad (2)$$

Integrated equation of Eq. 2 is called Brownian dynamics;[7]

$$\mathbf{r}_i(t+h) = \mathbf{r}_i(t) + \frac{\mathbf{F}_i(t)}{\zeta_i}h + \sqrt{\frac{2k_B T}{\zeta_i}h}\boldsymbol{\omega}_i, \qquad (3)$$

where h is a time step and $\boldsymbol{\omega}_i$ is a random noise vector obtained from Gaussian distribution.

Time step of 10 fs was used for single time step BD simulation. For multiple time step algorithm, short time step, $\Delta\tau$, of 5 fs and long time step, Δt, of 40 fs were used.[4] Cut-off method was not used. All bond lengths were constrained with LINCS algorithm.[8] Stokes radius of each atom was its van der Waals radius plus 1.4 Å. Coordinates and energies were recorded every 100 ps during the simulation. For analysis, the structures collected for first 10 ns were removed. All calculations were performed on a 2.8 GHz Pentium4 processor based on Linux.

2.2. Force field

We used the AMBER91 united-atom force field for amino acids[9] with some modifications as the followings. An angle-dependent, 12-10 hydrogen-bond potential, V_{hb}, was used for hydrogen-bonding atoms in combination with van der Waals potential, V_{vdW}:

$$V_{\text{hb}} = \sum_{i,j}(\frac{C_{ij}}{r_{ij}^{12}} - \frac{D_{ij}}{r_{ij}^{10}})F(\theta_{\text{A-H-D}}, \theta_{\text{AA-A-H}}), \quad (4)$$

$$V_{\text{vdW}} = \sum_{i,j}(\frac{A_{ij}}{r_{ij}^{12}} - \frac{B_{ij}}{r_{ij}^{6}})(1 - F(\theta_{\text{A-H-D}}, \theta_{\text{AA-A-H}})). \quad (5)$$

Here, A_{ij}, B_{ij}, C_{ij}, and D_{ij} are the parameters that depend on atom type i and j, and r_{ij} is the distance between atom i and j. The angle-dependent term, $F(\theta_{\text{A-H-D}}, \theta_{\text{AA-A-H}})$, varies depending on the type of hybridized orbital of the acceptor atom:

For sp^2 acceptor,
$F(\theta_{\text{A-H-D}}, \theta_{\text{AA-A-H}}) = \cos^4(\theta_{\text{A-H-D}})\cos^4(\theta_{\text{A-H-D}} - 155°)$
$(\theta_{\text{A-H-D}} > 90°, \theta_{\text{AA-A-H}} - 155° > 90°)$,
and for sp^3 acceptor, $\quad (6)$
$F(\theta_{\text{A-H-D}}, \theta_{\text{AA-A-H}}) = \cos^4(\theta_{\text{A-H-D}})\cos^4(\theta_{\text{A-H-D}} - 109.5°)$
$(\theta_{\text{A-H-D}} > 90°, \theta_{\text{AA-A-H}} - 109.5° > 90°)$,

where $\theta_{\text{A-H-D}}$ is the acceptor-hydrogen-donor angle and $\theta_{\text{AA-A-H}}$ is the base-acceptor-hydrogen angle (where the base is the atom that attaches to the acceptor).

To reproduce the solvation effects, three implicit solvent models were used: distance-dependent dielectric model (DD), solvent-accessible surface area model (SA), and effective charge model (EC).[5] In the DD model, $\epsilon = 2r_{ij}$ was used. The atomic solvation parameters used in the SA model were $\sigma(C) = 12$ cal/mol/Å^2, $\sigma(O, N) = -116$ cal/mol/Å^2, $\sigma(S) = -18$ cal/mol/Å^2, and $\sigma(O^-/N^+) = -280$ cal/mol/Å^2.[6]

The EC model was introduced by us to represent the shielding effect of oriented water molecules around a point charge,[5] in which atomic charge of atom i, q_i, is neutralized as a function of solvent-accessible surface area of the atom, $SA_i(\mathbf{r}^N)$, in a given atomic coordinate $\mathbf{r}^N$ ($\mathbf{r}^N$ is a position vector of Nth atom):

$$q'_i = q_i \left[\frac{1 - SA_i(\mathbf{r}^N)/S_i}{\alpha_{\text{int}}} + \frac{SA_i(\mathbf{r}^N)/S_i}{\alpha_{\text{ext}}} \right] \tag{7}$$

Here q'_i is the effective charge of atom i, S_i is the total solvent-accessible surface area of isolated atom i, α_{int} is a shielding parameter against interior of the solute (wherein α_{int} is set at unity), and α_{ext} is a shielding parameter for exterior water. In this study, $\alpha_{\text{ext}} = 5$ was used.

2.3. Umbrella sampling

Free energy along the chosen coordinate (called as reaction coordinate) is known as a potential of mean force (PMF). For calculating PMF, sampling of large conformational space separated by several multiples of $k_B T$ is necessary. Unfortunately, conventional simulation methods, such as Monte Carlo or molecular dynamics (MD), do not sample these regions adequately in biomolcuier system. Umbrella sampling method is a traditional technique that overcomes this sampling problem by modifying the potential function so that the unfavorable states are sampled sufficiently.[10] The modified potential function ($V'(\mathbf{r}^N)$) is written as follow:

$$V'(\mathbf{r}^N) = V(\mathbf{r}^N) + U(\mathbf{r}^N), \tag{8}$$

where ($V(\mathbf{r}^N)$) is a potential function of a protein-solvent system and ($U(\mathbf{r}^N)$) is a weighting function called as "umbrella potential". Typically, this umbrella potential takes a quadratic form:

$$U(\mathbf{r}^N) = K_{\text{umb}} (\mathbf{r}^N - \mathbf{r}_0^N)^2. \tag{9}$$

Here, K_{umb} is a force constant and $\mathbf{r}_0^N$ is an equilibrium coordinate. Simulation using the modified potential function will be biased toward the region of $\mathbf{r}_0^N$, resulting non-Boltzmann distribution. The corresponding Boltzmann average of any property A can be extracted from the non-Boltzmann distribution using a method introduce by Torrie and Valleau:[11]

$$\langle A \rangle = \frac{\langle A(\mathbf{r}^N) \exp \left[U(\mathbf{r}^N)/k_B T \right] \rangle_U}{\langle \exp \left[U(\mathbf{r}^N)/k_B T \right] \rangle_U}. \tag{10}$$

The subscript U indicates that an ensemble average is based on the probability determined by the modified potential function ($V'(\mathbf{r}^N)$).

Usually, multiple simulations using sequential values of $\mathbf{r}_0^N$ along the reaction coordinate are performed for efficient sampling and data obtained by these simulations are combined by WHAM (Weighted Histogram Analysis Method).[12]

2.4. Models

A designed $\beta\beta\alpha$ folded peptide named pda8d (PDB code 1psv) with the sequence KPYTARIKGRTFSNEKELRDFLETFTGR (28 residues)[13] was used to evaluate effectiveness of BD compared to MD. For folding simulation, two short peptides that fold into their three dimensional structures in aqueous solution were used. The first peptide is an analogue of the helical C-peptide of ribonuclease A termed peptide III, whose sequence is acetyl-AETAAAKFLRAHA-NH$_2$ (13 residues).[14] The second peptide is the designed β-hairpin peptide, BH8, whose sequence is RGITVNGKTYGR (12 residues).[15] For binding affinity calculation, ubiquitin ligase Ned4 WW-domain (50 residues) and Pro-rich peptide (17 residues) complex structure determined by NMR (PDB code 1i5h)[16] was used as an initial state.

2.5. Analysis

2.5.1. Native contacts

We defined nine backbone-backbone hydrogen bonds (the O$\cdots$H distance is smaller than 3 Å) between residue i and $i + 4$ as native contacts of peptide III. For BH8, four interstrand backbone-backbone hydrogen bonds (Ile3 NH-Tyr10 CO, Ile3 CO-Tyr10 NH, Val5 NH-Lys8 CO, and Val5 CO-Lys8 NH) and 3 interstrand side-chain - side-chain interactions (Ile3-Tyr10, Thr4-Thr9, and Val5-Lys8; distances between geometrical centers of side-chains are smaller than 7 Å) were used for native contacts.

2.5.2. Cluster analysis

The method of cluster analysis is based on structural similarity that was measured using distance-based root mean square deviation (dRMS).[6] The dRMS is evaluated for each pair of structures. For each conformation, the number of neighbors is calculated using a dRMS cutoff of 2 Å. The conformation with the highest number of neighbors (the most populated cluster) is defined as the center of the first cluster. All the neighbors of this conformation are removed from the ensemble of conformations. The center of the second cluster is determined in the same way as for the first cluster. This procedure is repeated until any one structure is assigned to one of such clusters.

3. Results and Discussion

3.1. *Computational time*

Table 1 lists the computational time required for 1 ns simulation of pda8d using BD, BD with multiple time step (MTS) algorithm and MD on a 2.8 GHz Pentium4 processor. BD simulation was faster than the conventional MD simulation by a factor of 50. When MTS algorithm was introduced in BD simulation, computational time was greatly reduced by a factor of 150 compared with the MD simulation. The BD simulations were stable and the artifacts often observed in simulations at vacuum condition were reduced (data not shown). These results indicate that this BD method would be effective for long time simulation.

Table 1. Computational time required for 1 ns simulation.

Algorithm	Number of atoms	Time (min)	Speedup factor
BD	304	38.8	53
MTS BD	304	12.8	161
MD	7,681	2,057	1

Note: [a]All calculations were performed using Pentium4 2.8 GHz processor. [b]Multiple time step algorithm was used with short time step of 5 fs and long time step of 40 fs. All covalent bonds were constrained with LINCS. [c] The simulation was performed using AMBER. The peptide was solvated using a box extending at least 10 Å in all directions. All covalent bonds were constrained with SHAKE. Cut-off radius was 9 Å. The time step was 2 fs.

3.2. *Folding simulations of α-helical and β-hairpin peptides*

To study the folding mechanism of the key structural elements of proteins, we performed long time simulations of α-helical, peptide III, and β-hairpin, BH8, peptides at 298 K using our BD with MTS algorithm and the implicit solvent DD/SA/EC models from the fully extended conformations. Folding simulations of the peptides were performed five times using different random seeds for each peptide.

3.2.1. *Folding trajectories*

Figure 2 shows the fraction of native contacts (Q) of the two peptides as a function of simulation time. For peptide III, although there were few states having $Q > 0.8$ due to lack of hydrogen bonds at C-terminus, the peptide reached the folded states from the extended states within 400 ns in all simulations. Because the formation of perfect helix ($Q = 1.0$) accompanies with large entropic cost of conformation, this state is not expected to exist in a significant amount. In the simulations of BH8, the peptide also folded from the extended structure in all trajectories.

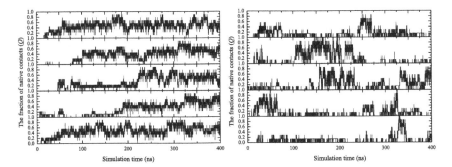

Fig. 2. Time evolutions of the fraction of native contacts during BD simulations of peptide III (left) and BH8 (right). Five trajectories obtained by the BD simulations using different random number seeds are shown.

3.2.2. *Energy components*

The average effective energy (effective energy is the intra-protein energy plus solvation free energy) and its components (van der Waals term, E_{vdW}, electrostatic term including the effects of DD/SA/EC implicit solvent models, E_{elec}') of the two peptides as a function of Q are shown in Figure 3. The total effective energy showed downhill profile for both peptides. The negative gradient of the total effective energy of peptide III was much larger than that of BH8. However, since variances of the total energies were too large, there were many non-native structures having lower effective energies than the energy of the native structure in both systems. This result indicates that it is impossible to predict the native states of the peptides based on the energy alone. For peptide III, the average values of E_{vdW} and E_{elec}' decreased with Q. For BH8, although the average value of E_{vdW}, decreased with Q, the slope of E_{elec}' is quite flat. These results indicate that the effective driving energy contributions to the folding of the peptides are concluded to be derived from both van der Waals and electrostatic terms for the α-helical peptide, peptide III, and from van der Waals term for β-hairpin peptide, BH8.

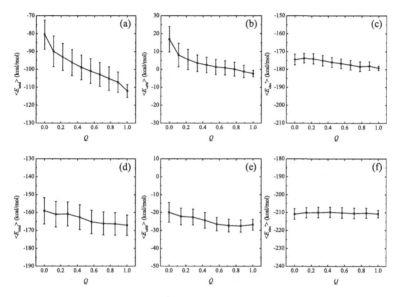

Fig. 3. Energy plots of the 2.0×10^4 conformations sampled during five simulations of (a-c) peptide III and (d-f) BH8 at 298 K. (a, d) Average of total effective energy, (b, e) average energy of van der Waals term and (c, f) average energy of effective electrostatic term as a function of Q are shown.

3.2.3. Cluster analysis

Next we performed a cluster analysis based on dRMS using about 20,000 structures obtained by the simulations of each peptide. The structures of the centers of the three most populated clusters for both peptides are shown in Figure 4. The most populated clusters of peptide III and BH8 contain 7% and 8% of all the conformations, respectively. Interestingly, the most populated cluster had higher average value of Q than that of other clusters and the folded structures belonged to these most populated clusters for both peptides. The central structure of cluster 1 of peptide III had a helical conformation throughout the peptide. For BH8, the central structure of the most populated cluster was a β-hairpin conformation that had side-chains of Ile3, Val5, Lys8, and Tyr10 protruding on the same side of plane of the strands, which is consistent with the NMR data. An important point is that the cluster analysis makes it possible to predict the native folded states from the structures obtained by the BD simulations.

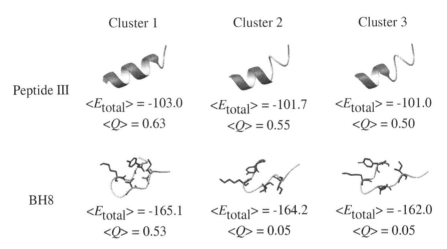

Fig. 4. Ribbon representations of the central structures of the three most populated clusters for peptide III (upper) and for BH8 (lower). From left to right, cluster 1, cluster 2 and cluster 3 are shown. The values of total effective energy (E_{total} in kcal/mol) and the fraction of native contacts, Q, averaged over the cluster are listed under each central structure. For BH8, residues of Ile3, Val5, Lys8 and Tyr10 are shown in sticks. The figures are generated with MOLMOL.[18]

3.3. *Binding affinity calculation*

Using MD method, we can follow only tens of nanoseconds of a protein dynamics. Hence the time is not enough to simulate the conformational change accompanying with the ligand binding reaction to a receptor. Since we have developed BD algorithm, we have extended it to enable the free energy estimation of a ligand binding using umbrella sampling.

For calculation of binding affinity of WW-domain and its binding peptide using BD/umbrella sampling method, the distance between Phe27 Cβ of WW-domain and Pro Cγ of the peptide was used as reaction coordinate;

$$U(r_{12}) = K_{\text{umb}}(r_{12} - r_0)^2. \tag{11}$$

Here, r_{12} is the distance between the atoms mentioned above and r_0 is the equilibrium distance. 36 sampling windows using $r_0 = 3.5, 4.0 \cdots, 21.0$ Å and $K_{\text{umb}} = 4$ kcal/mol/Å^2 were used. In each window, 1.0 ns and 1.1 ns BD simulations were performed for equilibration and sampling, respectively.

Figure 5 shows a PMF of WW-domain/peptide binding as a function of the reaction coordinate obtained by using the BD/umbrella sampling method. The binding free energy was estimated as 9 kcal/mol. In experiment, that value was estimated as 6 kcal/mol. The accordance is not good enough to be quantitative. This may be due to the inappropriate force parameters used in this study. We are trying to improve the parameters toward being quantitative by our method.

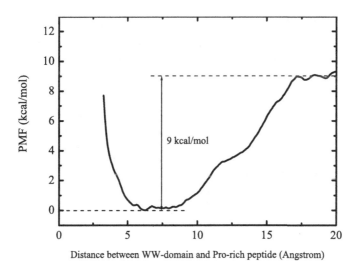

Fig. 5. Binding free energy estimation of WW-domain/Pro-rich peptide by BD/umbrella sampling simulation. In experiment, the binding free energy was estimated as 6 kcal/mol, and in calculation it was 9 kcal/mol. PMF means potential of mean force.

4. Conclusion

Now the BD method made it possible to simulate the folding of the key secondary structure elements in proteins, as well as to estimate binding affinity of a protein-protein interaction as described above. Adding to the conclusion of this report, I would like to state a future direction of our study. For understanding folding mechanism and predicting the folded structures of proteins from their amino acid sequences, the folding simulations of proteins having more than 50 amino acid residues should be essential. To tackle this problem, we have launched the protein structure prediction server named "TANPAKU", in which thousands of BD simulations are performed on distributed computing platform simultaneously. The tremendous computational power would enable us to search for an optimal parameter set more efficiently. This will lead the simulations to being more quantitative. Furthermore, we would like to develop a quantum algorithm of BD simulation in collaboration with physicists and mathematicians. We believe that these studies will become a key step to solving "the secret" of protein folding and binding.

Acknowledgments

This work was supported by the Academic Frontier Project from MEXT (Ministry of Education, Culture, Sports, Science and Technology of Japan).

References

1. Y. Sugita and Y. Okamoto, *Chem. Phys. Lett.* **314**, 141 (1999).
2. W. C. Still, A. Tempczyk, R. C. Hawley and T. Hendrickson, *J. Am. Chem. Soc.* **112**, 6127 (1990).
3. T. Ando, T. Meguro and I. Yamato, *J. Comput. Chem., Jpn.* **1**, 103 (2002).
4. T. Ando, T. Meguro and I. Yamato, *Mol. Simul.* **29**, 471 (2003).
5. T. Ando, T. Meguro, I. Yamato, *J. Comput. Chem., Jpn.* **3**, 129 (2004).
6. T. Ando and I. Yamato, *Mol. Simul.* **31**, 683 (2005).
7. D. L. Ermak and J. A. McCammon, *J.Chem.Phys.* **69**, 1352 (1978).
8. B. Hess, H. Bekker, H. J. C. Berendsen and J. G. E. M. Fraaije, *J. Comp. Chem.* **18**, 1463 (1997).
9. S. J. Weiner, P. A. Kollman, D. A. Case, U. C. Singh, C. Ghio, G. Alagona, S. Profeta, and P. Weiner, *J. Am. Chem. Soc.* **106**, 765 (1984).
10. G. M. Torrie and J. P. Valleau, *Chem. Phys. Lett.* **28**, 578 (1974).
11. G. M. Torrie and J. P. Valleau, *J. Comput. Phys.* **23**, 187 (1977).
12. S. Kumar, D. Bouzida, R. H. Swedsen, P. A. Kollman and J. M. Rosenberg, *J. Comp. Chem.* **13**, 1011 (1992).
13. B. I. Dahiyat, C. A. Sarisky and S. L. Mayo, *J. Mol. Biol.* **273**, 789 (1997).
14. K. R. Shoemaker, P. S. Kim, E. J. York, J. M. Stewart and R. L. Baldwin, *Nature* **326**, 563 (1987).
15. M. Ramirez-Alvarado, F. J. Blanco and L. Serrano, *Nat. Struct. Biol.* **3**, 604 (1996).
16. V. Kanelis, D. Rotin and J. D. Forman-Kay, *Nat. Struct. Biol.* **8**, 407 (2001).
17. W. L. DeLano, The PyMOL Molecular Graphics System (2002). http://pymol.sourceforge.net/
18. R. Koradi, M. Billeter and K. Wuthrich, *J. Mol. Graphics* **14**, 51 (1996).

QUANTUM FOKKER-PLANCK MODELS: THE LINDBLAD AND WIGNER APPROACHES

A. ARNOLD

Institute for Analysis and Scientific Computing, TU Wien,
Wiedner Hauptstr. 8, A-1040 Wien, Austria
E-mail: Anton.Arnold@tuwien.ac.at

F. FAGNOLA

Dipartimento di Matematica, Politecnico di Milano,
Piazza Leonardo da Vinci 32, I-20133 Milano, Italy
E-mail: franco.fagnola@polimi.it

L. NEUMANN

Institute for Analysis and Scientific Computing, TU Wien,
Wiedner Hauptstr. 8, A-1040 Wien, Austria
E-mail: Lukas.Neumann@tuwien.ac.at

In this article we try to bridge the gap between the quantum dynamical semigroup and Wigner function approaches to quantum open systems. In particular we study stationary states and the long time asymptotics for the quantum Fokker–Planck equation. Our new results apply to open quantum systems in a harmonic confinement potential, perturbed by a (large) sub-quadratic term.

Keywords: Quantum Markov Semigroups, Quantum Fokker Planck, steady state, large-time convergence.

1. Quantum Fokker–Planck model

This paper is concerned with the mathematical analysis of quantum Fokker–Planck (QFP) models, a special type of open quantum systems that models the quantum mechanical charge-transport including diffusive effects, as needed, *e.g.*, in the description of quantum Brownian motion,[1] quantum optics,[2] and semiconductor device simulations.[3] We shall consider two equivalent descriptions, the Wigner function formalism and the density matrix formalism.

In the quantum kinetic Wigner picture a quantum state is described by the real valued Wigner function $w(x,v,t)$, where $(x,v) \in \mathbb{R}^2$ denotes the position–velocity phase space. Its time evolution in a harmonic confinement potential $V_0(x) = \omega^2 \frac{x^2}{2}$ with $\omega > 0$ is given by the Wigner Fokker–Planck[4–6] (WFP) equation

$$\partial_t w = \omega^2 x \partial_v w - v \partial_x w + Qw, \quad (1)$$
$$Qw = 2\gamma \partial_v (vw) + D_{pp} \Delta_v w + D_{qq} \Delta_x w + 2 D_{pq} \partial_v \partial_x w.$$

The (real valued) diffusion constants D and the friction γ satisfy the Lindblad condition

$$\Delta := D_{pp} D_{qq} - D_{pq}^2 - \gamma^2/4 \geq 0, \quad (2)$$

and $D_{pp}, D_{qq} \geq 0$. Moreover we assume that the particle mass and $\hbar$ are scaled to 1.

WFP can be considered as a quantum mechanical generalization of the usual kinetic Fokker–Planck equation (or Kramer's equation), to which it is known to converge in the classical limit $\hbar \to 0$, after an appropriate rescaling of the appearing physical parameters.[7,8] The WFP equation has been partly derived in Ref. 9 as a rigorous scaling limit for a system of particles interacting with a heat bath of phonons.

In recent years, mathematical studies of WFP type equations mainly focused on the Cauchy problem (with or without self-consistent Poisson-coupling).[4,10–14] In the present work we shall be concerned with the steady state problem for the WFP equation and the large-time convergence to such steady states. Stationary equations for quantum systems, based on the Wigner formalism, seem to be rather difficult. For a purely quadratic confinement potential, this problem was dealt with in Ref. 6 using PDE–tools. The extension to harmonic potentials with a small, smooth perturbation was recently obtained in Ref. 15 using fixed point arguments and spectral theory. Here we consider large perturbations of the harmonic potential. To this end we shall work in the density matrix formalism, using tools from operator theory.

In the density matrix formalism a quantum state is described by a density matrix $\sigma \in \mathscr{T}_1^+(\mathsf{h})$, the cone of positive trace class operators on some Hilbert space h. The time evolution with the initial state $\rho_0 = \sigma$ is governed by the linear QFP equation or master equation

$$\frac{d\rho_t}{dt} = \mathcal{L}_*(\rho_t), \quad (3)$$

with the Lindbladian

$$\mathcal{L}_*(\rho) = -\frac{i}{2}\left[p^2 + \omega^2 q^2 + V(q), \rho\right] - i\gamma\left[q, \{p, \rho\}\right]$$
$$- D_{qq}[p,[p,\rho]] - D_{pp}[q,[q,\rho]] + 2D_{pq}[q,[p,\rho]], \qquad (4)$$

where $V(q)$ is the perturbation of the harmonic potential.

Global in time solutions to such master equations were established in Ref. 13 (nonlinear QFP–Poisson equation) starting from the construction of the associated minimal quantum dynamical semigroup (QDS).[16]

General methods for the study of quantum master equations and their large time behavior, including the existence of steady states and convergence towards them were developed in Quantum Probability.

Applicable sufficient conditions for proving uniqueness, i.e. trace preservation, of the solution obtained by the minimal semigroup method were given in Ref. 17 (see also Ref. 18). A criterion based on a non-commutative generalization of Liapounov functions for proving the existence of steady states was developed in Ref. 19. The support of steady states and decomposition of a quantum Markov semigroup into its transient and recurrent components were studied in Ref. 20 and Ref. 21. When the support of a steady state is full, i.e. it is faithful, uniqueness of steady states and convergence towards steady states can be deduced from simple algebraic conditions based on commutators of operators appearing in a Lindblad form representation of the master equation (see Ref. 22 for bounded and Ref. 23 for unbounded operators). Many of these methods generalize those of stochastic analysis in the study of classical Markov semigroups and processes. We refer to the lecture note Ref. 24 for a comprehensive account.

In this paper we study the master equation (3) by the above methods. We first prove the existence and trace preservation (i.e. uniqueness) of solutions and then the existence of a steady state. If the diffusion constants D_{pp}, D_{pq}, D_{qq}, and the friction γ satisfy the Lindblad condition (2) with the *strict* inequality, we prove that this quantum Markovian evolution is irreducible in the sense of Ref. 25. As a consequence, steady states must be faithful and one can apply simple commutator conditions on the operators in the GKSL representation to establish uniqueness of the steady state and large time convergence towards this state.

When $\Delta = 0$ we conjecture that (see Sect. 9), unless V is zero and the limiting conditions $D_{qp} = -\gamma D_{qq}$, $D_{pp} = \omega^2 D_{qq}$ are satisfied, the quantum Markov semigroup is still irreducible. But the invariant subspace problem that has to be solved for proving this becomes very difficult and we were not been able to solve it.

The paper is organized as follows: In Section 2 we review the equivalence of the kinetic Wigner formalism and the Lindblad approach to open quantum systems. Some technical preliminaries are presented in §3 and 4. In §5 we construct the minimal QDS for (3), (4) with external potentials that grow at most subquadratically. The markovianity of the semigroup is proved in §6. This yields uniqueness and mass–conservation of the solution to (3), (4). In §7 we establish the existence of a steady state and in §8 we prove that the solution converges to this unique steady state for arbitrary initial data provided the Lindblad condition (2) is fulfilled with strict inequality. The limiting case $\Delta = 0$ is studied in §9.

2. Passage from the Wigner equation to the master equation

In this section we show how to pass from the Wigner language to the GKLS (Gorini, Kossakowski, Sudarshan;[26] Lindblad[27]) language. In order to keep the notation simple, we shall confine our presentation to the one dimensional case. However, the results extend to higher dimensions. The underlying Hilbert space of our considerations is $\mathsf{h} = L^2(\mathbb{R})$. We denote by q and p the standard position and momentum operators ($p = -i\partial_x$). They satisfy the canonical commutation relation (CCR) $[q, p] = i\mathbb{1}$.

The Wigner function $w(x, v, t)$ of a state $\rho_t = \mathcal{T}_t^*(\sigma)$, is (up to normalization) the anti Fourier transform of

$$\varphi(\xi, \eta, t) = \operatorname{tr}\left(\rho_t\, e^{-i(\xi q + \eta p)}\right). \tag{5}$$

Using (5) we shall now transform the WFP equation (1) into an evolution equation for the corresponding density matrix ρ_t.

For notational simplicity we do not always denote the time dependence of density matrices or Wigner functions explicitly.

As a consequence of the CCR we have

$$e^{-i(\xi q + \eta p)} = e^{-i\xi q} e^{-i\eta p} e^{i\xi\eta/2}, \tag{6}$$

$$e^{-i(\xi q + \eta p)} = e^{-i\eta p} e^{-i\xi q} e^{-i\xi\eta/2}, \tag{7}$$

$$e^{-i(\xi q + \eta p)} = e^{-i\eta p/2} e^{-i\xi q} e^{-i\eta p/2}. \tag{8}$$

Assuming that ρ is sufficiently regular, by differentiating (6) and (7) and using the cyclic property of the trace, we find

$$\partial_\xi \varphi(\xi, \eta) = -i\operatorname{tr}\left(\rho\, q\, e^{-i(\xi q + \eta p)}\right) + \frac{i\eta}{2}\varphi(\xi, \eta),$$

$$\partial_\xi \varphi(\xi, \eta) = -i\operatorname{tr}\left(q\, \rho\, e^{-i(\xi q + \eta p)}\right) - \frac{i\eta}{2}\varphi(\xi, \eta).$$

Subtracting, and respectively, summing the above equations we have

$$\eta \varphi(\xi, \eta) = -\text{tr}\left([q, \rho]\, e^{-i(\xi q + \eta p)}\right),$$

$$\partial_\xi \varphi(\xi, \eta) = -\frac{i}{2} \text{tr}\left(\{q, \rho\}\, e^{-i(\xi q + \eta p)}\right).$$

In a similar way we obtain formulae for the products and derivatives with respect to ξ:

$$\xi \varphi(\xi, \eta) = \text{tr}\left([p, \rho]\, e^{-i(\xi q + \eta p)}\right),$$

$$\partial_\eta \varphi(\xi, \eta) = -\frac{i}{2} \text{tr}\left(\{p, \rho\}\, e^{-i(\xi q + \eta p)}\right).$$

The Wigner function is the anti Fourier transform of φ

$$w(x, v) = \left(\frac{1}{2\pi}\right)^2 \int_{\mathbb{R}^2} e^{i(\xi x + \eta v)} \varphi(\xi, \eta) d\xi d\eta.$$

The factor is chosen such that the total mass is given by

$$m = \text{tr}(\rho) = \varphi(0, 0) = \int_{\mathbb{R}^2} w(x, v) dx dv.$$

Hence, trace conservation is equivalent to mass conservation.
Integrating by parts (and again assuming sufficient regularity and decay) we obtain

$$xw(x, v) = \frac{1}{2\pi} \int_{\mathbb{R}^2} x e^{i(\xi x + \eta v)} \varphi(\xi, \eta) d\xi d\eta$$

$$= \frac{1}{2\pi} \int_{\mathbb{R}^2} -i \left(\partial_\xi e^{i(\xi x + \eta v)}\right) \varphi(\xi, \eta) d\xi d\eta$$

$$= \left[\frac{-i e^{i(\xi x + \eta v)} \varphi(\xi, \eta)}{2\pi}\right]_{-\infty}^{+\infty} + \frac{i}{2\pi} \int_{\mathbb{R}^2} e^{i(\xi x + \eta v)} (\partial_\xi \varphi(\xi, \eta)) d\xi d\eta$$

$$= \frac{1}{2\pi} \int_{\mathbb{R}^2} e^{i(\xi x + \eta v)} \left(\frac{1}{2} \text{tr}\left(\{q, \rho\}\, e^{-i(\xi q + \eta p)}\right)\right) d\xi d\eta.$$

In a similar way one can calculate

$$vw(x, v) = \frac{1}{2\pi} \int_{\mathbb{R}^2} e^{i(\xi x + \eta v)} \left(\frac{1}{2} \text{tr}\left(\{p, \rho\}\, e^{-i(\xi q + \eta p)}\right)\right) d\xi d\eta,$$

$$\partial_x w(x, v) = \frac{1}{2\pi} \int_{\mathbb{R}^2} e^{i(\xi x + \eta v)} \left(\text{tr}\left(i[p, \rho]\, e^{-i(\xi q + \eta p)}\right)\right) d\xi d\eta,$$

$$\partial_v w(x, v) = \frac{1}{2\pi} \int_{\mathbb{R}^2} e^{i(\xi x + \eta v)} \left(\text{tr}\left(-i[q, \rho]\, e^{-i(\xi q + \eta p)}\right)\right) d\xi d\eta.$$

The above formulae lead to the following dictionary for translating a master equation from the Wigner function language to the GKSL language:

transformation on w	transformation on ρ
xw	$\frac{1}{2}\{q, \rho\}$
vw	$\frac{1}{2}\{p, \rho\}$
$\partial_x w$	$i[p, \rho]$
$\partial_v w$	$-i[q, \rho]$

For the terms appearing in the WFP equation we have:

transformation on w	transformation on ρ
$x\partial_v w$	$-\frac{i}{2}\{q, [q, \rho]\} = -\frac{i}{2}[q^2, \rho]$
$v\partial_x w$	$\frac{i}{2}\{p, [p, \rho]\} = \frac{i}{2}[p^2, \rho]$
$\partial_v(vw)$	$-\frac{i}{2}[q, \{p, \rho\}] = -\frac{i}{2}\{p, [q, \rho]\} + \rho$
$\Delta_v w$	$-[q, [q, \rho]]$
$\Delta_x w$	$-[p, [p, \rho]]$
$\partial_x \partial_v w$	$[p, [q, \rho]]$

Using this dictionary we find the following GKSL form ($\frac{d\rho_t}{dt} = \mathcal{L}_*(\rho_t)$) of the linear QFP equation

$$\mathcal{L}_*(\rho) = -\frac{i}{2}\left[p^2 + \omega^2 q^2, \rho_t\right] - i\gamma\left[q, \{p, \rho_t\}\right] \qquad (9)$$
$$- D_{qq}[p, [p, \rho_t]] - D_{pp}[q, [q, \rho_t]] + 2D_{pq}[q, [p, \rho_t]].$$

This corresponds to choosing $\lambda = \mu = \gamma$ in (3.8) of Ref. 5 (see also Ref. 28).

The dual equation of (9) with an added perturbation potential V reads

$$\mathcal{L}(A) = \frac{i}{2}\left[p^2 + \omega^2 q^2 + 2V(q), A\right] + i\gamma\{p, [q, A]\}$$
$$- D_{qq}[p, [p, A]] - D_{pp}[q, [q, A]] + 2D_{pq}[q, [p, A]], \quad A \in \mathscr{B}(\mathsf{h}).$$

It can be written[4] in (generalised) GKSL form like

$$\mathcal{L}(A) = i[H, A] - \frac{1}{2}\sum_{\ell=1}^{2}(L_\ell^* L_\ell A - 2L_\ell^* A L_\ell + A L_\ell^* L_\ell) \qquad (10)$$

with the "adjusted" Hamiltonian

$$H = \frac{1}{2}\left(p^2 + \omega^2 q^2 + \gamma(pq + qp)\right) + V(q),$$

and the Lindblad operators L_1 and L_2 given by

$$L_1 = \frac{-2D_{pq} + i\gamma}{\sqrt{2D_{pp}}}p + \sqrt{2D_{pp}}\,q, \qquad L_2 = \frac{2\sqrt{\Delta}}{\sqrt{2D_{pp}}}p. \qquad (11)$$

3. Key inequalities for the existence of a steady state

We aim at applying the criterion for existence of a normal invariant state by Fagnola and Rebolledo.[19] To this end we have to find a positive operator X and an operator Y bounded from below, satisfying

$$\mathcal{L}(X) \leq -Y,$$

which in addition both have finite dimensional spectral projections associated with intervals $]-\infty, \Lambda]$. To illustrate the technique we first present the computation for the harmonic potential only. The perturbation potential $V(q)$ will be added later on.

Consider the Lindbladian

$$\mathcal{L}_*(\rho) = -\frac{i}{2}\left[p^2 + \omega^2 q^2, \rho\right] - i\gamma\left[q, \{p, \rho\}\right]$$
$$- D_{qq}[p, [p, \rho]] - D_{pp}[q, [q, \rho]] + 2D_{pq}[q, [p, \rho]]$$

with dual[28]

$$\mathcal{L}(X) = \frac{i}{2}\left[p^2 + \omega^2 q^2, X\right] + i\gamma\{p, [q, X]\}$$
$$- D_{qq}[p, [p, X]] - D_{pp}[q, [q, X]] + 2D_{pq}[q, [p, X]].$$

Straightforward computations with the CCR $[q, p] = i\mathbf{1}$ yield

Lemma 3.1. *The following formulae hold for f, g smooth:*

$$\mathcal{L}(f(p)) = -\frac{\omega^2}{2}\left(qf'(p) + f'(p)q\right) - 2\gamma p f'(p) + D_{pp}f''(p),$$
$$\mathcal{L}(g(q)) = \frac{1}{2}\left(pg'(q) + g'(q)p\right) + D_{qq}g''(q),$$
$$\mathcal{L}(pq + qp) = 2\left(p^2 - \omega^2 q^2\right) - 2\gamma(pq + qp) + 4D_{pq}.$$

This suggests looking for X, Y given by second order polynomials in p and q (i.e. $f(p) = p^2$, $g(q) = q^2$). Therefore we start studying some algebraic properties of these operators:

Lemma 3.2. *For all $r, s > 0$ such that $rs > 1$ the operators*

$$rp^2 - (pq + qp) + sq^2, \qquad rp^2 + (pq + qp) + sq^2$$

are strictly positive and have discrete spectrum. Moreover all spectral projections associated with bounded intervals are finite dimensional.

Proof. Let $r_0, s_0 > 0$ be such that $r_0 < r$, $s_0 < s$ and $r_0 s_0 = 1$. Then

$$\left|\sqrt{r_0}\, p - \sqrt{s_0}\, q\right|^2 = r_0 p^2 - (pq + qp) + s_0 q^2 \geq 0.$$

It follows that
$$rp^2 - (pq+qp) + sq^2 = |\sqrt{r_0}\,p - \sqrt{s_0}\,q|^2 + (r-r_0)p^2 + (s-s_0)q^2.$$

Therefore the resolvent of $rp^2 - (pq+qp) + sq^2$ is dominated by the resolvent of a multiple (indeed $\min\{(r-r_0), (s-s_0)\}$) of the number operator $\frac{1}{2}(p^2+q^2-1)$. Since the latter is compact, also the resolvent of $rp^2 - (pq+qp) + sq^2$ is compact and self-adjoint. Hence it has a discrete spectrum that might only accumulate at 0.

The proof for the second operator is the same. □

We choose X of the form
$$X = rp^2 + (pq+qp) + sq^2 \tag{12}$$

and compute
$$\mathcal{L}(rp^2 + (pq+qp) + sq^2) = -2(2\gamma r - 1)p^2 - 2\omega^2 q^2$$
$$+ (s - 2\gamma - \omega^2 r)(pq+qp) + 2rD_{pp} + 4D_{pq} + 2sD_{qq}.$$

The conditions required on X and Y in Theorem 7.1 (*i.e.* $X > 0$ and $\mathcal{L}(X) \leq -Y$) hold if
$$rs > 1,$$
$$4\omega^2(2\gamma r - 1) > \left|s - 2\gamma - \omega^2 r\right|^2.$$

Letting r and s go to infinity with $s - 2\gamma - \omega^2 r$ constant (that can be 0, for simplicity), it is clear that, when $\gamma > 0$, we can find r and s large enough satisfying the above condition. We take, *e.g.* any $r > (2\gamma)^{-1}$ and $s = 2\gamma + \omega^2 r$ since
$$rs = 2\gamma r + \omega^2 r^2 > 1 + \frac{\omega^2}{4\gamma^2} > 1.$$

Y will be chosen later in Theorem 7.1.

We now add the perturbation potential. Let $V : \mathbb{R} \to \mathbb{R}$ be a smooth function satisfying a growth condition like
$$|V'(x)| \leq g_V \left(1 + |x|^2\right)^{\alpha/2} \tag{13}$$

with $g_V > 0$ and $0 \leq \alpha < 1$. Hence, this perturbation potential is strictly sub-quadratic. It gives rise to one additional term in $\mathcal{L}(X)$, namely:
$$i\left[V(q), p^2\right] = -(pV'(q) + V'(q)p),$$
$$i\left[V(q), pq+qp\right] = -2qV'(q).$$

Therefore we find now

$$\mathcal{L}(rp^2 + (pq+qp) + sq^2) \qquad (14)$$
$$= -2(2\gamma r - 1)p^2 - 2\omega^2 q^2 + (s - 2\gamma - \omega^2 r)(pq+qp)$$
$$-r\left(pV'(q) + V'(q)p\right) - 2qV'(q) + 2rD_{pp} + 4D_{pq} + 2sD_{qq}.$$

Note that (due to the positivity of $|\epsilon^{1/2}p \pm \epsilon^{-1/2}V'(q)|^2$)

$$-\left(\epsilon p^2 + \frac{1}{\epsilon}(V'(q))^2\right) \leq pV'(q) + V'(q)p \leq \epsilon p^2 + \frac{1}{\epsilon}(V'(q))^2.$$

Therefore, playing on the ε and the bound on the derivative of V, we can find the needed inequality $\mathcal{L}(X) \leq -Y$. This will be used in §7 to prove the existence of a steady state.

4. Domain problems

First we define the number operator $N := \frac{1}{2}(p^2 + q^2 - 1)$ on h with

$$\mathrm{Dom}(N) = \left\{ u \in \mathsf{h} \,\middle|\, Nu \in \mathsf{h} \right\} = \left\{ u \in \mathsf{h} \,\middle|\, p^2 u, q^2 u \in \mathsf{h} \right\},$$

where the last equality follows easily from $\|Nu\|_{\mathsf{h}}^2 < \infty$ by an integration by parts. $C_c^\infty(\mathbb{R})$ is a core for N (cf. Ref. 30, e.g.). Let X be the self-adjoint extension of (12) (still denoted by X). $\mathrm{Dom}(X)$ is its maximum domain and $C_c^\infty(\mathbb{R})$ is a core for X.

The position and momentum operators are defined on $\mathrm{Dom}(N^{1/2})$. Both q and p have, by Nelson's analytic vector theorem, self-adjoint extensions that will be still denoted by q and p.

First we shall compare the domains of N and X. To this end we need

Lemma 4.1. *Let $r, s > 0$ with $rs > 1$ and define*

$$R := r^{1/2}p + r^{-1/2}q, \qquad S := s^{1/2}q + s^{-1/2}p.$$

Then, for all $u \in C_c^\infty(\mathbb{R})$ the following identities hold

$$\langle u, X^2 u \rangle = \left(s - \frac{1}{r}\right)^2 \langle u, q^4 u \rangle + 2\left(s - \frac{1}{r}\right)\langle u, qR^2 q u \rangle$$
$$+ \langle u, R^4 u \rangle - 2(rs-1)\|u\|^2,$$

$$\langle u, X^2 u \rangle = \left(r - \frac{1}{s}\right)^2 \langle u, p^4 u \rangle + 2\left(r - \frac{1}{s}\right)\langle u, pS^2 p u \rangle$$
$$+ \langle u, S^4 u \rangle - 2(rs-1)\|u\|^2.$$

Proof. Since u belongs to the domain of any monomial in p and q the proof can be reduced to the algebraic computation avoiding writing u's.

Starting from the identity $X = \theta q^2 + R^2$ with $\theta = s - 1/r$ we have
$$X^2 = \theta^2 q^4 + \theta \left(q^2 R^2 + R^2 q^2 \right) + R^4.$$
The mixed product term can be written in the form
$$\begin{aligned} q^2 R^2 + R^2 q^2 &= qR^2 q + q\left[q, R^2\right] + qR^2 q + \left[R^2, q\right] q \\ &= 2qR^2 q + qR\left[q, R\right] + q\left[q, R\right] R + \left[R, q\right] Rq + R\left[R, q\right] q \\ &= 2qR^2 q + 2ir^{1/2} qR - 2ir^{1/2} Rq \\ &= 2qR^2 q + 2ir^{1/2} \left[q, R\right] \\ &= 2qR^2 q - 2r. \end{aligned}$$
The conclusion is now immediate. □

The following lemma gives similar inequalities for the operators $rp^2 + sq^2$ (*i.e.* X without mixed products) that will be useful in the sequel

Lemma 4.2. *For all $r, s > 0$ and $u \in C_c^\infty(\mathbb{R})$ we have*
$$\left\langle u, \left(rp^2 + sq^2\right)^2 u \right\rangle \geq (r \wedge s)^2 \left\langle u, \left(p^2 + q^2\right)^2 u \right\rangle - 2\left(rs - (r \wedge s)^2\right) \|u\|^2,$$
$$\left\langle u, \left(rp^2 + sq^2\right)^2 u \right\rangle \leq (r \vee s)^2 \left\langle u, \left(p^2 + q^2\right)^2 u \right\rangle + 2\left((r \vee s)^2 - rs\right) \|u\|^2,$$
where $r \wedge s = \min\{r, s\}$ and $r \vee s = \max\{r, s\}$.

Proof. Indeed
$$\begin{aligned} \left(rp^2 + sq^2\right)^2 &= r^2 p^4 + rs \left(p^2 q^2 + q^2 p^2\right) + s^2 q^4 \\ &= r^2 p^4 + 2rs \, pq^2 p + s^2 q^4 - 2rs \\ &\geq (r \wedge s)^2 \left(p^4 + 2pq^2 p + q^4\right) - 2rs \\ &= (r \wedge s)^2 \left(p^2 + q^2\right)^2 - 2\left(rs - (r \wedge s)^2\right). \end{aligned}$$
Moreover
$$\begin{aligned} \left(rp^2 + sq^2\right)^2 &= r^2 p^4 + rs \left(p^2 q^2 + q^2 p^2\right) + s^2 q^4 \\ &= r^2 p^4 + 2rs \, pq^2 p + s^2 q^4 - 2rs \\ &\leq (r \vee s)^2 \left(p^4 + 2pq^2 p + q^4\right) - 2rs \\ &= (r \vee s)^2 \left(p^2 + q^2\right)^2 + 2\left((r \vee s)^2 - rs\right). \end{aligned}$$
This completes the proof. □

Proposition 4.1. *The domains of the operators N and X coincide.*

Proof. We first show that Dom(X)⊆Dom(N). By Lemma 4.1, for all $u \in C_c^\infty$ we have

$$\|Xu\|^2 + 2(rs-1)\|u\|^2 \geq \frac{1}{2} \min\left\{(s-r^{-1})^2, (r-s^{-1})^2\right\} \langle u, (p^4+q^4)u\rangle$$
$$\geq \frac{1}{4} \min\left\{(s-r^{-1})^2, (r-s^{-1})^2\right\} \langle u, (p^2+q^2)^2 u\rangle,$$

where we used the elementary inequality $(p^2+q^2)^2 \leq 2(p^4+q^4)$ and $r, s > 0$; $rs > 1$. Therefore we find a constant $c_1(r,s) > 0$ such that

$$\|Xu\|^2 + 2(rs-1)\|u\|^2 \geq c_1(r,s)\left\|(p^2+q^2)u\right\|^2. \tag{15}$$

Now, if $u \in \text{Dom}(X)$, then there exists a sequence $(u_n)_{n\geq 1}$ is C_c^∞ converging in norm to u such that $(Xu_n)_{n\geq 1}$ converges in norm to Xu (since X is closed). The above inequality shows then that the sequence $(Nu_n)_{n\geq 1}$ is also norm convergent and its limit (N is closed) is Nu.

This shows that Dom(X) ⊆ Dom(N) and the two domains coincide since the opposite inclusion holds true by the construction of X itself. □

For future reference we briefly recall the definition of the annihilation operator $a := \frac{1}{\sqrt{2}}(q+ip)$ and the creation operator $a^\dagger := \frac{1}{\sqrt{2}}(q-ip)$. Using the isomorphic identification of h with $l^2(\mathbb{N}_0)$ (via the eigenfunctions of N, i.e. the Hermite functions), these operators can also be represented as

$$ae_0 = 0, \quad ae_{j+1} = \sqrt{j+1}\, e_j, \quad a_j^\dagger = \sqrt{j+1}\, e_{j+1}, \quad \text{for } j \in \mathbb{N}_0.$$

Here, $(e_j)_{j\geq 0}$ is the canonical orthonormal basis of $l^2(\mathbb{N}_0)$.

5. Construction of the minimal quantum dynamical semigroup

In this section we shall establish the existence of the minimal quantum dynamical semigroup (QDS) for the Lindbladian (10). To this end we first consider the operator G, defined on Dom(N) by

$$G = -\frac{1}{2}(L_1^* L_1 + L_2^* L_2) - iH = -\left(D_{qq} + \frac{i}{2}\right)p^2 - \left(D_{pp} + \frac{i\omega^2}{2}\right)q^2$$
$$+ \left(D_{pq} - \frac{i\gamma}{2}\right)(pq+qp) + \frac{\gamma}{2} - iV(q). \tag{16}$$

We suppose that the potential V is twice differentiable and satisfies the growth condition (13).

Proposition 5.1. *The domains of the operators N, G, and G^* coincide.*

Proof. Since $\mathrm{Dom}(N) \subseteq \mathrm{Dom}(G)$ by construction, it suffices to check the opposite inclusion. To this end we proceed as before finding an estimate of $\|Nu\|$ by the graph norm of G. Putting

$$G_0 = -\frac{1}{2}\left(L_1^* L_1 + L_2^* L_2\right) = -D_{qq}p^2 + D_{pq}(pq+qp) - D_{pp}q^2 + \frac{\gamma}{2},$$

we have for all $u \in C_c^\infty(\mathbb{R})$

$$\|Gu\|^2 = \|G_0 u\|^2 + \langle u, i\,[H, G_0]\,u\rangle + \|Hu\|^2.$$

A straightforward computation yields

$$\begin{aligned}i\,[G_0, H] =\ & -2(\gamma D_{qq} + D_{pq})p^2 + 2(\gamma D_{pp} + \omega^2 D_{pq})q^2 + 2D_{pq}qV'(q) \\ & + \left(D_{pp} - \omega^2 D_{qq}\right)(pq+qp) - D_{qq}\left(pV'(q) + V'(q)p\right).\end{aligned}$$

The above commutator has quadratic monomials in p and q and other terms including $V'(q)$ whose growth is sublinear due to our hypothesis on the potential. It follows then that we can find a constant $c_3 > 0$ such that

$$|\langle u, i\,[G_0, H]\,u\rangle| \le c_3 \langle u, (p^2 + q^2)u\rangle.$$

Hence, for all $\epsilon > 0$, by the Schwarz inequality

$$|\langle u, i\,[G_0, H]\,u\rangle| \le \epsilon \left\|(p^2 + q^2)u\right\|^2 + c_3^2 \epsilon^{-1}\|u\|^2.$$

It follows that

$$\|Gu\|^2 \ge \|G_0 u\|^2 - \epsilon \left\|(p^2 + q^2)u\right\|^2 - c_3^2 \epsilon^{-1}\|u\|^2. \tag{17}$$

Case 1:
Let $D_{pq} \ne 0$ and $D_{pp}D_{qq} > D_{pq}^2$ (cp. to the Lindblad condition (2)). Then

$$\frac{D_{qq}}{D_{pq}}\frac{D_{pp}}{D_{pq}} = \frac{D_{pp}D_{qq}}{D_{pq}^2} > 1$$

and

$$G_0 = -D_{pq}\left(\frac{D_{qq}}{D_{pq}}p^2 - (pq+qp) + \frac{D_{pp}}{D_{pq}}q^2\right) + \frac{\gamma}{2}.$$

Therefore G_0 is a multiple of an operator like those of Lemma 3.2 (up to a constant). By Inequality (15) we can find constants $c_4, c_5 > 0$ such that $\|G_0 u\|^2 \ge c_4 \left\|(p^2 + q^2)u\right\|^2 - c_5 \|u\|^2$. Therefore we find the inequality

$$\|Gu\|^2 \ge (c_4 - \epsilon)\|(p^2 + q^2)u\|^2 - c_6\|u\|^2 \tag{18}$$

for all $u \in C_c^\infty(\mathbb{R})$. Choosing $\epsilon = c_4/2$, we find an inequality allowing us to repeat the above argument for a sequence $(u_n)_{n \ge 1}$ (see the proof of

Prop. 4.1) and prove the inclusion $\text{Dom}(G) \subseteq \text{Dom}(N)$.

Case 2:
Let $D_{pq} = 0$. The estimate (18) now follows from (17) and Lemma 4.2. We conclude again that $\text{Dom}(G) \subseteq \text{Dom}(N)$.

Case 3:
Let $D_{pq} \neq 0$ and $D_{pp}D_{qq} = D_{pq}^2$ (and hence $\gamma = 0$). To recover (18) we start from

$$\|Gu\|^2 \geq \|Hu\|^2 - \epsilon \left\|(p^2 + q^2)u\right\|^2 - c_3^2 \epsilon^{-1} \|u\|^2 \qquad (19)$$

(in analogy to (17)), where we have now $H = \frac{1}{2}(p^2 + \omega^2 q^2) + V(q)$. From Lemma 4.2 we obtain

$$\|(p^2 + q^2)u\|^2 \leq c_1 \|\frac{1}{2}(p^2 + \omega^2 q^2)u\|^2 + c_2 \|u\|^2$$
$$\leq 2c_1(\|Hu\|^2 + \|V(q)u\|^2) + c_2 \|u\|^2.$$

Since $V(q)$ is sub-quadratic we have

$$\|V(q)u\|^2 \leq \epsilon \|q^2 u\|^2 + c_3^2 \epsilon^{-1} \|u\|^2 \leq \epsilon \|(p^2 + q^2)u\|^2 + c_4 \|u\|^2.$$

Hence,

$$\|(p^2 + q^2)u\|^2 - c_5 \|u\|^2 \leq c_6 \|Hu\|^2, \qquad (20)$$

and we conclude by combining (19) and (20). □

Remark 5.1. From Propositions 4.1 and 5.1 we infer that the graph norms of G, X, N are equivalent. Hence G is relatively bounded by X.

The operator G is clearly dissipative because

$$\Re \langle u, Gu \rangle = -\frac{1}{2} \sum_{\ell=1}^{2} \|L_\ell u\|^2 \leq 0$$

for all $u \in \text{Dom}(N)$. Therefore, by Prop. 3.1.15 of Ref. 29 it is closable and its closure is dissipative. We denote by the same symbol G the closure. Analogously, G^* is also dissipative.

Hence, the Lumer–Phillips theorem (see Theorem 3.1.16 of Ref. 29, e.g.) yields:

Proposition 5.2. *The operator G generates a strongly continuous contraction semigroup $(P_t)_{t \geq 0}$ on* h.

Since $\text{Dom}(G) = \text{Dom}(N) \subset \text{Dom}(L_\ell)$, $\ell = 1, 2$, and since

$$\langle Gv, u\rangle + \sum_{\ell=1}^{2}\langle L_\ell v, L_\ell u\rangle + \langle v, Gu\rangle = 0 \quad \forall u, v \in \text{Dom}(N),$$

condition **(H)** of Ref. 19 holds and we can construct $\mathcal{T}$, the minimal QDS[16] associated with G and the L_ℓ's.

6. Markovianity of the Quantum Dynamical Semigroup

The hypotheses for constructing the minimal quantum dynamical semigroup with form generator

$$\pounds(A)[v, u] = \langle Gv, Au\rangle + \sum_{\ell=1}^{2}\langle L_\ell v, AL_\ell u\rangle + \langle v, AGu\rangle, \quad A \in \mathcal{B}(\mathsf{h}) \quad (21)$$

hold by Prop. 5.2.

Now we want to show that the minimal semigroup $\mathcal{T}$ is Markov and hence mass conserving. To this end we apply Theorem 4.4 in Ref. 17. The algebraic computations in Sec. 3 suggest to consider an operator X of the form (12) with $r > (2\gamma)^{-1}$ and $s = 2\gamma + \omega^2 r$ on the linear manifold $C_c^\infty(\mathbb{R})$. The algebraic computations of this section can be made in the quadratic form sense.

We now check that the operator X satisfies the fundamental hypothesis **C** of Ref. 17 starting from domain properties:

Proposition 6.1. *The following properties of G, L_ℓ, and N hold:*

(1) $\text{Dom}(G) = \text{Dom}(X) \subseteq \text{Dom}(X^{1/2})$ *and* $\text{Dom}(X) = \text{Dom}(G)$ *is a core for* $X^{1/2}$,
(2) $L_\ell(\text{Dom}(G^2)) \subseteq \text{Dom}(X^{1/2})$ *for* $\ell = 1, 2$.

Proof. Clearly, the first part of (1) follows from the Propositions 4.1, 5.1 and the second assertion is a well-known property of the square root of a positive operator (cf. Thm. V.3.24 of Ref. 31).

Property (2) follows from the inclusion $\text{Dom}(G^2) \subseteq \text{Dom}(G) = \text{Dom}(N)$ and $L_\ell(\text{Dom}(N)) \subseteq \text{Dom}(N^{1/2})$. □

We now apply the sufficient condition for conservativity taking as the operator C (cf. Ref. 17) the self-adjoint operator

$$Xu = rp^2 + (pq + qp) + (\omega^2 r + 2\gamma)q^2 \text{ on } \text{Dom}(X) = \text{Dom}(N)$$

with $r > (2\gamma)^{-1}$.

Proposition 6.2. *Suppose that V is twice differentiable and satisfies the growth condition (13). Then there exists a positive constant b such that*

$$2\Re\langle Xu, Gu\rangle + \sum_{\ell=1}^{2} \left\|X^{1/2} L_\ell u\right\|^2 \le b \left\|X^{1/2} u\right\|^2 \tag{22}$$

for all $u \in \text{Dom}(N)$.

Proof. We first check the above inequality for $u \in C_c^\infty(\mathbb{R})$. The vector u clearly belongs to the domain of the operators $XG, G^*X, L_\ell^* X L_\ell$. Therefore, the left-hand side is equal to $\langle u, \mathcal{L}(X)u\rangle$. From (14) we have then

$$\langle u, \mathcal{L}(X)u\rangle = -2\left\langle u, \left((2\gamma r - 1)p^2 + \omega^2 q^2\right)u\right\rangle \tag{23}$$
$$+ \left(2rD_{pp} + 4D_{pq} + 2(\omega^2 r + 2\gamma)D_{qq}\right)\|u\|^2 + i\langle u, [V(q), X]u\rangle.$$

Estimating the commutator as follows (cf. (14))

$$|i\langle u, [V(q), X]u\rangle| = |-\langle u, (r(pV'(q) + V'(q)p) + 2qV'(q))u\rangle|$$
$$\le 2r\|pu\| \cdot \|V'(q)u\| + 2\|qu\| \cdot \|V'(q)u\|$$
$$\le \langle u, (r^2 p^2 + q^2)u\rangle + 2\left\langle u, |V'(q)|^2 u\right\rangle$$
$$\le \langle u, (r^2 p^2 + (2g_V^2 + 1)q^2 + 2g_V^2)u\rangle,$$

and putting $c_5 = \max\left\{r^2, 2g_V^2 + 1, 2rD_{pp} + 4D_{pq} + 2(\omega^2 r + 2\gamma)D_{qq}\right\}$ we find

$$|\langle u, \mathcal{L}(X)u\rangle| \le c_5 \langle u, (p^2 + q^2 + 2)u\rangle$$
$$= c_5 \langle u, (2N + 3)u\rangle$$
$$\le b\langle u, Xu\rangle$$

as in the proof of Prop. 4.1. Hence, we have now proved the inequality (22) for $u \in C_c^\infty(\mathbb{R})$. The extension to arbitrary $u \in \text{Dom}(N)$ follows by a standard approximation argument. □

This result yields

Theorem 6.1. *Suppose that the potential V is twice differentiable and satisfies the growth condition (13). Then the minimal semigroup associated with the above operators G, L_1, L_2 is Markov.*

Proof. It suffices to apply Theorem 4.4 from Ref. 17, choosing the positive, self-adjoint operator $\Phi := -G_0$ introduced in the proof of Prop. 5.1. The hypothesis **C** holds by Propositions 6.1 and 6.2, and the hypothesis **A** by Prop. 5.2. Moreover, we choose the positive, self-adjoint operator C as a sufficiently large multiple of X to satisfy

$$\langle \Phi^{\frac{1}{2}} u,\, \Phi^{\frac{1}{2}} u \rangle \leq \langle C^{\frac{1}{2}} u,\, C^{\frac{1}{2}} u \rangle \quad \forall u \in \mathrm{Dom}(X).$$ □

7. Stationary state

In order to establish the existence of a steady state of QFP we now start to verify the conditions of Theorem VI.1 in Ref. 19.

Theorem 7.1. *Suppose that $\gamma > 0$ and the potential V is twice differentiable and satisfies the growth condition (13). Then the quantum Markov semigroup (QMS) with form generator (21) has a normal invariant state.*

Proof. We shall apply Theorem IV.1 from Ref. 19. Hypothesis **(H)** is satisfied due to Prop. 5.2. Now choose

$$X := rp^2 + (pq + qp) + \left(\omega^2 r + 2\gamma\right) q^2, \quad Y := c_6(p^2 + q^2) - c_7 \mathbb{1}$$

with $r > (2\gamma)^{-1}$ and $c_6, c_7 > 0$. Indeed X is clearly positive (cf. Lemma 3.2). Y is bounded below with finite dimensional spectral projections associated with intervals $]-\infty, \Lambda]$, since it is a translation of a multiple of the number operator.

In order to check the fundamental inequality

$$\mathcal{L}(X)[u, u] \leq -\langle u, Yu \rangle \tag{24}$$

for all $u \in \mathrm{Dom}(N)$ we start from the identity (23) in the proof of Prop. 6.2 and estimate the commutator as follows

$$\begin{aligned}
|i \langle u, [V(q), X]u \rangle| &= |- \langle u, (r(pV'(q) + V'(q)p) + 2qV'(q))\, u \rangle| \\
&\leq 2r\|pu\| \cdot \|V'(q)u\| + 2\|qu\| \cdot \|V'(q)u\| \\
&\leq \epsilon \langle u, (p^2 + q^2)u \rangle + \frac{r^2 + 1}{\epsilon} \langle u, |V'(q)|^2 u \rangle \\
&\leq \epsilon \langle u, (p^2 + q^2)u \rangle + g_V^2 \frac{r^2 + 1}{\epsilon} \langle u, (1 + |x|^{2\alpha})u \rangle,
\end{aligned}$$

where we used the elementary inequality

$$(1 + |x|^2)^\alpha \leq 1 + |x|^{2\alpha}$$

for $0 \leq \alpha \leq 1$. With the Young inequality we have
$$|x|^{2\alpha} = \eta^{-1} \cdot (\eta |x|^{2\alpha}) \leq \frac{(\eta |x|^{2\alpha})^{1/\alpha}}{1/\alpha} + \frac{(\eta^{-1})^{1/(1-\alpha)}}{1/(1-\alpha)} = \alpha \eta^{1/\alpha} |x|^2 + \frac{1-\alpha}{\eta^{1/(1-\alpha)}}$$
for all $\eta > 0$. Choosing $\eta = (\epsilon^2/r^2 + 1)^\alpha$ we obtain
$$|x|^{2\alpha} \leq \frac{\alpha \epsilon^2}{r^2+1} |x|^2 + \frac{(1-\alpha)(r^2+1)^{\alpha/(1-\alpha)}}{\epsilon^{2\alpha/(1-\alpha)}}.$$
Therefore we have
$$|i \langle u, [V(q), X]u \rangle| \leq \epsilon \langle u, (p^2+q^2)u \rangle + \alpha \epsilon g_V^2 \langle u, q^2 u \rangle$$
$$+ g_V^2 \frac{r^2+1}{\epsilon} \left(1 + (1-\alpha) \left(\frac{r^2+1}{\epsilon^2} \right)^{\alpha/(1-\alpha)} \right) \|u\|^2$$
$$\leq (1 + \alpha g_V^2) \epsilon \langle u, (p^2+q^2)u \rangle + g_V^2 \|u\|^2 \left(\frac{(1-\alpha)(r^2+1)^{1/(1-\alpha)}}{\epsilon^{(1+\alpha)/(1-\alpha)}} + \frac{r^2+1}{\epsilon} \right).$$
This inequality and (23) give
$$\langle u, \mathcal{L}(X)u \rangle \leq -2 \langle u, ((2\gamma r - 1)p^2 + \omega^2 q^2) u \rangle + (1 + \alpha g_V^2)\epsilon \langle u, (p^2+q^2)u \rangle$$
$$+ \left(2r D_{pp} + 4 D_{pq} + 2(\omega^2 r + 2\gamma) D_{qq} \right) \|u\|^2$$
$$+ g_V^2 \left(\frac{(1-\alpha)(r^2+1)^{1/(1-\alpha)}}{\epsilon^{(1+\alpha)/(1-\alpha)}} + \frac{r^2+1}{\epsilon} \right) \|u\|^2.$$
For all $r > (2\gamma)^{-1}$, we can take an ϵ small enough such that
$$c_6 := 2 \min \left\{ (2\gamma r - 1), \omega^2 \right\} - (1 + \alpha g_V^2)\epsilon > 0.$$
Putting
$$c_7 := 2r D_{pp} + 4 D_{pq} + 2(\omega^2 r + 2\gamma) D_{qq} + g_V^2 \frac{(1-\alpha)(r^2+1)^{1/(1-\alpha)}}{\epsilon^{(1+\alpha)/(1-\alpha)}} + g_V^2 \frac{r^2+1}{\epsilon}$$
we find the asserted inequality (24):
$$\langle u, \mathcal{L}(X)u \rangle \leq -c_6 \langle u, (p^2+q^2)u \rangle + c_7 \|u\|^2$$
for $u \in C_c^\infty(\mathbb{R})$. The extension to arbitrary $u \in \text{Dom}(N)$ follows by a standard approximation argument.

Clearly G is relatively bounded with respect to X by previous results (cf. §§4, 5). Moreover we have $(\lambda + X)^{-1}(\text{Dom}(N)) = \text{Dom}(N^2)$, and hence
$$L_\ell \left((\lambda + X)^{-1} \text{Dom}(N) \right) \subseteq L_\ell(\text{Dom}(N)) \subseteq \text{Dom}(N^{1/2}) = \text{Dom}(X^{1/2})$$
for $\ell = 1, 2$; $\lambda \geq 1$. Therefore Theorem IV.1[19] can be applied and it yields the existence of a normal invariant state. □

We remark that the operators X and Y do not commute here, in contrast to most examples in Ref. 19.

8. Irreducibility and large time behavior

A QMS $\mathcal{T}$ on $\mathcal{B}(\mathsf{h})$ is called *irreducible* if the only subharmonic projections[20] Π in h (*i.e.* projections satisfying $\mathcal{T}_t(\Pi) \geq \Pi$ for all $t \geq 0$) are the trivial ones 0 or $\mathbb{1}$.

If a projection Π is subharmonic, the total mass of any normal state σ with support in Π (*i.e.* such that $\Pi\sigma\Pi = \Pi\sigma = \sigma\Pi = \sigma$), remains concentrated in Π during the evolution. Indeed, the state $\mathcal{T}_t^*(\sigma)$ at time t then satisfies

$$1 = \mathrm{tr}\,(\mathcal{T}_t^*(\sigma)) \geq \mathrm{tr}\,(\mathcal{T}_t^*(\sigma)\Pi) = \mathrm{tr}\,(\sigma\mathcal{T}_t(\Pi)) \geq \mathrm{tr}\,(\sigma\Pi) = \mathrm{tr}\,(\sigma) = 1\,.$$

As an example, the support projection of a normal stationary state for a QMS is subharmonic (cf. Th. II.1 in Ref. 20). Thus if a QMS is irreducible and has a normal invariant state, then its support projection must be $\mathbb{1}$, *i.e.* it must be faithful.

In this section we shall prove that the QMS we constructed in §5-6 is irreducible if the strict inequality $\Delta > 0$ holds. The more delicate limiting case $\Delta = 0$ is postponed to the next section.

Subharmonic projections are characterized by the following theorem.[20]

Theorem 8.1. *A projection Π is subharmonic for the QMS associated with the operators G, L_ℓ if and only if its range $\mathcal{X}$ is an invariant subspace for all the operators P_t of the contraction semigroup generated by G (i.e. $\forall t \geq 0 : P_t \mathcal{X} \subseteq \mathcal{X}$) and*

$$L_\ell(\mathcal{X} \cap Dom(G)) \subseteq \mathcal{X}$$

for all ℓ's.

The application to our model yields the following

Theorem 8.2. *Suppose that $\Delta > 0$. Then the QMS $\mathcal{T}$ associated with (the closed extensions of) the operators G, L_ℓ given by (16) and (11) is irreducible.*

Proof. Let Π be a subharmonic projection with range $\mathcal{X} \subseteq \mathsf{h}$. We shall prove that either $\Pi = 0$ or $\Pi = \mathbb{1}$, and hence $\mathcal{T}$ is irreducible.

The domain of N is P_t invariant because it coincides with $\mathrm{Dom}(G)$ which is obviously P_t-invariant. Since both L_1, L_2 map $\mathrm{Dom}(N)$ into $\mathrm{Dom}(N^{1/2})$ and $\mathrm{Dom}(N^{1/2})$ into h, we have

$$L_\ell\left(\mathcal{X} \cap \mathrm{Dom}(N)\right) \subseteq \mathcal{X} \cap \mathrm{Dom}(N^{1/2}), \quad \text{and}$$
$$L_\ell\left(\mathcal{X} \cap \mathrm{Dom}(N^n)\right) \subseteq \mathcal{X} \cap \mathrm{Dom}(N^{n-1/2}). \qquad (25)$$

Then, by the linear independence of L_1 and L_2 (due to $\Delta > 0$)

$$p\left(\mathcal{X} \cap \mathrm{Dom}(N)\right) \subseteq \mathcal{X} \cap \mathrm{Dom}(N^{1/2}), \quad \text{and}$$
$$q\left(\mathcal{X} \cap \mathrm{Dom}(N)\right) \subseteq \mathcal{X} \cap \mathrm{Dom}(N^{1/2}),$$

and thus, since $N = (p^2 + q^2 - 1)/2$,

$$N\left(\mathcal{X} \cap \mathrm{Dom}(N)\right) \subseteq \mathcal{X}.$$

For all $n > 0$ the resolvent operator $R(n; G) = (n - G)^{-1}$ maps h in $\mathrm{Dom}(G) = \mathrm{Dom}(N)$, therefore the operator $N_n = nNR(n; G)$ is defined on h. It is also bounded, because N is relatively bounded with respect to G and the identity $GR(n; G)u = -u + nR(n; G)u$ yields the inequalities

$$\|N_n u\| \le c\,\|nGR(n; G)u\| + c\,\|nR(n; G)u\|$$
$$\le cn\,\|u\| + c(n+1)\,\|nR(n; G)u\|,$$

for all $u \in \mathsf{h}$.

The subspace $\mathcal{X}$ is clearly N_n-invariant because

$$R(n; G) = \int_0^\infty \exp(-nt) P_t\, dt$$

maps it into $\mathcal{X} \cap \mathrm{Dom}(G) = \mathcal{X} \cap \mathrm{Dom}(N)$ and $N(\mathcal{X} \cap \mathrm{Dom}(N)) \subseteq \mathcal{X}$. It follows that $\mathcal{X}$ is invariant for the operators $\mathrm{e}^{-tN_n} = \sum_{k \ge 0}(-tN_n)^k/k!$ ($t \ge 0$) of the (semi)group generated by N_n.

Notice that, for all $u \in \mathrm{Dom}(N) = \mathrm{Dom}(G)$, we have

$$\|N_n u - Nu\| = \|N\left(nR(n; G)u - u\right)\|$$
$$\le c\,\|G\left(nR(n; G)u - u\right)\| + c\,\|nR(n; G)u - u\|$$
$$= c\,\|nR(n; G)Gu - Gu\| + c\,\|nR(n; G)u - u\|.$$

Thus, by the well-known properties of the Yosida approximations, $N_n u$ converges towards Nu as n tends to infinity for all $u \in \mathrm{Dom}(N)$. It follows then from a well-known result in semigroup theory (cf. 31, Chapter IX) that the operators e^{-tN_n} converge towards the operators e^{-tN} of the semigroup generated by $-N$ in the strong operator topology on h uniformly for t in bounded subsets of $[0, +\infty[$. Therefore $\mathcal{X}$ is e^{-tN}-invariant for all $t \ge 0$.

Since the operators $e^{-tN}, t > 0$ are compact and self-adjoint, it follows that $\mathcal{X}$ is generated by eigenvectors of N,

$$\mathcal{X} = \overline{\text{Lin}\{e_j \mid j \in J\}},$$

for some $J \subseteq \mathbb{N}_0$. Here $(e_j)_{j \geq 0}$ is the canonical orthonormal basis of $\ell^2(\mathbb{N}_0)$. Moreover, using $\mathcal{X}_\infty := \mathcal{X} \cap_{n \geq 1} \text{Dom}(N^n)$, we have $\text{Lin}\{e_j \mid j \in J\} \subseteq \mathcal{X}_\infty \subseteq \mathcal{X}$, since e_j are the eigenvectors of N.

By (25), $\mathcal{X}_\infty$ is L_1, L_2-invariant. Since a and $a^\dagger$ are related to L_1, L_2 by an invertible linear transformation, $\mathcal{X}_\infty$ is also $a, a^\dagger$-invariant. Therefore, if J is not empty (i.e. $\mathcal{X} \neq \{0\}$) taking $m = \min J$ we find immediately that $e_0 = a^m e_m/\sqrt{m!}$ belongs to $\mathcal{X}_\infty \subseteq \mathcal{X}$. Thus, any e_k with $k > 0$ belongs to $\mathcal{X}_\infty$ because $a^{\dagger k} e_0 = \sqrt{k!} e_k$ and $\mathcal{X} = \mathcal{X}_\infty$ coincides with the whole of h. □

Corollary 8.1. *If $\Delta > 0$, then all normal invariant states are faithful.*

Proof. The support projection of an invariant state is subharmonic (and non-zero). Since the QMS is irreducible, all non-zero subharmonic projections must coincide with the identity operator $\mathbb{1}$. Hence, any normal invariant state must be faithful. □

Recall the following classical result.[22]

Theorem 8.3. *Let $\mathcal{T}$ be the unital minimal QMS associated with operators G, L_ℓ. Suppose that $\mathcal{T}$ has a faithful normal invariant state ρ. Then the vector space of fixed points*

$$\mathcal{F}(\mathcal{T}) = \{x \in \mathcal{B}(\mathsf{h}) \mid \mathcal{T}_t(x) = x, \; \forall t \geq 0\}$$

is an algebra and the invariant state ρ is unique if and only if $\mathcal{F}(\mathcal{T}) = \mathbb{C}\mathbb{1}$.

Remember that the QMS under consideration admits a steady state by Theorem 7.1 and by Corollary 8.1 this invariant state is faithful provided that we have strict inequalitiy in the Lindblad condition (2).

The next result, taken from Proposition 2.3 in Ref. 23 and the Correction in Ref. 32, allows us to apply immediately the above theorem and show that the QMS converges towards its unique invariant state. Let us introduce first

$$\mathcal{N}(\mathcal{T}) = \{x \in \mathcal{B}(\mathsf{h}) \mid \mathcal{T}_t(x^*x) = \mathcal{T}_t(x^*)\mathcal{T}_t(x), \mathcal{T}_t(xx^*) = \mathcal{T}_t(x)\mathcal{T}_t(x^*) \; \forall t \geq 0\}.$$

If $\mathcal{T}$ has a *faithful* normal invariant state, it is easy to show by the Schwarz property $\mathcal{T}_t(x^*)\mathcal{T}_t(x) \leq \mathcal{T}_t(x^*x)$, that $\mathcal{N}(\mathcal{T})$ is an algebra.

Theorem 8.4. [23,32] *Let T be the unital minimal QMS associated with the operators G, L_ℓ. Suppose that: i) there exists a domain D_c which is a core for both G and G^*, and ii) for all $u \in D_c$, its image $R(n;G)u$ belongs to $Dom(G^*)$ and the sequence $(nG^*R(n;G)u)_{n\geq 1}$ converges strongly. Then*

(1) $\mathcal{F}(T) = \{H, L_1, L_2\}'$,
(2) $\mathcal{N}(T) \subseteq \{L_1, L_2\}'$,
(3) if $\mathcal{F}(T) = \mathcal{N}(T)$ then for all initial states σ, $T_t^*(\sigma)$ converges as t goes to infinity towards an invariant state in the trace norm.

In the above Theorem the set $\{H, L_1, L_2\}'$ denotes the *commutant*, i.e. the set of all operators that commute with H as well as with L_1 and L_2 (analogously for $\{L_1, L_2\}'$) and R denotes the resolvent. Because L_1 and L_2 are linearly independent, as long as we have strict inequality in (2), we see that $\{L_1, L_2\}'$ consists only of operators commuting with q and p. This yields $\mathcal{N}(T) = \mathbb{C}\mathbf{1}$ and since $\mathbb{C}\mathbf{1} \subseteq \mathcal{F}(T) \subseteq \mathcal{N}(T)$ also $\mathbb{C}\mathbf{1} = \mathcal{F}(T) = \mathcal{N}(T)$.

To apply Theorem 8.4 and thus ensure convergence we need to verify the conditions (i) and (ii). Condition (i) is obvious from Prop. 5.1 (one might pick $C_c^\infty(\mathbb{R})$ as a core). Condition (ii) is also easily checked because the operators G, G^* and N have the same domain by Prop. 5.1. Moreover, their graph norms are equivalent (cf. Remark 5.1). Now, for all $u \in D_c = C_c^\infty(\mathbb{R})$, since $nGR(n;G)u = nR(n;G)Gu$, the sequence $(nR(n;G)u)_{n\geq 1}$ converges to u in the graph norm of G. By Remark 5.1 it is also convergent in the graph norm of G^*, i.e. $(nG^*R(n;G)u)_{n\geq 1}$ converges to G^*u.

Altogether we have proved the following

Corollary 8.2. *Let $\gamma > 0$ and $V \in C^2(\mathbb{R})$ satisfy (13). If $\Delta > 0$ the QMS associated with G and L_ℓ has a unique faithful normal invariant state ρ. Moreover, for all normal initial states σ, we have*

$$\lim_{t\to\infty} T_t^*(\sigma) = \rho$$

in the trace norm.

9. The limiting case $\Delta = 0$

Here we study the case when $\Delta = D_{pp}D_{qq} - D_{pq}^2 - \gamma^2/4 = 0$ and $\gamma > 0$. We start with the case $V = 0$. In this situation we can compare our result to the explicit formula for the (unique normalized) steady state in Ref. 6 for zero perturbing potential. The kernel of the density matrix of the steady state can be calculated by means of Fourier transform[6] and is given by

$$\rho_\infty(x,y) = \frac{\gamma\omega}{\pi\sqrt{\gamma Q_{22}}} \exp\left(-\frac{1}{4\gamma Q_{22}}\left[\gamma^2\omega^2(x+y)^2 + Q(x-y)^2\right]\right) \times \exp\left(-i\omega\frac{Q_{12}}{Q_{22}}\left(\frac{x^2-y^2}{2}\right)\right), \quad (26)$$

where the abbreviations are

$$Q_{11} = D_{pp} + \omega^2 D_{qq}$$
$$Q_{12} = 2\omega\gamma D_{qq}$$
$$Q_{22} = D_{pp} + \omega^2 D_{qq} + 4\gamma(D_{pq} + \gamma D_{qq}) \quad \text{and}$$
$$Q = Q_{11}Q_{22} - Q_{12}^2.$$

One can see that this becomes a *pure* state if and only if $Q = \gamma^2\omega^2$. We will now discuss its implications on the relation between diffusion and friction coefficients.

Lemma 9.1. *Let $V = 0$, $\gamma > 0$, and the Lindblad condition hold. The steady state given by (26) is a pure state, i.e. $Q = \gamma^2\omega^2$, if and only if $\gamma < \omega$ and*

$$0 = D_{pp}D_{qq} - D_{pq}^2 - \gamma^2/4 \quad (27)$$
$$D_{pq} = -\gamma D_{qq} \quad (28)$$
$$D_{pp} = \omega^2 D_{qq}. \quad (29)$$

In this case

$$D_{qq} = \frac{\gamma}{2\sqrt{\omega^2-\gamma^2}}, \quad \text{and}$$
$$\rho_\infty(x,y) = \frac{1}{\pi}\sqrt[4]{\omega^2-\gamma^2}\, e^{-(cx^2+\bar{c}y^2)} \quad \text{with} \quad (30)$$
$$c = \frac{1}{2}(\sqrt{\omega^2-\gamma^2} + i\gamma). \quad (31)$$

Proof. We rewrite

$$0 = Q - \gamma^2\omega^2 = \left(1 - \frac{\gamma^2}{\omega^2}\right)\left(D_{pp} - D_{qq}\omega^2\right)^2 \quad (32)$$
$$+ \frac{\gamma^2}{\omega^2}\left(D_{pp} + 2D_{pq}\frac{\omega^2}{\gamma} + D_{qq}\omega^2\right)^2 \quad (33)$$
$$+ 4\omega^2\left(D_{pp}D_{qq} - D_{pq}^2 - \frac{\gamma^2}{4}\right). \quad (34)$$

In the case $\gamma < \omega$ this directly implies (27)–(29), because of (2).
For $\gamma = \omega$ the term (33) is a quadratic polynomial in ω. But it has no real zero, since $D_{pp}D_{qq} - D_{pq}^2 > 0$ by (34).
For the case $\gamma > \omega$ we rewrite (32)–(34) as

$$0 = \left(D_{pp} + 2D_{pq}\gamma + D_{qq}\omega^2\right)^2 \tag{35}$$

$$+ 4\gamma^2 \left(D_{pp}D_{qq} - D_{pq}^2 - \frac{\gamma^2}{4}\right) \tag{36}$$

$$+ \gamma^2(\gamma^2 - \omega^2), \tag{37}$$

which is a contradiction since the summand (36) is non–negative by (2). The form of D_{qq} and the kernel of the density matrix of the steady state follow by straightforward calculations. □

Corollary 9.1. *Let $V = 0$, $\Delta = 0$, and $0 < \gamma < \omega$. Under the conditions (27), (28) and (29) of Lemma 9.1 the semigroup is not irreducible. It admits a steady state that is not faithful.*

We now want to interpret this result in terms of the generators. Condition (27) implies that the Lindblad operators L_1 and L_2 are no longer linearly independent since L_2 is zero. The operator L_1 has a nontrivial kernel containing functions of the type $\exp(-cx^2)$, with c from (31), which are clearly also in the kernel of $L_1^*L_1$. Hence, density matrices with kernel (30) are initially not affected by the dissipative part of the evolution. However the Hamiltonian evolution might move them away from the kernel. It is exactly conditions (28) and (29) that ensure that H is (up to an additive constant) a scalar multiple of $L_1^*L_1$ and thus $[H, L_1^*L_1] = 0$. Of course in this case functions of type (30) stay unaffected by the evolution.

Using the explicit formula (26) for the kernel of the density matrix in the case $V = 0$ one can check the non–faithfulness of the steady state also directly:

Remark 9.1. *For ρ not to be faithful we have to find a $u \in \mathsf{h}$ with $u \neq 0$, such that $\langle u, \rho u \rangle = 0$. By positivity of ρ this is equivalent to $\rho u = 0$. Using (26), such a u satisfies*

$$0 = \int_{\mathbb{R}} \exp\left(-\frac{1}{4\gamma Q_{22}} \left[\gamma^2 \omega^2 (x+y)^2 + Q(x-y)^2\right]\right)$$
$$\times \exp\left(-i\omega \frac{Q_{12}}{Q_{22}} \left(\frac{x^2 - y^2}{2}\right)\right) u(y)\, dy,$$

for all $x \in \mathbb{R}$. Looking pointwise in x we can drop the factors depending only on x and get

$$0 = \int_{\mathbb{R}} \exp\left(-\frac{\gamma^2\omega^2 - Q}{2\gamma Q_{22}} xy\right)$$
$$\times \exp\left(-\frac{1}{4\gamma Q_{22}} \left(\gamma^2\omega^2 + Q - 2i\omega\gamma Q_{12}\right) y^2\right) u(y) dy.$$

Provided the steady state is not pure, i.e. $Q \neq \gamma^2\omega^2$, one can interpret this as the Laplace–transform of the term in the second line. Due to the decay of the exponential factor it exists for all $x \in \mathbb{R}$ and it can not be zero for all x unless u is zero. This shows that the steady-state ρ is faithful except when $Q = \gamma^2\omega^2$. In the case of a pure steady state the first exponential becomes one and the above integral vanishes for all functions that are orthogonal to

$$\exp\left(-\frac{1}{4\gamma Q_{22}} \left(\gamma^2\omega^2 + Q - 2i\omega\gamma Q_{12}\right) y^2\right) = \exp(-\bar{c}y^2).$$

Note that the complex conjugate of this function is in the kernel of L_1, as noted earlier.

The above discussion on the explicit steady state for $V = 0$ and algebraic computations of the invariant subspaces for G and L_1 showing that they should be trivial, lead us to the following

Conjecture. When $\Delta = 0$ but $V \neq 0$ (with $V \in C^2(\mathbb{R})$ satisfying (13)) the semigroup has a unique faithful normal invariant state and the conclusion of Corollary 8.2 holds.

The full proof of this conjecture, however, entails a lot of analytical difficulties and technicalities on invariant subspaces for strongly continuous semigroups and their perturbations and will be postponed to a forthcoming paper.

Acknowledgement: A. A. acknowledges partial support from the DFG under Grant No. AR 277/3-3, the ESF in the project "Global and geometrical aspects of nonlinear partial differential equations", and the Wissenschaftskolleg *Differentialgleichungen* of the FWF.

F. F. acknowledges partial support from the GNAMPA group project 2008 "Quantum Markov Semigroups". He also acknowledges the Institute for Analysis and Scientific Computing, TU Wien, were a part of this paper was written, for warm hospitality.

The authors are grateful to the organizers of the 28th conference on "Quantum probability and related topics", Guanajuato, Mexico, and in particular to Roberto Quezada.

References

1. L. Diósi, *On high-temperature Markovian equations for quantum Brownian motion*, Europhys. Lett. **22** (1993), 1-3.
2. M. Elk and P. Lambropoulos, *Connection between approximate Fokker-Planck equations and the Wigner function applied to the micromaser*, Quantum Semiclass. Opt. **8** (1996), 23–37.
3. A. Donarini, T. Novotný, and A.P. Jauho, *Simple models suffice for the single-dot quantum shuttle*, New J. of Physics **7** (2005), 237–262.
4. A. Arnold, J.L. López, P.A. Markowich, and J. Soler, *An Analysis of Quantum Fokker-Planck Models: A Wigner Function Approach*. Rev. Mat. Iberoam. **20**(3) (2004), 771-814.
   Revised: http://www.math.tuwien.ac.at/~arnold/papers/wpfp.pdf.
5. A. Sandulescu and A. Scutaru, *Open quantum systems and the damping of collective modes in deep inelastic collisions*. Ann. Phys. **173** (1987), 277–317.
6. C. Sparber, J.A. Carrillo, J. Dolbeault, and P.A. Markowich, *On the long time behavior of the quantum Fokker-Planck equation*. Monatsh. Math. **141**(3) (2004), 237–257.
7. R. Bosi, *Classical limit for linear and nonlinear quantum Fokker-Planck systems*, preprint 2007.
8. P.L. Lions and T. Paul, *Sur les mesures de Wigner*. Rev. Mat. Iberoam., **9** (1993), 553–618.
9. F. Castella, L. Erdös, F. Frommlet, and P. Markowich, *Fokker-Planck equations as Scaling Limit of Reversible Quantum Systems*. J. Stat. Physics **100**(3/4) (2000), 543–601.
10. A. Arnold, E. Dhamo, and C. Manzini, *The Wigner-Poisson-Fokker-Planck system: global-in-time solutions and dispersive effects*. Annales de l'IHP (C) - Analyse non lineaire **24**(4) (2007), 645–676.
11. A. Arnold, E. Dhamo, and C. Manzini, *Dispersive effects in quantum kinetic equations*. Indiana Univ. Math. J. **56**(3) (2007), 1299–1332.
12. J.A. Cañizo, J.L. López, and J. Nieto, *Global L^1 theory and regularity for the 3D nonlinear Wigner-Poisson-Fokker-Planck system*. J. Diff. Equ. **198** (2004), 356–373.
13. A. Arnold, and C. Sparber, Quantum dynamical semigroups for diffusion models with Hartree interaction. Comm. Math. Phys. **251**(1) (2004), 179–207.
14. A. Arnold, *Mathematical Properties of Quantum Evolution Equations*. In: G. Allaire, A. Arnold, P. Degond, Th.Y. Hou, **Quantum Transport - Modelling, Analysis and Asymptotics**, Lecture Notes in Mathematics 1946, Springer, Berlin (2008), ISBN: 978-3-540-79573-5.
15. A. Arnold, I.M. Gamba, M.P. Gualdani, and C. Sparber, *The Wigner-Fokker-Planck Equation: Stationary States and Large Time Behavior*, submitted, 2007.
16. E.B. Davies, **Quantum Theory of Open Systems**, Academic Press (1976).
17. A.M. Chebotarev and F. Fagnola, *Sufficient conditions for conservativity of quantum dynamical semigroups*. J. Funct. Anal. **153** (1998), 382–404.
18. J.C. García and R. Quezada, *Hille-Yosida estimate and nonconservativity*

criteria for quantum dynamical semigroups. Infinite Dimens. Anal. Quantum Probab. Rel. Topics **7** no.3 (2004), 383–394.
19. F. Fagnola and R. Rebolledo, *On the existence of stationary states for quantum dynamical semigroups.* J. Math. Phys. **42**(3) (2001), 1296–1308.
20. F. Fagnola and R. Rebolledo, *Subharmonic projections for a quantum Markov semigroup.* J. Math. Phys. **43**(2) (2002), 1074–1082.
21. V. Umanità, *Classification and decomposition of Quantum Markov Semigroups.* Probab. Theory Relat. Fields **134** (2006), 603–623.
22. A. Frigerio *Quantum dynamical semigroups and approach to equilibrium.* Lett. Math. Phys. **2**(2) (1977/78), 79–87.
23. F. Fagnola and R. Rebolledo, *The approach to equilibrium of a class of quantum dynamical semigroups.* Infin. Dimens. Anal. Quantum Probab. Relat. Top. **1**(4) (1998), 561–572.
24. F. Fagnola and R. Rebolledo, *Notes on the Qualitative Behaviour of Quantum Markov Semigroups.* In: S. Attal, A. Joye, C.-A. Pillet (eds.) **Open Quantum Systems III - Recent Developments**. Lecture Notes in Mathematics 1882 p. 161–206. Springer Berlin, Heidelberg (2006). ISBN 978-3-540-30993-2.
25. F. Fagnola and R. Rebolledo, *Transience and Recurrence of Quantum Markov Semigroups.* Probab. Theory Related Fields **126** n.2 (2003), 289–306.
26. V. Gorini, A. Kossakowski, and E.C.G. Sudarshan, *Completely positive dynamical semigroups of N-level systems.* J. Math. Phys. **17** (1976), 821–825.
27. G. Lindblad, *On the generators of quantum dynamical semigroups.* Commun. Math. Phys. **48** (1976), 119–130.
28. A. Isar, A. Sandulescu, H. Scutaru, E. Stefanescu, and W. Scheid, *Open quantum Systems.* Int. J. Mod. Phys. E, **3**(2) (1994), 635–719.
29. O. Bratteli and D. W. Robinson, **Operator Algebras and Quantum Statistical Mechanics, Vol I**. Springer (1987).
30. M. Reed and B. Simon, **Methods of Modern Mathematical Physics II**. Academic Press, 6th edition (1986).
31. T. Kato, **Perturbation Theory for Linear Operators**, Springer (1995).
32. F. Fagnola and R. Rebolledo, Algebraic Conditions for Convergence to a Steady State. Preprint (2008).

HILBERT SPACE OF ANALYTIC FUNCTIONS ASSOCIATED WITH A ROTATION INVARIANT MEASURE

N. ASAI

Department of Mathematics
Aichi University of Education
Kariya, 448-8542, Japan
nasai@auecc.aichi-edu.ac.jp

This paper contains not only a review of Ref.[4,6] but also new results. Let $\tilde{\mu}$ be a probability measure on $\mathbb{C}$ derived from the complex moment problem associated with the Jacobi-Szegö parameter. First, we show that $\tilde{\mu}$ has a rotation invariance property. It means that a one-mode interacting Fock space can be realized as a Hilbert space of analytic L^2-functions with respect to a rotation invariant measure on $\mathbb{C}$. As typical examples, the Gaussian and Bessel kernel measures on $\mathbb{C}$ are presented. Secondly, the relationship among the golden ratio, Bessel kernel measure and classical random variables is discussed. Furthermore, the distribution of the Lévy's stochastic area is also treated. In the end, we give a short remark on the binomial distribution.

1. Preliminaries.

Let μ be a probability measure on $I \subset \mathbb{R}$ with finite moments of all orders such that the linear span of the monomials $x^n, n \geq 0$, is dense in $L^2(I, \mu)$. Then it is known[11,25] that there exists a complete orthogonal system $\{P_n(x)\}_{n=0}^{\infty}$ of polynomials with leading coefficient 1 for $L^2(I, \mu)$ with $P_0 = 1$, a sequence $\{\omega_n\}_{n=0}^{\infty}$ of nonnegative real numbers, and a sequence $\{\alpha_n\}_{n=0}^{\infty}$ of real numbers such that the following recurrence formula holds:

$$(x - \alpha_n)P_n(x) = P_{n+1}(x) + \omega_n P_{n-1}(x), \quad n \geq 0,$$

where $\omega_0 = 1$ and $P_{-1} = 0$ by convention. The numbers ω_n, α_n are called the *Jacobi-Szegö parameters* of μ. It can be shown that the sequence $\alpha_n = 0$ for all n if and only if μ is symmetric.[11] For a probability measure μ with the associated sequence $\{\omega_n\}_{n=0}^{\infty}$, we define a sequence $\{\lambda_n\}_{n=0}^{\infty}$ by

$$\lambda_n = \omega_0 \omega_1 \cdots \omega_n, \quad n \geq 0. \tag{1}$$

Assume that $G_\lambda(z) := \sum_{n=0}^{\infty} \frac{1}{\lambda_n} z^n$ has a positive radious of convergence, denoted by $r_\lambda > 0$. For $f \in L^2(I, \mu)$, the author[2] has introduced a function $S_\mu f$ by

$$(S_\mu f)(z) = \langle E_\lambda(\cdot, \overline{z}), f \rangle_{L^2(\mu)} = \int_I E_\lambda(x, z) f(x) \, d\mu(x), \quad z \in \Omega_\lambda, \qquad (2)$$

where

$$E_\lambda(x, z) = \sum_{n=0}^{\infty} \frac{P_n(x)}{\lambda_n} z^n$$

and $\Omega_\lambda = \{z \in \mathbb{C} : |z| < \sqrt{r_\lambda}\}$. Note that $\|E_\lambda(\cdot, z)\|_{L^2(\mu)} = G_\lambda(|z|^2) < \infty$. The set $\{E_\lambda(\cdot, z) : z \in \Omega_\lambda\}$ is linearly independent and spans a dense subspace of $L^2(I, \mu)$. The S_μ-transform in Eq.(2) is a non-Gaussian analogue of the well-known Segal-Bargmann transform discussed in Ref.[10,14,23]

2. Hilbert space of analytic L^2-functions with respect to a rotation invariant measure.

Suppose that a probability measure $\widetilde{\mu}$ on Ω_λ is determined by the moment problem for a given sequence $\{\lambda_n\}_{n=0}^{\infty}$ in Eq.(1),

$$\int_{\Omega_\lambda} \overline{z}^n z^m \, d\widetilde{\mu}(z) = \delta_{n,m} \lambda_n, \qquad (3)$$

where $\delta_{n,m}$ is the Kronecker delta. We remark here that $\widetilde{\mu}$ is computed independently of a parameter α_n.

It is not difficult to generalize Proposition 4.4 in Ref.[6] as follows.

Proposition 2.1. *Suppose $\{\lambda_n\}_{n=0}^{\infty}$ satisfies the condition*

$$0 \le \lim_{n \to \infty} \frac{\lambda_n^{1/n}}{n^2} < \infty. \qquad (4)$$

Then the measure $\widetilde{\mu}$ satisfying (3) is unique.

The Hilbert space of analytic L^2-functions with respect to $\widetilde{\mu}$, namely $\mathcal{H}L^2(\Omega_\lambda, \widetilde{\mu})$, is a closed subspace of $L^2(\Omega_\lambda, \widetilde{\mu})$. Hence, there is an orthogonal projection operator $P : L^2(\Omega_\lambda, \widetilde{\mu}) \to \mathcal{H}L^2(\Omega_\lambda, \widetilde{\mu})$. This projection is given by

$$PF(z) = \int_{\Omega_\lambda} K_\lambda(\overline{z}, w) F(w) \, d\widetilde{\mu}(w)$$

with the reproducing kernel function

$$K_\lambda(z, w) := (S_\mu E_\lambda(x, w))(z) = \sum_{n=0}^{\infty} \frac{(zw)^n}{\lambda_n}$$

for all $F \in L^2(\Omega_\lambda, \widetilde{\mu})$. Note that $\{z^n\}_{n=0}^\infty$ is considered as an orthogonal basis for $\mathcal{H}L^2(\Omega_\lambda, \widetilde{\mu})$ and satisfies $(S_\mu P_n(x))(z) = z^n$. This was one of the key points to introduce the integral kernel function $E_\lambda(x, z)$ in Ref.[2] The norm on $\mathcal{H}L^2(\Omega_\lambda, \widetilde{\mu})$ is the $L^2(\Omega_\lambda, \widetilde{\mu})$-norm. It can be shown that S_μ is a unitary operator from $L^2(I, \mu)$ onto $\mathcal{H}L^2(\Omega_\lambda, \widetilde{\mu})$.

Let us denote the bosonic annihilation and creation operators, b and b^*, respectively, which satisfy the commutation relation $[b, b^*] = I$. For example, we can realize these operators by $b = \frac{d}{dz}$ and $b^* = z$, $z \in \mathbb{C}$. With b and b^*, we have introduced in Ref.[4] the *deformed annihilation operator* B^-, *deformed creation operator* B^+, and *conservation (neutral, preservation) operator* B° as

$$B^- = c_{0,1}b + c_{1,2}b^*b^2, \tag{5}$$

$$B^+ = b^*, \tag{6}$$

$$B^\circ = c_{0,0}I + c_{1,1}b^*b, \tag{7}$$

where $c_{0,1} > 0, c_{1,2} \geq 0$ and $c_{0,0}, c_{1,1} \in \mathbb{R}$. The case of $c_{1,2} < 0$ will be mentioned in Section 3. See Ref.[4] for those who are interested in how we thought of these deformed operators and why they should be considered.

By Accardi-Bożejko isomorphism[1,4] with Eqs.(5)(6)(7), it is easy to see that the Jacobi-Szegö parameters are represented as

$$\omega_0 = 1, \quad \omega_n = n(c_{0,1} + c_{1,2}(n-1)), \quad n \geq 1 \tag{8}$$

$$\alpha_n = c_{0,0} + c_{1,1}n, \quad n \geq 0, \tag{9}$$

where $c_{0,1} > 0, c_{1,2} \geq 0$ and $c_{0,0}, c_{1,1} \in \mathbb{R}$. Thus, we have $\Omega_\lambda = \mathbb{C}$ and $r_\lambda = \infty$. Then the Hilbert space $\mathcal{H}L^2(\mathbb{C}, \widetilde{\mu})$ equipped with $\{B^-, B^+, B^\circ\}$ is a realization of the *one-mode interacting Fock space* in terms of the space of analytic L^2-functions with respect to $\widetilde{\mu}$. Moreover, the S_μ-transform is a unitary operator from $L^2(I, \mu)$ onto $\mathcal{H}L^2(\mathbb{C}, \widetilde{\mu})$ with the intertwining property,

$$S_\mu^{-1}\left(c_{0,0} + (c_{0,1}b + b^*) + c_{1,1}b^*b + c_{1,2}b^*b^2\right)S_\mu = Q_x,$$

where Q_x is the multiplication operator by x on $L^2(I, \mu)$.

We are able to give the probabilistic meaning to the multiplication operator Q_x by x on $L^2(I, \mu)$. The multiplication operator Q_x can be considered as the classical random variable x and decomposed into the sum of operators B^-, B^+, B° in Eqs.(5)(6)(7), which are the generators of finite dimensional Lie algebras as treated in Section 4. This is a typical example of the classical-quantum correspondence in quantum probability theory.

In addition, constants $c_{0,0}$ and $c_{0,1}$ describe the mean and variance of μ, respectively (Table in Section 3).

Suppose that a probability measure $\widetilde{\mu}$ on $\mathbb{C}$ is given in the form of

$$d\widetilde{\mu}(z) = \frac{1}{2\pi}\rho(r)drd\theta \tag{10}$$

for $z = re^{i\theta}$ ($0 \leq \theta < 2\pi$, $0 < r < \infty$). Then it is easy to see that the moment problem (3) is reduced to the form of

$$\int_0^\infty r^{2n}\rho(r)dr = \lambda_n. \tag{11}$$

If one can construct a density function $\rho(r)$ uniquely due to the Mellin transform, then the existence of the measure $\widetilde{\mu}$ satisfying Eq.(3) is guaranteed. Hence, if λ_n satisfies the condition in Eq.(4), then $\widetilde{\mu}$ has the form uniquely as in Eq.(10).

Definition 2.1. Let $\mathcal{B}(\mathbb{R}^2)$ be the Borel σ-algebra on $\mathbb{R}^2$. A measure ν on $\mathbb{R}^2$ is rotation invariant if $\nu(B) = \nu(T^{-1}B)$ for every orthogonal matrix T and $B \in \mathcal{B}(\mathbb{R}^2)$, where $T^{-1}B = \{T^{-1}\boldsymbol{x} : \boldsymbol{x} \in B\}$.

Theorem 2.1. *Suppose that the Jacobi-Szegö parameter ω_n is given by Eq.(8). Then $\widetilde{\mu}$ is a unique rotation invariant solution of the moment problem in Eq.(3).*

Proof. Let $z = re^{i\theta}$ and $B \in \mathcal{B}(\mathbb{R}^2)$. For every orthogonal matrix T and $\boldsymbol{x} \in B$, $r = |\boldsymbol{x}| = |T^{-1}\boldsymbol{x}|$ holds. Due to $\rho(r) = \rho(|\boldsymbol{x}|) = \rho(|T^{-1}\boldsymbol{x}|)$, we get

$$\widetilde{\mu}(z) = \widetilde{\mu}(|\boldsymbol{x}|) = \widetilde{\mu}(|T^{-1}\boldsymbol{x}|).$$

Therefore, $\widetilde{\mu}$ is rotation invariant on $\mathbb{R}^2 \cong \mathbb{C}$.

Next, we prove uniqueness of $\widetilde{\mu}$. Since the case of $c_{1,2} = 0$ has been discussed in Ref.,[6] let us assume $c_{1,2} \neq 0$. For n sufficiently large, one can get

$$\frac{\lambda_n^{1/n}}{n^2} \sim \frac{c_{1,2}}{n^2}\left\{\sqrt{2\pi n}\left(\frac{n}{e}\right)^n \frac{\Gamma(n+\alpha)}{\Gamma(\alpha)}\right\}^{\frac{1}{n}} \leq \frac{c_{1,2}}{e}(\sqrt{2\pi n})^{\frac{1}{n}}\left(1 + \frac{\alpha-1}{n}\right) \tag{12}$$

$\alpha = c_{0,1}/c_{1,2}$ with the help of the Stirling formula. Hence the Jacobi-Szegö parameter ω_n given by Eq.(8) satisfies the condition in Eq.(4) in Proposition 2.1. Moreover, we get by Eq.(12)

$$\sum_{n=1}^\infty (\lambda_n)^{-\frac{1}{2n}} = \infty. \tag{13}$$

Thus, $\rho(r)$ is a unique measure satisfying Eq.(11) (The criterion in Eq.(13) is in Ref.[24] See also Ref.[6]). Therefore, $\widetilde{\mu}$ is uniquely determined by the moment problem in Eq.(3) with the form of Eq.(10). □

Hence, $\mathcal{H}L^2(\mathbb{C}, \widetilde{\mu})$ is a rotation invariant interacting Fock space. Important examples of $\widetilde{\mu}$ are the Gaussian and Bessel kernel measures on $\mathbb{C}$, which we present as follows. A constant $c_{1,2}$ plays key roles so as to construct $\widetilde{\mu}$ by solving the moment problem (3) (or (11)).

First of all, we consider the case of $c_{1,2} = 0$ as follows.

Theorem 2.2. (*Asai-Kubo-Kuo*[6]) *Suppose that the Jacobi-Szegö parameters have the forms in Eqs.(8)(9) under $c_{1,2} = 0$. Then we have the following results:*
(1) $\widetilde{\mu}$ *is a Gaussian measure $h_{c_{0,1}}$ on $\mathbb{C}$ expressed in the form of*

$$dh_{c_{0,1}}(z) := \frac{1}{\pi c_{0,1}} \exp\left(-\frac{|z|^2}{c_{0,1}}\right) d^2 z, \qquad (14)$$

where $d^2 z = dx_1 dx_2$, $x_1, x_2 \in \mathbb{R}$.
(2) S_μ *is a unitary operator from $L^2(I, \mu)$ onto $\mathcal{H}L^2(\mathbb{C}, h_{c_{0,1}})$ satisfying*

$$S_\mu^{-1}\left(c_{0,0} + (c_{0,1}b + b^*) + c_{1,1}b^*b\right) S_\mu = Q_x,$$

where Q_x is the multiplication operator by x on $L^2(I, \mu)$.

It is easy to see from Eq.(14) that $h_{c_{0,1}}$ is rotation invariant on $\mathbb{R}^2 \cong \mathbb{C}$ and known as the standard bivariable normal law whose covariance matrix is I.

Example 2.1. We have two examples of μ. If the Gaussian measure on $\mathbb{R}$ and Poisson measure on $\mathbb{N}_0 := \{0\} \cup \mathbb{N}$ are considered, $c_{1,2} = 0$ appears. Note that $c_{1,1} = 0$ in the case of Gaussian. See Table in Section 3. It is important to notice that if $c_{0,1} = \sigma^2 = \lambda$ holds, then the classical Gaussian and Poisson random variables are simultaneously realized on the same space $\mathcal{H}L^2(\mathbb{C}, h_{\sigma^2})$. Consult our paper[6] for details.

Secondly, the case of $c_{1,2} > 0$ is as follows:

Theorem 2.3. (*Asai*[4]) *Assume that the Jacobi-Szegö parameters have the forms in Eqs.(8)(9) under $c_{1,2} > 0$. Then we have the following results:*
(1) $\widetilde{\mu}$ *is a Bessel kernel measure $\gamma_{\alpha, c_{1,2}}$ on $\mathbb{C}$ expressed in the form of*

$$d\gamma_{\alpha, c_{1,2}}(z) := \frac{2 c_{1,2}^{-\frac{1}{2}(1+\alpha)}}{\pi \Gamma(\alpha)} |z|^{\alpha - 1} K_{1-\alpha}\left(2 c_{1,2}^{-\frac{1}{2}} |z|\right) d^2 z, \quad \alpha = c_{0,1}/c_{1,2},$$

where $d^2z = dx_1 dx_2$, $x_1, x_2 \in \mathbb{R}$. Note that Γ is the Gamma function and K_ν is so-called the modified Bessel function given by

$$K_\nu(x) = \frac{\pi}{2\sin(\nu\pi)}(I_{-\nu}(x) - I_\nu(x)),$$

where

$$I_\nu(x) = \left(\frac{x}{2}\right)^\nu \sum_{n=0}^\infty \frac{(x/2)^{2n}}{n!\Gamma(n+\nu+1)}.$$

(2) S_μ is a unitary operator from $L^2(I,\mu)$ onto $\mathcal{H}L^2(\mathbb{C}, \gamma_{\alpha,c_{1,2}})$ satisfying

$$S_\mu^{-1}\left(c_{0,0} + (c_{0,1}b + b^*) + c_{1,1}b^*b + c_{1,2}b^*b^2\right)S_\mu = Q_x,$$

where $\alpha = c_{0,1}/c_{1,2}$ and Q_x is the multiplication operator by x on $L^2(I,\mu)$.
(3) The measure $\gamma_{\alpha,c_{1,2}}$ has the following integral representation

$$d\gamma_{\alpha,c_{1,2}}(z) = \frac{1}{\Gamma(\alpha)}\left(\int_0^\infty h_{c_{1,2}t}(z)e^{-t}t^{\alpha-1}dt\right)dz, \tag{15}$$

where $\alpha = c_{0,1}/c_{1,2}$ and

$$h_{c_{1,2}t}(z) := \frac{1}{\pi c_{1,2}t}\exp\left(-\frac{|z|^2}{c_{1,2}t}\right).$$

A relationship between $\mathcal{H}L^2(\mathbb{C}, \gamma_{\alpha,c_{1,2}})$ and Riesz potentials has been discussed in Ref.[5] from the viewpoint of one-mode interacting Fock space approach.

Theorem 2.4. *The Bessel kernel measure $\gamma_{\alpha,c_{1,2}}$ is rotation invariant on $\mathbb{R}^2 \cong \mathbb{C}$.*

Proof. Let $z = x_1 + ix_2 \in \mathbb{C}$, $x_1, x_2 \in \mathbb{R}$ and $\boldsymbol{x} = (x_1, x_2)$. Since $|\boldsymbol{x}| = |T^{-1}\boldsymbol{x}|$ for an orthogonal matrix T, $|z| = |\boldsymbol{x}| = |T^{-1}\boldsymbol{x}|$. Since $h_{c_{1,2}t}$ is rotation invariant on $\mathbb{R}^2 \cong \mathbb{C}$ and Eq.(15) holds, the Bessel kernel measure $\gamma_{\alpha,c_{1,2}}$ is rotation invariant on $\mathbb{R}^2 \cong \mathbb{C}$. □

Our approach can cover the following examples.[4,12,16–20]

Example 2.2. For $c_{1,2} > 0$ case, we have three examples of μ classified by the signs of $c_{1,1}^2 - 4c_{1,2}$:
(1) If μ is the Gamma distribution on $\mathbb{R}_+$, then $c_{1,1} \neq 0$ and $c_{1,1}^2 = 4c_{1,2}$ are satisfied.
(2) If μ is the Pascal (Negative binomial) distribution on $\mathbb{N}_0 = \{0\} \cup \mathbb{N}$, then $c_{1,1}^2 > 4c_{1,2}$ is satisfied.
(3) If μ is the Meixner distribution on $\mathbb{R}$, then $c_{1,1}^2 < 4c_{1,2}$.

The author derived the discriminant $c_{1,1}^2 - 4c_{1,2}$ from the second order differential equation

$$\left\{c_{1,2}\frac{d^2}{dz^2} + c_{1,1}\frac{d}{dz} + 1\right\}(S_\mu E_\lambda(x,z))(w) = 0$$

near $z = \infty$ in Eq.(2.10) on Ref.[4] There are other relevant approaches in Remark 2.1, later. The reader can consult Table in Section 3 for explicit forms of distributions, Jacobi-Szegö parameters and $c_{0,0}, c_{0,1}, c_{1,1}, c_{1,2}$ and so on.

These examples have interesting features as follows.

(I) Suppose further that the Jacobi-Szegö parameters have the forms in Eqs.(8)(9) with $c_{0,1} = k = r = 2a$ and $c_{1,2} = 1 = qp^{-2} = (2\sin\phi)^{-2}$ in Table of Section 3. That is, $p = \tau^{-1}, q = 2 - \tau$ and $\phi = \frac{1}{6}\pi, \frac{5}{6}\pi$, where τ represents the golden ratio,

$$\tau = \frac{1 + \sqrt{5}}{2},$$

coming from a quadratic equation $\tau^2 - \tau - 1 = 0$. Then all of the classical Gamma, Pascal and Meixner random variables are simultaneously realized on the same space $\mathcal{H}L^2(\mathbb{C}, \gamma_{k,1})$ by Theorem 2.3. Moreover, it is easy to see that

$$(c_{0,0}, c_{0,1}, c_{1,1}, c_{1,2}) = (\tau^{-1}k, k, 2\tau - 1, 1)$$

in Pascal case. Such a case could be considered as an equilibrium state in some sense. It is not clear whether the golden ratio τ provides other probabilistic meanings. To our best knowledge, there may not exist papers which have pointed out relationships among three examples above, τ and a Hilbert space of analytic L^2 functions, and so on.

(II) Let us point out that the Meixner distribution,[17]

$$\frac{(2\sin\phi)^{2a}}{2\pi\Gamma(2a)}e^{(2\phi-\pi)x}|\Gamma(a+ix)|^2, \ a > 0, \ 0 < \phi < \pi, \ x \in \mathbb{R}, \qquad (16)$$

contains an interesting example with $a = 1/2$ and $\phi = \pi/2$. Since $\Gamma(1) = 1$ and

$$\left|\Gamma\left(\frac{1}{2} + ix\right)\right|^2 = \frac{\pi}{\cosh(\pi x)},$$

the probability density function in Eq.(16) with $a = 1/2$ and $\phi = \pi/2$ becomes

$$\frac{1}{\pi}\left|\Gamma\left(\frac{1}{2}+ix\right)\right|^2 = \frac{1}{\cosh(\pi x)} = \frac{2}{e^{\pi x}+e^{-\pi x}}.$$

This is called the distribution of the *Lévy's stochastic area* and a typical example of *self-reciprocal functions* just like a Gaussian distribution. See Ref.[13,15,16,20,22] and references therein. Moreover, $(c_{0,1}, c_{1,2}) = \left(\frac{1}{4}, \frac{1}{4}\right)$ and $(c_{0,0}, c_{1,1}) = (0,0)$ in this case. Consult Table in Section 3.

Remark 2.1. Other several derivations of a discriminant $c_{1,1}^2 - 4c_{1,2}$ are known as follows. In Ref.,[19] Meixner derived it from generating functions of exponential type. In Ref.,[20] Morris considered the natural exponential family of probability distributions with the variance being at most a quadratic function of the mean μ,

$$V(\mu) = v_0 + v_1\mu + v_2\mu^2.$$

In Ref.,[16] Ismail-May examined essentially the following problem: when the left hand side of

$$\exp\left(\lambda \int_c^{g(x)} \frac{tdt}{at^2+bt+c}\right) = \int_{-\infty}^{\infty} e^{\lambda ux} C(\lambda, u) du$$

is given under certain conditions, what kinds of probability density functions $C(\lambda, u)$ can be obtained to construct approximation operators?. In Ref.,[18] Kubo applied the multiplicative renormalization method[7-9] to the exponential family of generating functions,

$$\varphi(t, x) = \frac{e^{\rho(t)x}}{E_x[e^{\rho(t)x}]},$$

and showed that $\rho(t)$ satisfies a differential equation $\rho'(t) = (\beta t^2 + \gamma t + 1)^{-1}$. In conclusion, each approach originates from different motivations and purposes, but the common key word among them is "quadratic".

3. Remark on $c_{1,2} < 0$ and Table of Examples for $c_{1,2} \geq 0$.

In this section, we shall make a short remark on the case of $c_{1,2} < 0$. Up to here, we have only considered the case of $c_{1,2} \geq 0$. However, there is an important example which does not satisfy $c_{1,2} \geq 0$. It is the case of the binomial distribution $B(N, p)$,

$$\binom{N}{n} p^n (1-p)^{N-n}, \quad 0 < p < 1, \quad n = 0, 1, 2, \cdots N.$$

It is known[11,25] that the orthogonal polynomials associated with $B(N,p)$ are called the *Krawtchouk polynomials* with the Jacobi-Szegö parameters given by

$$\omega_0 = 1, \quad \omega_n = np(1-p)(N-(n-1)), \quad 1 \leq n \leq N \qquad (17)$$
$$\alpha_n = Np + (1-2p)n, \quad 0 \leq n \leq N. \qquad (18)$$

Hence, we can see $c_{1,2} = -p(1-p) < 0$ for $p \in (0,1)$. Other quantities are $c_{0,1} = Np(1-p)$, $c_{0,0} = Np$ and $c_{1,1} = 1-2p$. Since $\omega_n \equiv 0$ for $n \geq N+1$, an associated one-mode interacting Fock space is finite dimensional. All other cases under $c_{1,2} \geq 0$ are infinite dimensional. Hence, a corresponding measure $\widetilde{\mu}$ for $B(N,p)$ is different from either the Gaussian measure $h_{c_{0,1}}$ or Bessel kernel measure $\gamma_{\alpha,c_{1,2}}$. Due to these reasons, $B(N,p)$ is excluded from Table.

If taking limits as $p \to 0$, $N \to \infty$ under $Np = \lambda = \text{const} > 0$, then we get $c_{1,2} \to 0$ and

$$\omega_n = np(1-p)(N-(n-1)) \longrightarrow \lambda n,$$
$$\alpha_n = Np + (1-2p)n \longrightarrow \lambda + n.$$

Then, the limit parameters λn and $\lambda + n$ of Eqs.(17)(18), respectively, are the Jacobi-Szegö parameters associated with the Charlier polynomials, which is the orthogonal polynomials with respect to the Poisson measure of a parameter λ (See Table in Section 3). Correspondingly, the oscillator algebra discussed in Corollary A.2 can be derived as a contraction of $\mathfrak{su}(2)$ in Corollary A.4 as $p \to 0$, $N \to \infty$ under $Np = \lambda > 0$. This is a counterpart of the Poisson's law of small numbers in classical probability theory.

In the next page, we give a table for our convenience to make a comparison among various features related to Gaussian, Poisson, Gamma, Pascal (Negative binomial), Meixner distributions and corresponding $P_n, \omega_n, \alpha_n, E_\lambda(x,z)$ and others.

μ	Gauss	Poisson	Gamma	Pascal	Meixner
I	$\mathbb{R}$	$\mathbb{N}_0$	$\mathbb{R}_+$	$\mathbb{N}_0$	$\mathbb{R}$
μ on I	$\frac{1}{\sqrt{2\pi\sigma^2}}\exp\left(-\frac{(x-m)^2}{2\sigma^2}\right)$	$\frac{\lambda^k}{k!}e^{-\lambda}$	$\frac{1}{\Gamma(k)}x^{k-1}e^{-x}$	$p^r\frac{\Gamma(x+r)}{\Gamma(r)\Gamma(x+1)}q^x$	$\frac{(2\sin\phi)^{2a}}{2\pi\Gamma(2a)}e^{(2\phi-\pi)x}\lvert\Gamma(a+ix)\rvert^2$
Parameters	$\sigma>0,\,m$	$\lambda>0$	$k>0$	$r>0, p+q=1, 0<p,q<1$	$a>0, 0<\phi<\pi$
P_n	Hermite	Charlier	Laguerre	Meixner	Meixner-Pollaczek
ω_n	$\sigma^2 n$	λn	$n(k+n-1)$	$\frac{q}{p^2}n(r+n-1)$	$\frac{n(2a+n-1)}{(2\sin\phi)^2}$
$(c_{0,1}, c_{1,2})$	$(\sigma^2, 0)$	$(\lambda, 0)$	$(k, 1)$	$\left(r\dfrac{q}{p^2}, \dfrac{q}{p^2}\right)$	$\left(\dfrac{2a}{(2\sin\phi)^2}, \dfrac{(2\sin\phi)^2}{2a}\right)$
$c_{0,1}/c_{1,2}$	—	—	k	r	1
λ_n	$\sigma^{2n} n!$	$\lambda^n n!$	$\frac{n!\Gamma(n+k)}{\Gamma(k)}$	$\frac{q^n n!\Gamma(n+r)}{p^{2n}\Gamma(r)}$	$\frac{n!\Gamma(n+2a)}{(2\sin\phi)^{2n}\Gamma(2a)}$
α_n	m	$\lambda+n$	$k+2n$	$\frac{1}{p}(qr+(1+q)n)$	$\dfrac{a+n}{\tan\phi}$
$(c_{0,0}, c_{1,1})$	$(m, 0)$	$(\lambda, 1)$	$(k, 2)$	$\left(r\dfrac{q}{p}, \dfrac{1+q}{p}\right)$	$\left(\dfrac{a}{\tan\phi}, \dfrac{1}{\tan\phi}\right)$
$c_{1,1}^2 - 4c_{1,2}$	0	1	0	1	-1
$E_\lambda(x,z)$	$\exp\left(\frac{z}{\sigma^2}(x-m)-\frac{z^2}{2\sigma^2}\right)$	$e^{-z}\left(1+\frac{z}{\lambda}\right)^x$	$(1+z)^{-k}e^{\frac{z}{1+z}x}$	$(1+z)^x(1+qz)^{-(x+r)}$	See *
$\tilde{\mu}$ on $\mathbb{C}$	Gauss	Gauss	Bessel	Bessel	Bessel

Note: * See formulas in Theorem 2 (iv) in Ref. 18 with replacing a, β, b, γ by $c_{0,1}, c_{1,2}, c_{0,0}, c_{1,1}$ for μ being the Meixner case, respectively. The Meixner distribution with $a=1/2$, $\phi=\pi/2$ is given by Eq.(2.2). To calculate P_n, ω_n, α_n and $E_\lambda(x;z)$, we can apply the multiplicative renormalization method proposed in Ref.[7–9]

Appendix A.

4. Connections with Lie algebras.

Let us examine operators B^-, B^+, B° from the algebraic point of view.[3,12,21] Suppose that $\beta = c_{0,1}I + 2c_{1,2}b^*b$. By direct computations, one can derive the following results.

Theorem A.1. *(1) B^-, B^+, B°, β generate a Lie algebra $\mathfrak{g}_1$ satisfying the relations,*

$$[B^-, B^+] = \beta,$$
$$[\beta, B^-] = -2c_{1,2}B^-, \quad [\beta, B^+] = 2c_{1,2}B^+,$$
$$[B^\circ, B^-] = -c_{1,1}B^-, \quad [B^\circ, B^+] = c_{1,1}B^+.$$

(2) B^-, B^+, β generate a Lie subalgebra $\mathfrak{g}_2$ of $\mathfrak{g}_1$ satisfying the following commutation relations,

$$[B^-, B^+] = \beta,$$
$$[\beta, B^-] = -2c_{1,2}B^-, \quad [\beta, B^+] = 2c_{1,2}B^+.$$

(3)

$$\mathfrak{g}_1 = \mathfrak{h} \oplus \mathfrak{g}_2,$$

where $\mathfrak{h}$ is the center of $\mathfrak{g}_1$.

Remark 4.1. One of referees brings an early paper by P. Foinsilver[12] to our attention. $B^- = L, B^+ = R, B^\circ = N, \beta = \rho, b = V, b^* = R$ and $c_{0,0} = a_0, c_{0,1} = c, c_{1,1} = a, c_{1,2} = b$ are the correspondences between our and his notations, respectively. Our derivation of these operators is slightly different from him. Moreover, our original motivation is to realize different kinds of classical random variables by $B^-, B^+, B^\circ, b, b^*$ acting on the same Hilbert space of analytic L^2-functions with respect to a probability measure on $\mathbb{C}$. For this purpose, we have clarified in Ref.[4,6] that S_μ defined in Section 2 plays important roles. See Section 2 and our papers[4,6] for details.

By Theorem A.1, we get related Lie algebras with $h_{c_{0,1}}$ as follows. See also Ref.[12]

Corollary A.1. *Suppose that $c_{1,1} = c_{1,2} = 0$. Then $\mathfrak{g}_1 = \mathfrak{g}_2$ holds and is isomorphic to the Heisenberg-Weyl algebra. In fact, we have*

$$[B^-, B^+] = c_{0,1}I = \beta,$$
$$[\beta, B^-] = 0, \qquad [\beta, B^+] = 0,$$
$$[B^\circ, B^-] = 0, \qquad [B^\circ, B^+] = 0.$$

Corollary A.2. *Suppose that $c_{1,2} = 0$, but $c_{1,1} \neq 0$. Then $\mathfrak{g}_1$ is isomorphic to the oscillator algebra and $\mathfrak{g}_2$ is isomorphic to the Heisenberg-Weyl algebra. In fact, we have*

$$[B^-, B^+] = c_{0,1} I = \beta,$$
$$[\beta, B^-] = 0, \qquad [\beta, B^*] = 0,$$
$$[A^\circ, B^-] = -B^-, \qquad [A^\circ, B^+] = B^+,$$

where $A^\circ = \dfrac{B^\circ}{c_{1,1}}$.

From Theorem A.1, we get associated Lie algebras with $\gamma_{\alpha, c_{1,2}}$ as follows.

Corollary A.3. *If $c_{1,2} > 0$, then we have*

$$[A^-, A^+] = 2\widetilde{\beta},$$
$$[\widetilde{\beta}, A^-] = -A^-, \; [\widetilde{\beta}, A^+] = A^+,$$

where $A^- = \dfrac{B^-}{\sqrt{c_{1,2}}}$, $A^+ = \dfrac{B^+}{\sqrt{c_{1,2}}}$ and $\widetilde{\beta} = \dfrac{\beta}{2c_{1,2}}$. That is, $\mathfrak{g}_2$ is isomorphic to $\mathfrak{sl}(2, \mathbb{R})$ or $\mathfrak{su}(1,1)$ or $\mathfrak{so}(2,1)$.

By Theorem A.1, we obtain

Corollary A.4. *Assume $c_{1,2} < 0$. Then we have*

$$[A^-, A^+] = -2\widetilde{\beta},$$
$$[\widetilde{\beta}, A^-] = -A^-, \; [\widetilde{\beta}, A^+] = A^+,$$

where $A^- = \dfrac{B^-}{\sqrt{-c_{1,2}}}$, $A^+ = \dfrac{B^+}{\sqrt{-c_{1,2}}}$ and $\widetilde{\beta} = -\dfrac{\beta}{2c_{1,2}}$. That is, $\mathfrak{g}_2$ is isomorphic to $\mathfrak{su}(2)$.

Acknowledgments.

The author wants to give his deepest appreciation to all organizers of "the 28th conference on Quantum Probability and Related Topics" held in CIMAT (Guanajuato, Mexico, September 2–9, 2007). He also thanks the referees for making useful comments and showing relevant papers.[12,16] This is an extensively revised version of the original manuscript.

References

1. L. Accardi and M. Bożejko, *Interacting Fock space and Gaussianization of probability measures.* Infinite Dimensional Analysis, Quantum Probability and Related Topics. **1** (1998) 663–670.
2. N. Asai, *Analytic characterization of one-mode interacting Fock space.* Infinite Dimensional Analysis, Quantum Probability and Related Topics. **4** (2001) 409–415.
3. N. Asai, *Bosonic realization of operators in one-mode interacting Fock space.* manuscript. (2002).
4. N. Asai, *Hilbert space of analytic functions associated with the modified Bessel function and related orthogonal polynomials.* Infinite Dimensional Analysis, Quantum Probability and Related Topics. **8** (2005) 505–514.
5. N. Asai, *Riesz potentials derived by one-mode interacting Fock space approach.* Colloquium Mathematicum. **109** (2007) 101–106.
6. N. Asai, I. Kubo, and H.-H. Kuo, *Segal-Bargmann transforms of one-mode interacting Fock spaces associated with Gaussian and Poisson measures.* Proc. Amer. Math. Soc. **131** (2003), no. 3, 815–823.
7. N. Asai, I. Kubo, and H.-H. Kuo, *Multiplicative renormalization and generating functions I.* Taiwanese J. Math. **7** (2003) 89–101.
8. N. Asai, I. Kubo, and H.-H. Kuo, *Generating functions of orthogonal polynomials and Szegö-Jacobi parameters.* Prob. Math. Stats. **23** (2003) 273–291.
9. N. Asai, I. Kubo, and H.-H. Kuo, *Multiplicative renormalization and generating functions II.* Taiwanese J. Math. **8** (2004) 593–628.
10. V. Bargmann, *On a Hilbert space of analytic functions and an associated integral transform, I.* Comm. Pure Appl. Math. **14** (1961) 187–214.
11. T. S. Chihara, *An Introduction to Orthogonal Polynomials.* (Gordon and Breach, 1978).
12. P. Feinsilver, *Lie algebras and recurrence relations I.* Acta Appl. Math. **13** (1988) 291–333.
13. W. Feller, *An Introduction to Probability Theory and Its Applications, Vol.2,* 2nd ed. (Wiley, 1971)
14. L. Gross and P. Malliavin, *Hall's transform and the Segal-Bargmann map.* in: Itô Stochastic Calculus and Probability Theory, N. Ikeda et al. (eds.) (Springer, 1996) pp. 73–116.
15. T. Hida, *Brownian Motion.* (Springer, 1980).
16. M. E. H. Ismail and C. P. May, *On a family of approximation operators.* J. Math. Anal. Appl. **63** (1978) 446–462.
17. R. Koekoek and R. F. Swarttouw, *The Askey-scheme of hypergeometric orthogonal polynomials and its q-analogue.* Delft Univ. of Tech., Report No. 98-17 (1998) http://aw.twi.tudelft.nl/ koekoek/askey/index.html
18. I. Kubo, *Generating functions of exponential type for orthogonal polynomials.* Infinite Dimensional Analysis, Quantum Probability and Related Topics. **7** (2004) 155–159.
19. J. Meixner, *Orthgonale Polynomsysteme mit einem besonderen Gestalt der erzeugenden Funktion.* J. London Math. Soc. **9** (1934) 6–13.
20. C. N. Morris, *Natural exponential families with quadratic variance functions.*

Ann. Stats. **10** (1982) 68–80.
21. I. Ojima, *private communication.* (2002).
22. K. Sato, *Lévy Processes and Infinitely Divisible Distributions.* (Cambridge, 1999).
23. I. E. Segal, *The complex wave representation of the free Boson field.* in: Essays Dedicated to M. G. Krein on the Occassion of His 70th Birthday, Advances in Math.: Supplementary Studies Vol.3, I. Goldberg and M. Kac (eds.) (Academic, 1978) pp. 321–344.
24. J. Shohat and J. Tamarkin, *The Problem of Moments.* Math Surveys, **1**, (AMS, 1943).
25. M. Szegö, *Orthogonal Polynomials.* Coll. Publ. **23** (AMS, 1975).

QUANTUM CONTINUOUS MEASUREMENTS: THE SPECTRUM OF THE OUTPUT

A. BARCHIELLI and M. GREGORATTI

Department of Mathematics, Politecnico di Milano,
Piazza Leonardo da Vinci, I-20133 Milano, Italy.
Also: Istituto Nazionale di Fisica Nucleare, Sezione di Milano.
Alberto.Barchielli@polimi.it   Matteo.Gregoratti@polimi.it
www.mate.polimi/QP

When a quantum system is monitored in continuous time, the result of the measurement is a stochastic process. When the output process is stationary, at least in the long run, the spectrum of the process can be introduced and its properties studied. A typical continuous measurement for quantum optical systems is the so called homodyne detection. In this paper we show how the Heisenberg uncertainty relations give rise to characteristic bounds on the possible homodyne spectra and we discuss how this is related to the typical quantum phenomenon of squeezing.

Keywords: Quantum continuous measurements; Homodyne detection; Autocorrelation function; Spectrum; Squeezing.

1. Quantum continuous measurements

A big achievement in the 70's-80's was to show that, inside the axiomatic formulation of quantum mechanics, based on *positive operator valued measures* and *instruments*,[1,2] a consistent formulation of the theory of measurements continuous in time (*quantum continuous measurements*) was possible.[2-8] The main applications of quantum continuous measurements are in the photon detection theory in quantum optics (*direct, heterodyne, homodyne detection*).[9-15] A very flexible and powerful formulation of continuous measurement theory was based on stochastic differential equations, of classical type (commuting noises, Itô calculus) and of quantum type (non commuting noises, Hudson-Parthasarathy equation).[5-16]

In this paper we start by giving a short presentation of continuous measurement theory based on quantum SDE's. We consider only the type of observables relevant for the description of homodyne detection and we make the mathematical simplification of introducing only bounded operators on the Hilbert space of the quantum system and a finite number of noises. Then, we introduce the spectrum of the classical stochastic process which represents the output and we study the general properties of the spectra of such classical processes by proving characteristic bounds due to the Heisenberg uncertainty principle. Finally, we present the case of a two-level atom, where the spectral analysis of the output can reveal the phenomenon of squeezing of the fluorescence light, a phenomenon related to the Heisenberg uncertainty relations.

1.1. *Hudson Parthasarathy equation*

Let $\mathcal{H}$ be the *system space*, the complex separable Hilbert space associated to the observed quantum system, which we call system S. Quantum stochastic calculus and the Hudson-Parthasarathy equation[17] allow to represent the continuous measurement process as an interaction of system S with some quantum fields combined with an observation in continuous time of these fields. Let us start by introducing such fields. We denote by Γ the Hilbert space associated with d boson fields, that is the symmetric *Fock space* over the "one–particle space" $L^2(\mathbb{R}_+) \otimes \mathbb{C}^d \simeq L^2(\mathbb{R}_+; \mathbb{C}^d)$, and we denote by $e(f)$, $f \in L^2(\mathbb{R}_+; \mathbb{C}^d)$, the *coherent vectors*, whose components in the $0, 1, \ldots, n, \ldots$ particle spaces are $e(f) := \exp\left(-\frac{1}{2}\|f\|^2\right)\left(1, f, (2!)^{-1/2} f \otimes f, \ldots, (n!)^{-1/2} f^{\otimes n}, \ldots\right)$.

Let $\{z_k, \ k \geq 1\}$ be the canonical basis in $\mathbb{C}^d$ and for any $f \in L^2(\mathbb{R}_+; \mathbb{C}^d)$ let us set $f_k(t) := \langle z_k | f(t) \rangle$. We denote by $A_k(t)$, $A_k^\dagger(t)$, $\Lambda_{kl}(t)$ the *annihilation, creation and conservation processes*:

$$A_k(t) \, e(f) = \int_0^t f_k(s) \, \mathrm{d}s \, e(f),$$

$$\langle e(g) | A_k^\dagger(t) e(f) \rangle = \int_0^t \overline{g_k(s)} \, \mathrm{d}s \, \langle e(g) | e(f) \rangle,$$

$$\langle e(g) | \Lambda_{kl}(t) e(f) \rangle = \int_0^t \overline{g_k(s)} \, f_l(s) \, \mathrm{d}s \, \langle e(g) | e(f) \rangle.$$

The annihilation and creation processes satisfy the *canonical commutation rules* (CCR); formally, $[A_k(t), A_l^\dagger(s)] = t \wedge s$, $[A_k(t), A_l(s)] = 0$, $[A_k^\dagger(t), A_l^\dagger(s)] = 0$.

Let H, R_k, S_{kl}, $k, l = 1, \ldots, d$, be bounded operators on $\mathcal{H}$ such that $H^* = H$ and $\sum_j S_{jk}^* S_{jl} = \sum_j S_{kj} S_{lj}^* = \delta_{kl}$. We set also $K := -\mathrm{i}H -$

$\frac{1}{2}\sum_k R_k^* R_k$. Then, the quantum stochastic differential equation[17]

$$dU(t) = \left\{ \sum_k R_k\, dA_k^\dagger(t) + \sum_{kl}(S_{kl} - \delta_{kl})\,d\Lambda_{kl}(t) \right.$$
$$\left. - \sum_{kl} R_k^* S_{kl}\,dA_l(t) + K\,dt \right\} U(t), \quad (1)$$

with the initial condition $U(0) = \mathbb{1}$, has a unique solution, which is a strongly continuous family of unitary operators on $\mathcal{H}\otimes\Gamma$, representing the system-field dynamics in the interaction picture with respect to the free field evolution.[18]

1.2. *The reduced dynamics of the system*

The states of a quantum system are represented by statistical operators, positive trace-class operators with trace one; let us denote by $\mathcal{S}(\mathcal{H})$ the set of statistical operators on $\mathcal{H}$. For every composed state σ in $\mathcal{S}(\mathcal{H}\otimes\Gamma)$, the partial trace Tr_Γ (resp. $\mathrm{Tr}_\mathcal{H}$) with respect to the field (resp. system) Hilbert space gives the reduced system (resp. field) state $\mathrm{Tr}_\Gamma\,\sigma$ in $\mathcal{S}(\mathcal{H})$ (resp. $\mathrm{Tr}_\mathcal{H}\,\sigma$ in $\mathcal{S}(\Gamma)$). As initial state of the composed system "system S plus fields" we take $\rho\otimes\varrho_\Gamma(f)\in\mathcal{S}(\mathcal{H}\otimes\Gamma)$, where $\rho\in\mathcal{S}(\mathcal{H})$ is generic and $\varrho_\Gamma(f)$ is a coherent state, $\varrho_\Gamma(f):=|e(f)\rangle\langle e(f)|$. One of the main properties of the Hudson-Parthasarathy equation is that, with such an initial state, the reduced dynamics of system S obeys a quantum master equation.[16,17] Indeed, we get

$$\frac{d}{dt}\eta_t = \mathcal{L}(t)[\eta_t], \qquad \eta_t := \mathrm{Tr}_\Gamma\left\{U(t)\bigl(\rho\otimes\varrho_\Gamma(f)\bigr)U(t)^*\right\}, \quad (2)$$

where the Liouville operator $\mathcal{L}(t)$ turns out to be given by

$$\mathcal{L}(t)[\rho] = \left(K - \sum_{kl} R_k^* S_{kl} f_l(t)\right)\rho + \rho\left(K^* - \sum_{kj}\overline{f_j(t)} S_{kj}^* R_k\right)$$
$$+ \sum_k \left(R_k - \sum_l S_{kl} f_l(t)\right)\rho\left(R_k^* - \sum_l S_{kl}^* \overline{f_l(t)}\right) - \|f(t)\|^2\rho. \quad (3)$$

A particularly important case is $S_{kl} = \delta_{kl}$, when $\mathcal{L}(t)$ reduces to

$$\mathcal{L}(t)[\rho] = -\mathrm{i}\left[H - \mathrm{i}\sum_k f_k(t)R_k^* + \mathrm{i}\sum_k \overline{f_k(t)}R_k,\, \rho\right]$$
$$+ \sum_k \left(R_k \rho R_k^* - \frac{1}{2}R_k^* R_k \rho - \frac{1}{2}\rho R_k^* R_k\right). \quad (4)$$

It is useful to introduce also the evolution operator from s to t by

$$\frac{\mathrm{d}}{\mathrm{d}t}\Upsilon(t,s) = \mathcal{L}(t) \circ \Upsilon(t,s), \qquad \Upsilon(s,s) = \mathbf{1}. \quad (5)$$

With this notation we have $\eta_t = \Upsilon(t,0)[\rho]$.

1.3. *The field observables*

The key point of the theory of continuous measurements is to consider field observables represented by time dependent, commuting families of selfadjoint operators in the Heisenberg picture.[16] Being commuting at different times, these observables represent outputs produced at different times which can be obtained in the same experiment. Here we present a very special case of family of observables, a field quadrature. Let us start by introducing the operators

$$Q(t;\vartheta,\nu) = \int_0^t e^{-\mathrm{i}(\nu s+\vartheta)}\mathrm{d}A_1^\dagger(s) + \int_0^t e^{\mathrm{i}(\nu s+\vartheta)}\mathrm{d}A_1(s), \qquad t \geq 0; \quad (6)$$

$\vartheta \in (-\pi,\pi]$ and $\nu > 0$ are fixed. The operators $Q(t;\vartheta,\nu)$ are selfadjoint (they are essentially selfadjoint on the linear span of the exponential vectors). By using CCR's, one can check that they commute: $[Q(t;\vartheta,\nu), Q(s;\vartheta,\nu)] = 0$ (better: the unitary groups generated by $Q(t;\vartheta,\nu)$ and $Q(s;\vartheta,\nu)$ commute). The operators (6) have to be interpreted as linear combinations of the formal increments $\mathrm{d}A_1^\dagger(s)$, $\mathrm{d}A_1(s)$ which represent field operators evolving with the free-field dynamics; therefore, they have to be intended as operators in the interaction picture. The important point is that these operators commute for different times also in the Heisenberg picture, because

$$Q^{\mathrm{out}}(t;\vartheta,\nu) := U(t)^* Q(t;\vartheta,\nu)U(t) = U(T)^* Q(t;\vartheta,\nu)U(T), \quad \forall T \geq t; \quad (7)$$

this is due to the factorization properties of the Fock space and to the properties of the solution of the Hudson-Parthasarathy equation. These "output" quadratures are our observables. They regard those bosons in "field 1" which eventually have interacted with S between time 0 and time

t. Commuting selfadjoint operators can be jointly diagonalized and the usual postulates of quantum mechanics give the probabilities for the joint measurement of the observables represented by the selfadjoint operators $Q^{\text{out}}(t; \vartheta, \nu)$, $t \geq 0$. Let us stress that operators of type (6) with different angles and frequencies represent incompatible observables, because they do not commute but satisfy

$$[Q(t;\theta,\nu), Q(s;\phi,\mu)] = \frac{4\mathrm{i}\sin\left(\frac{t\wedge s}{2}(\nu-\mu)\right)\sin\left(\theta-\phi+\frac{t\wedge s}{2}(\nu-\mu)\right)}{\nu-\mu}.$$

When "field 1" represents the electromagnetic field, a physical realization of a measurement of the observables (7) is implemented by what is called an heterodyne/homodyne scheme. The light emitted by the system in the "channel 1" interferes with an intense coherent monochromatic laser beam of frequency ν. The mathematical description of the apparatus is given in Section 3.5 of Ref. 16.

1.4. *Characteristic operator, probabilities and moments*

The commuting selfadjoint operators (6) have a joint pvm (projection valued measure) which gives the distribution of probability for the measurement. Anyway, at least the finite-dimensional distributions of the output can be obtained via an explicit and easier object, the *characteristic operator* $\widehat{\Phi}_t(k; \vartheta, \nu)$, a kind of Fourier transform of this pvm. For any real test function $k \in L^\infty(\mathbb{R}_+)$ and any time $t > 0$ we define the unitary Weyl operator

$$\widehat{\Phi}_t(k;\vartheta,\nu) = \exp\left\{\mathrm{i}\int_0^t k(s)\,\mathrm{d}Q(s;\vartheta,\nu)\right\}$$

$$= \exp\left\{\mathrm{i}\int_0^t k(s)\mathrm{e}^{-\mathrm{i}(\nu s+\vartheta)}\mathrm{d}A_1^\dagger(s) + \mathrm{i}\int_0^t k(s)\mathrm{e}^{\mathrm{i}(\nu s+\vartheta)}\mathrm{d}A_1(s)\right\}. \quad (8)$$

Then, there exists a measurable space $(\Omega, \mathcal{F})$, a pvm ξ_ϑ^ν (acting on Γ) with value space $(\Omega, \mathcal{F})$, a family of real valued measurable functions $\{X(t;\cdot),\ t\geq 0\}$ on Ω, such that $X(0;\omega) = 0$, and, for any choice of n, $0 = t_0 < t_1 < \cdots < t_n \leq t$, $\kappa_j \in \mathbb{R}$,

$$\widehat{\Phi}_t(k;\vartheta,\nu) = \exp\left\{\mathrm{i}\sum_{j=1}^n \kappa_j\left[Q(t_j;\vartheta,\nu) - Q(t_{j-1};\vartheta,\nu)\right]\right\}$$

$$= \int_\Omega \exp\left\{\mathrm{i}\sum_{j=1}^n \kappa_j\left[X(t_j;\omega) - X(t_{j-1};\omega)\right]\right\}\xi_\vartheta^\nu(\mathrm{d}\omega), \quad (9)$$

where $k(s) = \sum_{j=1}^{n} 1_{(t_{j-1},t_j)}(s) \kappa_j$. Let us stress that the pvm depends on the observables and, so, on the parameters ϑ and ν, while the choices of the trajectory space (the measurable space $(\Omega, \mathcal{F})$) and of the process $X(t)$ are independent of ϑ and ν.

Then, we introduce the characteristic functional

$$\Phi_t(k; \vartheta, \nu) = \operatorname{Tr} \left\{ \exp\left\{ i \int_0^t k(s) \, dQ^{\text{out}}(s; \vartheta, \nu) \right\} \rho \otimes \varrho_\Gamma(f) \right\}$$
$$= \operatorname{Tr} \left\{ \widehat{\Phi}_t(k; \vartheta, \nu) U(t) \left(\rho \otimes \eta(f) \right) U(t)^* \right\}. \quad (10)$$

All the finite-dimensional probabilities of the increments of the process $X(t)$ are determined by

$$\Phi_t(k; \vartheta, \nu) = \int_{\mathbb{R}^n} \left(\prod_{j=1}^{n} e^{i\kappa_j \cdot x_j} \right)$$
$$\times \mathbb{P}_\rho^{\vartheta, \nu} \left[\Delta X(t_0, t_1) \in dx_1, \ldots, \Delta X(t_{n-1}, t_n) \in dx_n \right], \quad (11)$$

where we have introduced the test function $k(s) = \sum_{j=1}^{n} 1_{(t_{j-1},t_j)}(s) \kappa_j$, with $0 = t_0 < t_1 < \cdots < t_n \le t$, $\kappa_j \in \mathbb{R}$.

The fact that the theory gives in a simple direct way the distribution for the increments of the process $X(t)$, rather than its finite-dimensional distributions, is related also to the interpretation: the output $X(t)$ actually is obtained by a continuous observation of the generalized process $I(t) = dX(t)/dt$ followed by post-measurement processing.

Starting from the characteristic functional it is possible to obtain the moments of the output process $I(t)$ and to express them by means of quantities concerning only system S.[16] Let us denote by $\mathbb{E}_\rho^{\vartheta, \nu}$ the expectation with respect to $\mathbb{P}_\rho^{\vartheta, \nu}$; for the first two moments we obtain the expressions

$$\mathbb{E}_\rho^{\vartheta, \nu}[I(t)] = \operatorname{Tr}\left\{ (Z(t) + Z(t)^*) \eta_t \right\}, \quad (12a)$$

$$\mathbb{E}_\rho^{\vartheta, \nu}[I(t)I(s)] = \delta(t - s)$$
$$+ \operatorname{Tr}\left\{ (Z(t_2) + Z(t_2)^*) \Upsilon(t_2, t_1) [Z(t_1)\eta_{t_1} + \eta_{t_1} Z(t_1)^*] \right\}, \quad (12b)$$

where $t_1 = t \wedge s$, $t_2 = t \vee s$ and

$$Z(t) := e^{i(\nu t + \vartheta)} \left(R_1 + \sum_k S_{1k} f_k(t) \right). \quad (12c)$$

2. The spectrum of the output

2.1. *The spectrum of a stationary process*

In the classical theory of stochastic processes, the spectrum is related to the Fourier transform of the autocorrelation function. Let Y be a stationary real stochastic process with finite moments; then, the mean is independent of time $\mathbb{E}[Y(t)] = \mathbb{E}[Y(0)] =: m_Y$, $\forall t \in \mathbb{R}$, and the second moment is invariant under time translations

$$\mathbb{E}[Y(t)Y(s)] = \mathbb{E}[Y(t-s)Y(0)] =: R_Y(t-s), \qquad \forall t, s \in \mathbb{R}. \tag{13}$$

The function $R_Y(\tau)$, $\tau \in \mathbb{R}$, is called the *autocorrelation function* of the process. Obviously, we have $\operatorname{Cov}[Y(t), Y(s)] = R_Y(t-s) - m_Y^2$.

The *spectrum* of the stationary stochastic process Y is the Fourier transform of its autocorrelation function:

$$S_Y(\mu) := \int_{-\infty}^{+\infty} e^{i\mu\tau} R_Y(\tau) \, d\tau. \tag{14}$$

This formula has to be intended in the sense of distributions. If $\operatorname{Cov}[Y(\tau), Y(0)] \in L^1(\mathbb{R})$, we can write

$$S_Y(\mu) := 2\pi m_Y^2 \delta(\mu) + \int_{-\infty}^{+\infty} e^{i\mu\tau} \operatorname{Cov}[Y(\tau), Y(0)] \, d\tau. \tag{15}$$

By the properties of the covariance, the function $\operatorname{Cov}[Y(\tau), Y(0)]$ is positive definite and, by the properties of positive definite functions, this implies $\int_{-\infty}^{+\infty} e^{i\mu\tau} \operatorname{Cov}[Y(\tau), Y(0)] \, d\tau \geq 0$; then, also $S_Y(\mu) \geq 0$.

By using the stationarity and some tricks on multiple integrals, one can check that an alternative expression of the spectrum is

$$S_Y(\mu) = \lim_{T \to +\infty} \frac{1}{T} \mathbb{E}\left[\left|\int_0^T e^{i\mu t} Y(t) \, dt\right|^2\right]. \tag{16}$$

The advantage now is that positivity appears explicitly and only positive times are involved. Expression (16) can be generalized also to processes which are stationary only in some asymptotic sense and to singular processes as our $I(t)$.

2.2. The spectrum of the output in a finite time horizon

Let us consider our output $I(t) = \mathrm{d}X(t)/\mathrm{d}t$ under the physical probability $\mathbb{P}_\rho^{\vartheta,\nu}$. We call "spectrum up to time T" of $I(t)$ the quantity

$$S_T(\mu;\vartheta,\nu) = \frac{1}{T}\mathbb{E}_\rho^{\vartheta,\nu}\left[\left|\int_0^T e^{\mathrm{i}\mu t}\,\mathrm{d}X(t)\right|^2\right]. \tag{17}$$

When the limit $T \to +\infty$ exists, we can speak of *spectrum of the output*, but this existence depends on the specific properties of the concrete model.

By writing the second moment defining the spectrum as the square of the mean plus the variance, the spectrum splits in an elastic or coherent part and in an inelastic or incoherent one:

$$S_T(\mu;\vartheta,\nu) = S_T^{\mathrm{el}}(\mu;\vartheta,\nu) + S_T^{\mathrm{inel}}(\mu;\vartheta,\nu), \tag{18}$$

$$S_T^{\mathrm{el}}(\mu;\vartheta,\nu) = \frac{1}{T}\left|\mathbb{E}_\rho^{\vartheta,\nu}\left[\int_0^T e^{\mathrm{i}\mu t}\,\mathrm{d}X(t)\right]\right|^2, \tag{19}$$

$$S_T^{\mathrm{inel}}(\mu;\vartheta,\nu) = \frac{1}{T}\mathrm{Var}_\rho^{\vartheta,\nu}\left[\int_0^T \cos\mu t\,\mathrm{d}X(t)\right]$$
$$+ \frac{1}{T}\mathrm{Var}_\rho^{\vartheta,\nu}\left[\int_0^T \sin\mu t\,\mathrm{d}X(t)\right]. \tag{20}$$

Let us note that

$$S_T^{\mathrm{el}}(\mu;\vartheta,\nu) = S_T^{\mathrm{el}}(-\mu;\vartheta,\nu), \qquad S_T^{\mathrm{inel}}(\mu;\vartheta,\nu) = S_T^{\mathrm{inel}}(-\mu;\vartheta,\nu). \tag{21}$$

By using the expressions (12) for the first two moments we get the spectrum in a form which involves only system operators:

$$S_T^{\mathrm{el}}(\mu;\vartheta,\nu) = \frac{1}{T}\left|\int_0^T e^{\mathrm{i}\mu t}\,\mathrm{Tr}\left\{(Z(t)+Z(t)^*)\eta_t\right\}\mathrm{d}t\right|^2, \tag{22a}$$

$$S_T^{\mathrm{inel}}(\mu;\vartheta,\nu) = 1 + \frac{2}{T}\int_0^T \mathrm{d}t\int_0^t \mathrm{d}s\,\cos\mu(t-s)$$
$$\times \mathrm{Tr}\left\{\left(\tilde{Z}(t)+\tilde{Z}(t)^*\right)\Upsilon(t,s)\left[\tilde{Z}(s)\eta_s + \eta_s\tilde{Z}(s)^*\right]\right\}, \tag{22b}$$

$$\tilde{Z}(t) = Z(t) - \mathrm{Tr}\left\{Z(t)\eta_t\right\}. \tag{22c}$$

2.3. Properties of the spectrum and the Heisenberg uncertainty relations

Equations (22) give the spectrum in terms of the reduced description of system S (the fields are traced out); this is useful for concrete computations. But the general properties of the spectrum are more easily obtained by working with the fields; so, here we trace out first system S. Let us define the reduced field state

$$\Pi_T(f) := \text{Tr}_{\mathcal{H}}\{U(T)(\rho \otimes \varrho_\Gamma(f))U(T)^*\} \tag{23}$$

and the field operators

$$Q_T(\mu;\vartheta,\nu) = \frac{1}{\sqrt{T}}\int_0^T e^{i\mu t}\,\mathrm{d}Q(t;\vartheta,\nu), \tag{24a}$$

$$\tilde{Q}_T(\mu;\vartheta,\nu) = Q_T(\mu;\vartheta,\nu) - \text{Tr}\{\Pi_T(f)Q_T(\mu;\vartheta,\nu)\}. \tag{24b}$$

Let us stress that $Q_T(\mu;\vartheta,\nu)$ commutes with its adjoint and that $Q_T(\mu;\vartheta,\nu)^* = Q_T(-\mu;\vartheta,\nu)$. By using Eqs. (10) and (11) and taking first the trace over $\mathcal{H}$, we get

$$S_T(\mu;\vartheta,\nu) = \text{Tr}\{\Pi_T(f)Q_T(\mu;\vartheta,\nu)^*Q_T(\mu;\vartheta,\nu)\} \geq 0, \tag{25a}$$

$$S_T^{\text{el}}(\mu;\vartheta,\nu) = |\text{Tr}\{\Pi_T(f)Q_T(\mu;\vartheta,\nu)\}|^2 \geq 0, \tag{25b}$$

$$S_T^{\text{inel}}(\mu;\vartheta,\nu) = \text{Tr}\{\Pi_T(f)\tilde{Q}_T(\mu;\vartheta,\nu)^*\tilde{Q}_T(\mu;\vartheta,\nu)\} \geq 0. \tag{25c}$$

To elaborate the previous expressions it is useful to introduce annihilation and creation operators for bosonic modes, which are only approximately orthogonal for finite T:

$$a_T(\omega) := \frac{1}{\sqrt{T}}\int_0^T e^{i\omega t}\,\mathrm{d}A_1(t) = \frac{e^{\frac{1}{2}\omega T}}{\sqrt{T}}\int_{-\frac{T}{2}}^{\frac{T}{2}} e^{i\omega t}\,\mathrm{d}A_1(t+T/2), \tag{26a}$$

$$[a_T(\omega), a_T(\omega')] = [a_T^\dagger(\omega), a_T^\dagger(\omega')] = 0, \tag{26b}$$

$$[a_T(\omega), a_T^\dagger(\omega')] = \begin{cases} 1 & \text{for } \omega' = \omega, \\ \frac{e^{i(\omega-\omega')T}-1}{i(\omega-\omega')T} & \text{for } \omega' \neq \omega. \end{cases} \tag{26c}$$

Then, we have easily

$$Q_T(\mu;\vartheta,\nu) = e^{i\vartheta}a_T(\nu+\mu) + e^{-i\vartheta}a_T^\dagger(\nu-\mu), \tag{27}$$

$$S_T(\mu;\vartheta,\nu) = 1 + \text{Tr}\Big\{\Pi_T(f)\Big(a_T^\dagger(\nu+\mu)a_T(\nu+\mu) + a_T^\dagger(\nu-\mu)a_T(\nu-\mu) \\ + e^{-2i\vartheta}a_T^\dagger(\nu+\mu)a_T^\dagger(\nu-\mu) + e^{2i\vartheta}a_T(\nu-\mu)a_T(\nu+\mu)\Big)\Big\}, \tag{28a}$$

$$S_T^{\text{el}}(\mu;\vartheta,\nu) = \left| e^{i\vartheta}\operatorname{Tr}\{\Pi_T(f)a_T(\nu+\mu)\} + e^{-i\vartheta}\operatorname{Tr}\{\Pi_T(f)a_T^\dagger(\nu-\mu)\}\right|^2. \tag{28b}$$

Theorem 2.1. *Independently of the system state ρ, of the field state $\varrho_\Gamma(f)$ and of the Hudson-Parthasarathy evolution U, for every ϑ and ν we have the two bounds*

$$\frac{1}{2}\left(S_T^{\text{inel}}(\mu;\vartheta,\nu) + S_T^{\text{inel}}(\mu;\vartheta\pm\tfrac{\pi}{2},\nu)\right) \geq 1, \tag{29}$$

$$S_T^{\text{inel}}(\mu;\vartheta,\nu) S_T^{\text{inel}}(\mu;\vartheta\pm\tfrac{\pi}{2},\nu) \geq 1. \tag{30}$$

Proof. The first bound comes easily from $S_T^{\text{inel}}(\mu;\vartheta,\nu) = S_T(\mu;\vartheta,\nu) - S_T^{\text{el}}(\mu;\vartheta,\nu)$ and Eqs. (28).

To prove the second bound, let us introduce the operator

$$B_T(\omega) := a_T(\nu-\omega) - \operatorname{Tr}\{\Pi_T(f)a_T(\nu-\omega)\}, \tag{31a}$$

which satisfy the CCR

$$[B_T(\omega), B_T^\dagger(\omega)] = 1, \qquad [B_T(\omega), B_T(\omega)] = [B_T^\dagger(\omega), B_T^\dagger(\omega)] = 0. \tag{31b}$$

Then, we can write

$$S_T^{\text{inel}}(\mu;\vartheta,\nu) = \operatorname{Tr}\Big\{\left(e^{-i\vartheta}B_T^\dagger(-\mu) + e^{i\vartheta}B_T(\mu)\right)\Pi_T(f) \\ \times \left(e^{-i\vartheta}B_T^\dagger(\mu) + e^{i\vartheta}B_T(-\mu)\right)\Big\}.$$

The usual tricks to derive the Heisenberg-Scrödinger-Robertson uncertainty relations can be generalized also to non-selfadjoint operators.[3,19] For any choice of the state ϱ and of the operators X_1, X_2 (with finite second moments with respect to ϱ) the 2×2 matrix with elements $\operatorname{Tr}\{X_i\varrho X_j^*\}$ is positive definite and, in particular, its determinant is not negative. Then, we have

$$\operatorname{Tr}\{X_1\varrho X_1^*\}\operatorname{Tr}\{X_2\varrho X_2^*\} \geq |\operatorname{Tr}\{X_1\varrho X_2^*\}|^2$$
$$\geq |\operatorname{Im}\operatorname{Tr}\{X_1\varrho X_2^*\}|^2 = \frac{1}{4}|\operatorname{Tr}\{\varrho(X_2^*X_1 - X_1^*X_2)\}|^2.$$

By taking $\varrho = \Pi_T(f)$, $X_1 = e^{-i\vartheta}B_T^\dagger(\mu) + e^{i\vartheta}B_T(-\mu)$, $X_2 = \mp i\left(e^{-i\vartheta}B_T^\dagger(\mu) - e^{i\vartheta}B_T(-\mu)\right)$, we get

$$S_T^{\text{inel}}(\mu;\vartheta,\nu) S_T^{\text{inel}}(\mu;\vartheta\pm\tfrac{\pi}{2},\nu) \\ \geq \left|1 + \operatorname{Tr}\{\Pi_T(f)\left(B_T^\dagger(\mu)B_T(\mu) - B_T^\dagger(-\mu)B_T(-\mu)\right)\}\right|^2.$$

But we can change μ in $-\mu$ and we have also

$$S_T^{\text{inel}}(\mu;\vartheta,\nu)S_T^{\text{inel}}(\mu;\vartheta\pm\tfrac{\pi}{2},\nu) = S_T^{\text{inel}}(-\mu;\vartheta,\nu)S_T^{\text{inel}}(-\mu;\vartheta\pm\tfrac{\pi}{2},\nu)$$
$$\geq \left|1+\text{Tr}\left\{\Pi_T(f)\left(B_T^\dagger(-\mu)B_T(-\mu)-B_T^\dagger(\mu)B_T(\mu)\right)\right\}\right|^2.$$

The two inequalities together give

$$S_T^{\text{inel}}(\mu;\vartheta,\nu)S_T^{\text{inel}}(\mu;\vartheta\pm\tfrac{\pi}{2},\nu)$$
$$\geq \left(1+\left|\text{Tr}\left\{\Pi_T(f)\left(B_T^\dagger(\mu)B_T(\mu)-B_T^\dagger(-\mu)B_T(-\mu)\right)\right\}\right|\right)^2 \geq 1, \quad (32)$$

which is what we wanted. □

Ref. 19 introduces a class of operators for the electromagnetic field, called *two-mode quadrature-phase amplitudes*, which have the structure (27) of our operators $Q_T(\mu;\vartheta,\nu)$. Anyway only two modes are involved, as if we fixed μ and ν. Let us denote here those operators by Q_{pa}. The paper explicitly constructs a class of quasi-free (or Gaussian) field states ϱ_{sq} for which $\text{Tr}\{\varrho_{\text{sq}}Q_{\text{pa}}^*Q_{\text{pa}}\} - |\text{Tr}\{\varrho_{\text{sq}}Q_{\text{pa}}\}|^2 < 1$. Such states are called *two-mode squeezed states*.

More generally, one speaks of *squeezed field* if, at least in a region of the μ line, for some ϑ one has $S_T^{\text{inel}}(\mu;\vartheta,\nu) < 1$. If this happens, the Heisenberg-type relation (30) says that necessarily $S_T^{\text{inel}}(\mu;\vartheta+\tfrac{\pi}{2},\nu) > 1$ in such a way that the product is bigger than one.

3. Squeezing of the fluorescence light of a two-level atom

Let us take as system S a two-level atom, which means $\mathcal{H} = \mathbb{C}^2$, $H = \frac{\omega_0}{2}\sigma_z$; $\omega_0 > 0$ is the *resonance frequency* of the atom. We denote by σ_- and σ_+ the lowering and rising operators and by $\sigma_x = \sigma_- + \sigma_+$, $\sigma_y = \text{i}(\sigma_- - \sigma_+)$, $\sigma_z = \sigma_+\sigma_- - \sigma_-\sigma_+$ the Pauli matrices; we set also $\sigma_\vartheta = \text{e}^{\text{i}\vartheta}\sigma_- + \text{e}^{-\text{i}\vartheta}\sigma_+$. We stimulate the atom with a coherent monochromatic laser and consider homodyne detection of the fluorescence light. The quantum fields Γ model the whole environment. The electromagnetic field is split in two fields, according to the direction of propagation: one field for the photons in the forward direction ($k = 2$), that of the stimulating laser and of the lost light, one field for the photons collected to the detector ($k = 1$). Assume that the interaction with the atom is dominated by absorption/emission and that the direct scattering is negligible:

$$S_{kl} = \delta_{kl}, \qquad R_1 = \sqrt{\gamma p}\,\sigma_-\,, \qquad R_2 = \sqrt{\gamma(1-p)}\,\sigma_-\,.$$

The coefficient $\gamma > 0$ is the natural *line-width* of the atom, p is the fraction of fluorescence light which reaches the detector and $1 - p$ is the fraction of lost light ($0 < p < 1$).[10,15,16,20] We introduce also the interaction with a thermal bath,

$$R_3 = \sqrt{\gamma \bar{n}}\, \sigma_-, \qquad R_4 = \sqrt{\gamma \bar{n}}\, \sigma_+, \qquad \bar{n} \geq 0,$$

and a term responsible of *dephasing* (or decoherence),

$$R_5 = \sqrt{\gamma k_d}\, \sigma_z, \qquad k_d \geq 0.$$

To represent a coherent monochromatic laser of frequency $\omega > 0$, we take $f_k(t) = \delta_{k2} \frac{i\Omega}{2\sqrt{\gamma(1-p)}} e^{-i\omega t} 1_{[0,T]}(t)$; T is a time larger than any other time in the theory and the limit $T \to +\infty$ is taken in all the physical quantities. The quantity $\Omega \geq 0$ is called *Rabi frequency* and $\Delta\omega = \omega_0 - \omega$ is called *detuning*. The squeezing in the fluorescence light is revealed by homodyne detection, which needs to maintain phase coherence between the laser stimulating the atom and the laser in the detection apparatus which determines the observables $Q(t; \vartheta, \nu)$; this in particular means that necessarily we must take $\nu = \omega$.

The limit $T \to +\infty$ can be taken in Eqs. (22) and it is independent of the atomic initial state.[20] The result is

$$S^{\text{el}}(\mu; \vartheta) := \lim_{T \to +\infty} S_T^{\text{el}}(\mu; \vartheta, \omega) = 2\pi\gamma p \left|\text{Tr}\left\{\sigma_\vartheta \rho_{\text{eq}}\right\}\right|^2 \delta(\mu), \qquad (33)$$

$$S^{\text{inel}}(\mu; \vartheta) := \lim_{T \to +\infty} S_T^{\text{inel}}(\mu; \vartheta, \omega) = 1 + 2\gamma p \left(\frac{A}{A^2 + \mu^2} \vec{t}\right) \cdot \vec{s}, \qquad (34)$$

where

$$\vec{t} = \text{Tr}\left[\left(e^{i\vartheta}\sigma_- \rho_{\text{eq}} + \rho_{\text{eq}} e^{-i\vartheta}\sigma_+ - \text{Tr}[\sigma_\vartheta \rho_{\text{eq}}] \rho_{\text{eq}}\right) \vec{\sigma}\right], \qquad \vec{s} = \begin{pmatrix} \cos \vartheta \\ \sin \vartheta \\ 0 \end{pmatrix},$$

$$\rho_{\text{eq}} = \frac{1}{2}\left(1 + \vec{x}_{\text{eq}} \cdot \vec{\sigma}\right), \qquad \vec{x}_{\text{eq}} = -\gamma A^{-1} \begin{pmatrix} 0 \\ 0 \\ 1 \end{pmatrix},$$

$$A = \begin{pmatrix} \gamma\left(\frac{1}{2} + \bar{n} + 2k_d\right) & \Delta\omega & 0 \\ -\Delta\omega & \gamma\left(\frac{1}{2} + \bar{n} + 2k_d\right) & \Omega \\ 0 & -\Omega & \gamma(1 + 2\bar{n}) \end{pmatrix}.$$

Examples of inelastic spectra are plotted for $\gamma = 1$, $\bar{n} = k_d = 0$, $p = 4/5$, and two different values of $\Delta\omega$. The Rabi frequency Ω and θ are chosen in both cases to get good visible minima of S^{inel} below 1. Thus in this case the

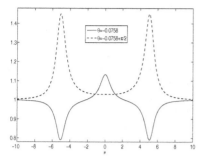
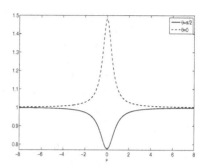

Fig. 1. $S^{\text{inel}}(\mu;\vartheta)$ with $\Delta\omega = 3.5$, $\Omega = 3.7021$.

Fig. 2. $S^{\text{inel}}(\mu;\vartheta)$ with $\Delta\omega = 0$, $\Omega = 0.2976$.

analysis of the homodyne spectrum reveals the squeezing of the detected light. Also complementary spectra are shown to verify Theorem 2.1. One could also compare the homodyne spectrum with and without $\bar{n}$ and k_{d}, thus verifying that the squeezing is very sensitive to any small perturbation.

References

1. K. Kraus, *States, Effects and Operations*, Lecture Notes in Physics **190** (Springer, Berlin, 1980).
2. E. B. Davies, *Quantum Theory of Open Systems* (Academic Press, London, 1976).
3. A. S. Holevo, *Statistical Structure of Quantum Theory*, Lecture Notes in Physics **m 67** (Springer, Berlin, 2001).
4. A. Barchielli, L. Lanz, G. M. Prosperi, *A model for the macroscopic description and continuous observation in quantum mechanics*, Nuovo Cimento **72 B** (1982) 79-121.
5. A. Barchielli, G. Lupieri, *Quantum stochastic calculus, operation valued stochastic processes and continuous measurements in quantum mechanics*, J. Math. Phys. **26** (1985) 2222–2230.
6. A. Barchielli, *Measurement theory and stochastic differential equations in quantum mechanics*, Phys. Rev. A **34** (1986) 1642–1649.
7. V. P. Belavkin, *Nondemolition measurements, nonlinear filtering and dynamic programming of quantum stochastic processes*. In A. Blaquière (ed.), *Modelling and Control of Systems*, Lecture Notes in Control and Information Sciences **121** (Springer, Berlin, 1988) pp. 245–265.
8. A. Barchielli, V. P. Belavkin, *Measurements continuous in time and a posteriori states in quantum mechanics*, J. Phys. A: Math. Gen. **24** (1991) 1495–1514.
9. A. Barchielli, *Direct and heterodyne detection and other applications of quantum stochastic calculus to quantum optics*, Quantum Opt. **2** (1990) 423–441.

10. H. J. Carmichael, *An Open System Approach to Quantum Optics*, Lect. Notes Phys. **m 18** (Springer, Berlin, 1993).
11. H. M. Wiseman, G. J. Milburn, *Interpretation of quantum jump and diffusion processes illustrated on the Bloch sphere*, Phys. Rev. A **47** (1993) 1652–1666.
12. H. M. Wiseman, G. J. Milburn, *Quantum theory of optical feedback via homodyne detection*, Phys. Rev. Lett. **70** (1993) 548–551.
13. H. M. Wiseman, *Quantum trajectories and quantum measurement theory*, Quantum Semiclass. Opt. **8** (1996) 205–222.
14. G. J. Milburn, *Classical and quantum conditional statistical dynamics*, Quantum Semiclass. Opt. **8** (1996) 269–276.
15. C. W. Gardiner, P. Zoller, *Quantum Noise* (Springer, Berlin, 2000).
16. A. Barchielli, *Continual Measurements in Quantum Mechanics and Quantum Stochastic Calculus*. In S. Attal, A. Joye, C.-A. Pillet (eds.), *Open Quantum Systems III*, Lecture Notes in Mathematics **1882** (Springer, Berlin, 2006), pp. 207–291.
17. K. R. Parthasarathy, *An Introduction to Quantum Stochastic Calculus* (Birkhäuser, Basel, 1992).
18. A. Frigerio, *Covariant Markov dilations of quantum dynamical semigroups*, Publ. RIMS Kyoto Univ. **21** (1985) 657–675.
19. C. M. Caves, B. L. Schumaker, *New formalism for two-photon quantum optics. I. Quadrature phases and squeezed states*, Phys. rev. A **31** (1985) 3068–3092.
20. A. Barchielli, M. Gregoratti, M. Licciardo, *Quantum trajectories, feedback and squeezing*, arXiv:0801.4710v1.

CHARACTERIZATION THEOREMS IN GEGENBAUER WHITE NOISE THEORY

A. BARHOUMI
Department of Mathematics
Higher School of Sci. and Tech. of Hammam-Sousse
Sousse University, Sousse, Tunisia
E-mail: abdessatar.barhoumi@ipein.rnu.tn

A. RIAHI and H. OUERDIANE
Department of Mathematics
Faculty of sciences of Tunis
University of Tunis El-Manar
1060 Tunis, Tunisia
E-mail: habib.ouerdiane@fst.rnu.tn
E-mail: a1riahi@yahoo.fr

The main purpose of this paper is to derive a general structure of Gegenbauer white noise analysis as a counterpart class of non-Lévy white noise. Namely, we consider, on an appropriate space of distributions, $\mathcal{N}'_\beta$, a Gegenbauer white noise measure, $\mathcal{G}_\beta$, and construct a nuclear triple $(\mathcal{N}_\beta) \subset L^2(\mathcal{N}'_\beta, \mathcal{G}_\beta) \subset (\mathcal{N}_\beta)^*$ of test and generalized functions. A basic role is played by the chaos expansion. By using the S_β-transform we prove a general characterization theorems for Gegenbauer white noise distributions, white noise test functions in terms of analytical functions.

Keywords: Fock space, Gateaux-entire function, Gegenbauer white noise, S_β-transform, Gegenbauer isometry.

1. Introduction

The paper is intended as a continuation of the recently developed Gegenbauer white noise functionals.[4] The main purpose is to present the construction and description of a new nuclear triple, which is obtained as image under some isomorphism of the corresponding triple centered at a suitable Fock space. This new development aims at realizing Hida's idea of white noise at a non-Lévy level and at investigating applications to infinite dimensional harmonic analysis and quantum physics.

The white noise theory for Brownian motion was first introduced by T. Hida, in the Gaussian case, in his celebrated lecture notes[9,10]. Later, Kubo and Takenaka reformulated Hida's theory by taking different test function spaces and using the S-transform as machinery. For non-Gaussian white noise analysis, Y. Itô[12] constructed a Poissonian counterpart of Hida's theory and Kondratiev et al.[1,13,20] established a purely non-Gaussian distribution theory in infinite dimensional analysis by means of a normalized Laplace transform; (see also[7]). In[6,15] a theory of the Lévy white noise analysis for the general case of Lévy processes is developed. See also the generalization and extension to various aspects in[16,17,19].

In view of these different developments, it is natural to ask if a white noise theory can also be developed for other processes. In particular, since Gegenbauer processes are becoming increasingly important in many applications, it would be of interest to have a white noise theory for such processes. In fact, Gegenbauer processes are long memory processes and are characterized by an unbounded power spectral density at zero. From this last singularity property, one can observe that a natural tool to analyze such processes appears to be a generalization of the wavelet transform in quite different way from the Lévy processes. This was the motivation of our previous paper[4] where a study on white noise functionals of general non-Lévy class is started.

The paper is organized as follows: Section 2 is devoted to a quick review of Gegenbauer white noise functionals with special emphasis on the chaos decomposition of the space $L^2(\mathcal{N}_\beta', \mathcal{B}(\mathcal{N}_\beta'), \mathcal{G}_\beta)$ by means of an orthogonal system of infinite dimensional Gegenbauer polynomials and an appropriate tensor product. In Section 3, we construct a nuclear triple of test and generalized functions. In Section 4, we define the S_β-transform as a main tool in working in these spaces. The S_β-transform serves to prove characterization theorems for test and generalized functions in terms of analytical functionals with hypergeometric growth condition.

2. Gegenbauer White Noise Space

Let $\mu_{\beta,\sigma}$ be the beta-type distributions with parameters $\beta > -\frac{1}{2}$ and $\sigma \in \mathbb{R}$ given by

$$\begin{cases} d\mu_{\beta,\sigma}(x) = \dfrac{1}{|\sigma|\sqrt{\pi}} \dfrac{\Gamma(\beta+1)}{\Gamma(\beta+\frac{1}{2})} \left(1 - \dfrac{x^2}{\sigma^2}\right)^{\beta-\frac{1}{2}} \chi_{]-|\sigma|,|\sigma|[}(x)\, dx \\ d\mu_{\beta,0}(x) = d\delta_0(x) \end{cases} \quad (1)$$

where δ_0 is the Dirac measure concentrated on the point 0 and $\Gamma(\cdot)$ is the Gamma function. For $\sigma = 1$, we have the standard beta-type distribution with parameter β :

$$d\mu_\beta(x) := d\mu_{\beta,1}(x) = \frac{1}{\sqrt{\pi}} \frac{\Gamma(\beta+1)}{\Gamma(\beta+\frac{1}{2})} (1-x^2)^{\beta-\frac{1}{2}} \chi_{]-1,1[}(x)\, dx. \qquad (2)$$

From the paper[3], we recall the following useful background. Apply the Gram-Schmidt orthogonalization process to the sequence $\{1, x, x^2, \cdots, x^n, \cdots\}$ to get a sequence $\{P_{\beta,n};\ n = 0, 1, 2, \cdots\}$ of orthogonal polynomials in $L^2(\mu_\beta)$. Here $P_{\beta,0}(x) = 1$ and for $\beta > 0$, $P_{\beta,n}$ is the Gegenbauer polynomial given by

$$P_{\beta,n}(x) = \frac{n!}{2^n (\beta)_n} \sum_{k=0}^{[n/2]} (-1)^n 2^{n-2k} \binom{-\beta}{n-k} \binom{n-k}{k} x^{n-2k},$$

where the shifted factorials are given by

$$(a)_0 = 1\ ;\ (a)_k = a(a+1)\cdots(a+k-1) = \frac{\Gamma(a+k)}{\Gamma(a)}, \quad \forall k \in \mathbb{N},$$

and for $\delta \in \mathbb{R}$, $p \in \mathbb{N}$, we have

$$\binom{\delta}{p} = \frac{\delta(\delta-1)\cdots(\delta-p+1)}{p!}.$$

For the Szegö-Jacobi parameters, since the measure μ_β is symmetric we have

$$\alpha_n = 0 \quad \text{and} \quad w_n = \frac{n(n-1+2\beta)}{4(n+\beta)(n-1+\beta)}, \quad n \geq 0.$$

It is well-known that these polynomials $P_{\beta,n}$ satisfy the recursion formula

$$x\, P_{\beta,n}(x) = P_{\beta,n+1}(x) + w_n P_{\beta,n-1}(x).$$

We recall that the generating function ψ_{μ_β} associated to the measure μ_β is given by

$$\psi_{\mu_\beta}(t;x) = \frac{1}{(1-2xt+t^2)^\beta} = \sum_{n=0}^{\infty} \frac{2^n (\beta)_n}{n!} t^n P_{\beta,n}(x).$$

The Bessel function of the first kind of order $\alpha > -\frac{1}{2}$ can be defined by

$$J_\alpha(x) = \left(\frac{x}{2}\right)^\alpha \sum_{k=0}^{+\infty} \frac{(-1)^k}{k!\,\Gamma(\alpha+k+1)} \left(\frac{x}{2}\right)^{2k}, \quad x > 0.$$

Moreover, we have the following Poisson-Mehler integral representation

$$J_\alpha(x) = \frac{1}{\sqrt{\pi}\,\Gamma(\alpha+\frac{1}{2})} \left(\frac{x}{2}\right)^\alpha \int_{-1}^{1} (1-t^2)^{\alpha-\frac{1}{2}} e^{ixt} dt. \qquad (3)$$

The normalized Bessel function of order $\alpha > -\frac{1}{2}$ is given by

$$j_\alpha(x) = \begin{cases} 2^\alpha\,\Gamma(\alpha+1)\,\frac{J_\alpha(x)}{x^\alpha} & \text{if } x \neq 0 \\ 1 & \text{if } x = 0 \end{cases} \qquad (4)$$

Using (3) and (4), the Fourier transform of the beta-type distribution, in Eq. (1), is given by

$$\widehat{\mu}_{\beta,\sigma}(x) = \int_{\mathbb{R}} e^{ixt} d\mu_{\beta,\sigma}(t) = j_\beta\left(|\sigma|\,x\right), \quad x \in \mathbb{R}. \qquad (5)$$

For simplicity, put $I =]-1,1[$. From the Favard theorem,[5] one can easily obtain

$$\|P_{\beta,n}\|^2_{L^2(I,\mu_\beta)} = \frac{n!}{4^n} \frac{(2\beta)_n}{(\beta)_n(\beta+1)_n} =: M_{\beta,n}.$$

Therefore, we define the corresponding Gegenbauer functions $\mathcal{H}_{\beta,n}$ by

$$\mathcal{H}_{\beta,n}(x) = \left(\frac{\Gamma(\beta+1)}{\sqrt{\pi}\,\Gamma(\beta+\frac{1}{2})\,M_{\beta,n}}\right)^{\frac{1}{2}} P_{\beta,n}(x)(1-x^2)^{\frac{2\beta-1}{4}}.$$

This gives an orthonormal basis $\{\mathcal{H}_{\beta,n};\ n = 0,1,2,\cdots\}$ for $H := L^2(I,dx)$.
Define the operator A_β, on H, by

$$A_\beta = (x^2-1)\frac{d^2}{dx^2} + 2x\frac{d}{dx} - \left(\beta-\frac{1}{2}\right)^2 \frac{x^2}{x^2-1} + 2.$$

Then the Gegenbauer functions $\mathcal{H}_{\beta,n}$ are eigenvectors of A_β, namely,

$$A_\beta \mathcal{H}_{\beta,n} = \lambda_{\beta,n} \mathcal{H}_{\beta,n}$$

with

$$\lambda_{\beta,n} = \beta + \frac{3}{2} + n(n+2\beta), \quad n = 0,1,2,\cdots.$$

Moreover, for any $p > \frac{1}{4}$, A_β^{-p} is a Hilbert-Schmidt operator satisfying

$$\|A_\beta^{-p}\|^2_{HS} = \sum_{n=0}^{\infty} \lambda_{\beta,n}^{-2p} < \infty.$$

Now, for each $p \in \mathbb{R}$, define a norm $|\cdot|_p$ on H by

$$|f|_p = \left|A_\beta^p |f|\right|_0 = \left(\sum_{n=0}^\infty \lambda_{\beta,n}^{2p} \langle |f|, \mathcal{H}_{\beta,n}\rangle^2\right)^{1/2}, \quad f \in H, \qquad (6)$$

where $|\cdot|_0$ and $\langle \cdot, \cdot \rangle$ are, respectively, the norm and the inner product of H. For $p \geq 0$, let $\mathcal{N}_{\beta,p}$ be the Hilbert space consisting of all $f \in H$ with $|f|_p < \infty$, and $\mathcal{N}_{\beta,-p}$ the completion of H with respect to $|\cdot|_{-p}$. Since A_β^{-1} is of Hilbert-Schmidt type, identifying H with its dual space we come to the real standard nuclear triple

$$\mathcal{N}_\beta := \bigcap_{p \geq 0} \mathcal{N}_{\beta,p} \subset H \subset \bigcup_{p \geq 0} \mathcal{N}_{\beta,-p} =: \mathcal{N}_\beta'.$$

Being compatible to the inner product of H, the canonical bilinear form on $\mathcal{N}_\beta' \times \mathcal{N}_\beta$ is denoted by $\langle \cdot, \cdot \rangle$ again. For $n \in \mathbb{N}$, we denote by $\mathcal{N}_\beta^{\hat{\otimes} n}$ the n-fold symmetric tensor product of $\mathcal{N}_\beta$ equipped with the π-topology and by $\mathcal{N}_{\beta,p}^{\hat{\otimes} n}$ the n-fold symmetric hilbertian tensor product of $\mathcal{N}_{\beta,p}$. We will preserve the notation $|\cdot|_p$ and $|\cdot|_{-p}$ for the norms on $\mathcal{N}_{\beta,p}^{\hat{\otimes} n}$ and $\mathcal{N}_{\beta,-p}^{\hat{\otimes} n}$, respectively.

Definition 2.1.[4] The probability measure $\mathcal{G}_\beta$ on $\mathcal{N}_\beta'$, of which characteristic function is given by

$$\int_{\mathcal{N}_\beta'} e^{i\langle w, \varphi \rangle} d\mathcal{G}_\beta(w) = j_\beta(\langle \varphi \rangle), \quad \varphi \in \mathcal{N}_\beta, \qquad (7)$$

where $\langle \varphi \rangle = \int_I \varphi(t) dt$, is called *the Gegenbauer white noise measure with parameter β*. The probability space $(\mathcal{N}_\beta', \mathcal{B}(\mathcal{N}_\beta'), \mathcal{G}_\beta)$, where $\mathcal{B}(\mathcal{N}_\beta')$ is the cylinder σ-algebra on $\mathcal{N}_\beta'$, is called *the Gegenbauer white noise space*.

Remark 2.1. For $\xi \in \mathcal{N}_\beta$ such that $\langle \xi \rangle \geq 0$, let X_ξ be the random variable defined, on $(\mathcal{N}_\beta', \mathcal{B}(\mathcal{N}_\beta'), \mathcal{G}_\beta)$, by

$$X_\xi(w) := \langle w, \xi \rangle.$$

Then X_ξ has a beta-type distribution with parameters β and $\langle \xi \rangle$. Notice that, from (6), the nuclear space $\mathcal{N}_\beta$ is closed under the absolute value. Then, for $\varphi \in \mathcal{N}_\beta$ such that $\langle |\varphi| \rangle \neq 0$, we recover the beta-type distribution μ_β through the random variable $X_{\widetilde{\varphi}}$, where $\widetilde{\varphi} := \frac{|\varphi|}{\langle |\varphi| \rangle}$. This gives an essence to our approach for defining the Gegenbauer white noise functionals. Moreover, our choice for the above normalization is suitable for the proof of the orthogonality property in the chaotic decomposition of the Hilbert space of square $\mathcal{G}_\beta$-integrable functionals. For more details we refer to the paper[4].

Definition 2.2.[4] For $w \in \mathcal{N}'_\beta$ and $n = 0, 1, 2, \cdots$, we define *the β-type tensor product* $: w^{\otimes n} :_\beta$ as the continuous linear functional on $\mathcal{N}_\beta^{\widehat{\otimes} n}$ characterized by

$$\langle : w^{\otimes n} :_\beta, \varphi^{\otimes n} \rangle = |\varphi|_0^n P_{\beta,n}\left(\frac{\langle w, |\varphi| \rangle}{\langle |\varphi| \rangle}\right), \quad \varphi \in \mathcal{N}_\beta, \tag{8}$$

and for any orthogonal vectors $\xi_1, \cdots, \xi_k \in \mathcal{N}_\beta$ and nonnegative integers n_j's such that $n_1 + \cdots + n_k = n$,

$$\langle : w^{\otimes n} :_\beta, \xi_1^{\otimes n_1} \widehat{\otimes} \cdots \widehat{\otimes} \xi_k^{\otimes n_k} \rangle = \langle : w^{\otimes n_1} :_\beta, \xi_1^{\otimes n_1} \rangle \cdots \langle : w^{\otimes n_k} :_\beta, \xi_k^{\otimes n_k} \rangle. \tag{9}$$

The β-Fock space $\mathcal{F}_\beta(H)$ over H is defined as the weighted direct sum of the n-th symmetric tensor powers $H^{\widehat{\otimes} n}$,

$$\mathcal{F}_\beta(H) := \bigoplus_{n=0}^{+\infty} M_{\beta,n} H^{\widehat{\otimes} n}.$$

Thus $\mathcal{F}_\beta(H)$ consists of sequences $\overrightarrow{f} = (f^{(0)}, f^{(1)}, \cdots)$ such that, for any $n \in \mathbb{N}$, $f^{(n)} \in H^{\widehat{\otimes} n}$ and

$$\|\overrightarrow{f}\|^2_{\mathcal{F}_\beta(H)} = \sum_{n=0}^{+\infty} M_{\beta,n} \|f^{(n)}\|^2_{H^{\widehat{\otimes} n}} < \infty.$$

Theorem 2.1.[4] *For each $F \in L^2(\mathcal{N}'_\beta, \mathcal{G}_\beta)$, there exists a unique sequences $\overrightarrow{f} = (f^{(n)})_{n=0}^\infty \in \mathcal{F}_\beta(H)$ such that*

$$F = \sum_{n=0}^{+\infty} \langle : .^{\otimes n} :_\beta, f^{(n)} \rangle \tag{10}$$

in the L^2-sense. Conversely, for any $\overrightarrow{f} = (f^{(n)})_{n=0}^\infty \in \mathcal{F}_\beta(H)$, (10) defines a function in $L^2(\mathcal{N}'_\beta, \mathcal{G}_\beta)$. In that case,

$$\|F\|^2_{L^2(\mathcal{N}'_\beta, \mathcal{G}_\beta)} = \sum_{n=0}^{+\infty} M_{\beta,n} \|f^{(n)}\|^2_{H^{\widehat{\otimes} n}} = \|\overrightarrow{f}\|^2_{\mathcal{F}_\beta(H)}.$$

The following unitary operator is called the Gegenbauer isometry:

$$\begin{aligned} I_\beta : \mathcal{F}_\beta(H) &\longrightarrow L^2(\mathcal{N}'_\beta, \mathcal{G}_\beta) \\ (f^{(n)})_{n=0}^\infty &\longmapsto F \end{aligned} \tag{11}$$

3. Gegenbauer Test and Generalized Functions

For simplicity, we will use $(L^2)_\beta$ to denote the space $L^2(\mathcal{N}'_\beta, \mathcal{B}(\mathcal{N}'_\beta), \mathcal{G}_\beta)$. We need to construct a space $(\mathcal{N}_\beta)$ of test functions and its dual space $(\mathcal{N}_\beta)^*$ of generalized functions such that the following is a nuclear triple

$$(\mathcal{N}_\beta) \subset (L^2)_\beta \subset (\mathcal{N}_\beta)^*.$$

Consider the space $\mathcal{P}(\mathcal{N}'_\beta)$ of continuous polynomials on $\mathcal{N}'_\beta$:

$$\mathcal{P}(\mathcal{N}'_\beta) = \left\{ \varphi,\ \varphi(w) = \sum_{k=0}^n \left\langle :w^{\otimes k}:_\beta, \varphi^{(k)} \right\rangle;\ \varphi^{(k)} \in \mathcal{N}_\beta^{\hat\otimes k},\ w \in \mathcal{N}'_\beta,\ n \in \mathbb{N} \right\}.$$

This space may endowed with a topology such that $\mathcal{P}(\mathcal{N}'_\beta)$ becomes a nuclear space. Therefore, if we denote by $\mathcal{P}'(\mathcal{N}'_\beta)$ the dual space of $\mathcal{P}(\mathcal{N}'_\beta)$ with respect to $(L^2)_\beta$, we have the following nuclear triple

$$\mathcal{P}(\mathcal{N}'_\beta) \subset (L^2)_\beta \subset \mathcal{P}'(\mathcal{N}'_\beta).$$

The bilinear dual pairing $\langle\!\langle \cdot, \cdot \rangle\!\rangle$ between $\mathcal{P}'(\mathcal{N}'_\beta)$ and $\mathcal{P}(\mathcal{N}'_\beta)$ is then related to the inner product on $(L^2)_\beta$ by :

$$\langle\!\langle F, \varphi \rangle\!\rangle = (F, \overline{\varphi})_{(L^2)_\beta}, \quad F \in (L^2)_\beta,\ \varphi \in \mathcal{P}(\mathcal{N}'_\beta)$$

where $\overline{\varphi}$ denotes the complex conjugate function of φ.

Now, for each $p, q \in \mathbb{N}$, we define the following norm

$$\|\varphi\|^2_{\beta,p,q} = \sum_{n=0}^\infty 2^{nq} M_{\beta,n} |\varphi^{(n)}|^2_p,$$

where $\varphi(w) = \sum_{n=0}^{+\infty} \left\langle :w^{\otimes n}:_\beta, \varphi^{(n)} \right\rangle \in \mathcal{P}(\mathcal{N}'_\beta)$. Then we define the space $(\mathcal{N}_{\beta,p})_q$ to be the completion of $\mathcal{P}(\mathcal{N}'_\beta)$ with respect to $\|\cdot\|_{\beta,p,q}$. Finally, the space of test functions $(\mathcal{N}_\beta)$ is defined to be the projective limit of the spaces $(\mathcal{N}_{\beta,p})_q$

$$(\mathcal{N}_\beta) = \operatorname*{proj\,lim}_{p,q \in \mathbb{N}} (\mathcal{N}_{\beta,p})_q.$$

Let $(\mathcal{N}_{\beta,-p})_{-q}$ be the dual, with respect to $(L^2)_\beta$, of $(\mathcal{N}_{\beta,p})_q$ and let $(\mathcal{N}_\beta)^*$ be the dual with respect to $(L^2)_\beta$ of $(\mathcal{N}_\beta)$. We denote by $\langle\!\langle \cdot, \cdot \rangle\!\rangle$ the corresponding dual pairing which is given by the extension of the scalar product on $(L^2)_\beta$. We know from the general duality theorem that

$$(\mathcal{N}_\beta)^* = \operatorname*{ind\,lim}_{p,q \in \mathbb{N}} (\mathcal{N}_{\beta,-p})_{-q}.$$

The space $(\mathcal{N}_\beta)^*$ is the space of generalized functions and we obtain the nuclear triple

$$(\mathcal{N}_\beta) \subset (L^2)_\beta \subset (\mathcal{N}_\beta)^*.$$

The chaos decomposition gives a natural expansion of $\Phi \in (\mathcal{N}_\beta)^*$ into generalized kernels $\Phi^{(n)} \in \mathcal{N}_\beta^{'\otimes n}$. Let $\Phi^{(n)} \in \mathcal{N}_\beta^{'\otimes n}$ be given. Then there exist a distribution $\langle : \cdot^{\otimes n} :_\beta, \Phi^{(n)} \rangle$ in $(\mathcal{N}_\beta)^*$ acting on $\varphi \in (\mathcal{N}_\beta)$ as

$$\langle\!\langle \langle : \cdot^{\otimes n} :_\beta, \Phi^{(n)} \rangle, \varphi \rangle\!\rangle := M_{\beta,n} \langle \Phi^{(n)}, \varphi^{(n)} \rangle.$$

Any $\Phi \in (\mathcal{N}_\beta)^*$ then has a unique decomposition

$$\Phi = \sum_{n=0}^{+\infty} \langle : w^{\otimes n} :_\beta, \Phi^{(n)} \rangle,$$

where the sum converges in $(\mathcal{N}_\beta)^*$ and we have

$$\langle\!\langle \Phi, \varphi \rangle\!\rangle = \sum_{n=0}^{+\infty} M_{\beta,n} \langle \Phi^{(n)}, \varphi^{(n)} \rangle, \quad \varphi \in (\mathcal{N}_\beta). \tag{12}$$

From the definition, one can prove that $(\mathcal{N}_{\beta,-p})_{-q}$ is the Hilbert space with norm

$$\|\Phi\|_{\beta,-p,-q}^2 = \sum_{n=0}^{\infty} 2^{-nq} M_{\beta,n} |\Phi^{(n)}|_{-p}^2.$$

4. Characterization Theorems for Test and Generalized Functions

In this section, we start by the definition of the S_β-transform in order to characterize test and generalized functionals in terms of analytical functions with appropriate growth condition.

4.1. The S_β-transform

Let $\mathcal{N}_{\beta,c} = \mathcal{N}_\beta + i\mathcal{N}_\beta$ and $\mathcal{N}_{\beta,p,c} = \mathcal{N}_{\beta,p} + i\mathcal{N}_{\beta,p}$ be the complexifications of $\mathcal{N}_\beta$ and $\mathcal{N}_{\beta,p}$, respectively. Let us remember, that a generalized hypergeometric series with r numerator parameters $a_1, \cdots, a_r$ and s denominator parameters $b_1, \cdots, b_r$ is defined by

$${}_rF_s(a_1, \cdots, a_r; b_1, \cdots, b_r; z) = \sum_{n=0}^{\infty} \frac{(a_1)_n \cdots (a_r)_n}{(b_1)_n \cdots (b_s)_n} \frac{z^n}{n!}. \tag{13}$$

It's showed that the series converges absolutely for $|z| < 1$. For more details on hypergeometric series we refer to[8] and references therein.

Now, for $f \in \mathcal{N}_\beta$, we define the β-exponential function $e_\beta(\cdot, f)$ by

$$e_\beta(w, f) = \sum_{n=0}^{+\infty} \frac{2^n}{n!} (\beta)_n \langle : w^{\otimes n} :_\beta, f^{\otimes n} \rangle, \quad w \in \mathcal{N}'_\beta. \tag{14}$$

Calculating its (p, q)-norm we find

$$\|e_\beta(\cdot, f)\|_{\beta, p, q}^2 = \sum_{n=0}^{+\infty} 2^{nq} M_{\beta, n} (\beta)_n^2 \frac{4^n}{(n!)^2} |f^{\otimes n}|_p^2 = G_\beta(2^q |f|_p^2),$$

where G_β is the following entire holomorphic function:

$$G_\beta(z) := {}_2F_1(2\beta, \beta; \beta+1; z) = \sum_{n=0}^{+\infty} \frac{(2\beta)_n (\beta)_n}{(\beta+1)_n} \frac{z^n}{n!}, \quad |z| < 1. \tag{15}$$

It follows that $\|e_\beta(\cdot, f)\|_{\beta, p, q}^2 < \infty$ if and only if $2^q |f|_p^2 < 1$.

In contrast to usual white noise analysis, the β-exponential function are not test functions, they are only in those $(\mathcal{N}_{\beta,p})_q$ for which $2^q |f|_p^2 < 1$. Let $\Phi \in (\mathcal{N}_\beta)^*$, then there exist $p, q \in \mathbb{N}$ such that $\Phi \in (\mathcal{N}_{\beta,-p})_{-q}$. For all $f \in \mathcal{N}_\beta$ with $2^q |f|_p^2 < 1$, we can define the S_β-transform of Φ as

$$S_\beta \Phi(f) := \langle\!\langle \Phi, e_\beta(\cdot, f) \rangle\!\rangle.$$

Using the duality form (12), we conclude that the S_β-transform of generalized function $\Phi = \sum_{n=0}^{\infty} \langle : \cdot^{\otimes n} :_\beta, \Phi^{(n)} \rangle$ is given by

$$S_\beta \Phi(f) = \sum_{n=0}^{\infty} \frac{(2\beta)_n}{2^n (\beta+1)_n} \langle \Phi^{(n)}, f^{\otimes n} \rangle.$$

This definition extends immediately to complex vectors $f \in \mathcal{N}_{\beta,c}$ with $2^q |f|_p^2 < 1$

$$S_\beta \Phi(f) := \langle\!\langle \Phi, e_\beta(\cdot, f) \rangle\!\rangle = \sum_{n=0}^{\infty} \frac{(2\beta)_n}{2^n (\beta+1)_n} \langle \Phi^{(n)}, f^{\otimes n} \rangle \tag{16}$$

Hence, for $\Phi \in (\mathcal{N}_{\beta,-p})_{-q}$, (16) defines the S_β-transform for all f from the following open neighborhood of zero

$$\mathcal{U}_{p,q} = \{ f \in \mathcal{N}_{\beta,c}, \; 2^q |f|_p^2 < 1 \}.$$

It is noteworthy that $\{e_\beta(\cdot, f), \; f \in \mathcal{U}_{p,q}\}$ is a dense subset of $(\mathcal{N}_{\beta,p})_q$. Then the S_β-transform is well defined on $(\mathcal{N}_{\beta,-p})_{-q}$.

4.2. Characterization of test and generalized functions

Let $U \subset \mathcal{N}_{\beta,c}$ be open and consider a function $F : U \longrightarrow \mathbb{C}$. F is said to be Gâteaux-holomorphic if for all $\xi_0 \in U$ and for all $\xi \in \mathcal{N}_{\beta,c}$ the mapping from $\mathbb{C}$ to $\mathbb{C} : z \to F(\xi_0 + z\xi)$ is holomorphic in some neighborhood of zero in $\mathbb{C}$. By the general theory, the n-th Gâteaux derivative

$$F_n(\xi_1, \cdots, \xi_n) = \frac{1}{n!} D_{\xi_1} \cdots D_{\xi_n} F(0) \qquad (17)$$

becomes a continuous n-linear form on $\mathcal{N}_\beta$. If F is Gâteaux-holomorphic, then there exists for every $\eta \in U$ a sequence of homogeneous polynomials $\frac{1}{n!} D_\eta^n F(0)$ such that

$$F(\eta + \xi) = \sum_{n=0}^{\infty} \frac{1}{n!} D_\eta^n F(\xi)$$

for all ξ from some open set $V \subset U$. F is said to be holomorphic on U, if for all η in U there exists an open set $V \subset U$ such that $\sum_{n=0}^{\infty} \frac{1}{n!} D_\eta^n F(\xi)$ converges uniformly on V to a continuous function. We say that F is holomorphic at ξ_0 if there is an open set U containing ξ_0 such that F is holomorphic on U. We consider germs of holomorphic functions, i.e. we identify F and G if there exists an open neighborhood of zero U such that $F(\xi) = G(\xi)$ for all $\xi \in U$. Thus, we define $\text{Hol}_0(\mathcal{N}_{\beta,c})$ as the algebra of germs of functions holomorphic at zero equipped with the inductive topology given by the following family of norms

$$\|F\|_{\beta,p,q} := \sup_{\xi \in \mathcal{U}_{p,q}} \left\{ |F(\xi)| \, G_\beta^{-1/2} \left(2^q |\xi|_p^2\right) \right\},$$

where G_β is the map given by the hypergeometric series in (15).

With these notations and definition, we have the following characterization theorem for generalized functions.

Theorem 4.1. *The S_β-transform defines a topological isomorphism of the space of generalized function $(\mathcal{N}_\beta)^*$ onto the space $\text{Hol}_0(\mathcal{N}_{\beta,c})$.*

Proof. Let $\Phi = \sum_{n=0}^{+\infty} \left\langle : w^{\otimes n} :_\beta, \Phi^{(n)} \right\rangle \in (\mathcal{N}_\beta)^*$ and $F = S_\beta \Phi$. Then there exist $p, q \in \mathbb{N}$ such that $\Phi \in (\mathcal{N}_{\beta,-p})_{-q}$. For $\xi \in \mathcal{U}_{p,q}$, by Cauchy's inequality

$$|S_\beta \Phi(\xi)| \leq \sum_{n=0}^{+\infty} \frac{(2\beta)_n}{2^n (\beta+1)_n} |\Phi^{(n)}|_{-p} |\xi|_p^n$$

$$\leq \left(\sum_{n=0}^{+\infty} 2^{-nq} M_{\beta,n} |\Phi^{(n)}|_{-p}^2 \right)^{1/2} \left(\sum_{n=0}^{+\infty} \frac{2^{nq} (2\beta)_n^2}{2^{2n} (\beta+1)_n^2 M_{\beta,n}} |\xi|_p^{2n} \right)^{1/2}$$

$$\leq \|\Phi\|_{\beta,-p,-q} \left[G_\beta(2^q |\xi|_p^2) \right]^{1/2}.$$

This shows uniform convergence of (16) on $\mathcal{U}_{p,q}$. Hence, $F \in \text{Hol}_0(\mathcal{N}_{\beta,c})$ and we have

$$\|S_\beta \Phi\|_{\beta,p,q} \leq \|\Phi\|_{\beta,-p,-q}.$$

This implies that S_β is injective and continuous from $(\mathcal{N}_\beta)^*$ to $\text{Hol}_0(\mathcal{N}_{\beta,c})$.
Conversely, let $F \in \text{Hol}_0(\mathcal{N}_{\beta,c})$ be given. There exist $p, q \in \mathbb{N}$ such that

$$F(\xi) = \sum_{n=0}^{+\infty} \langle F_n, \xi^{\otimes n} \rangle = \sum_{n=0}^{+\infty} D_\xi^n F(0),$$

converges uniformly and is bounded on $\mathcal{U}_{p,q}$. For $\xi \in \mathcal{N}_{\beta,c}$ with $|\xi|_p = 1$ and $z \in \mathbb{C}$ with $|z|^2 2^q < 1$, since F is bounded by a constant K on $\mathcal{U}_{p,q}$, Cauchy's integral formula yields

$$|\langle F_n, \xi^{\otimes n} \rangle| = \left| \frac{1}{2i\pi} \int_{|z|=2^{-\frac{(q+1)}{2}}} \frac{F(z\xi)}{z^{n+1}} dz \right| \leq K 2^{\frac{n(q+1)}{2}}.$$

Then, by using the polarization identity, we derive

$$\sup \left\{ |\langle F_n, \xi_1 \widehat{\otimes} \cdots \widehat{\otimes} \xi_n \rangle|, \xi_1, \cdots, \xi_n \in \mathcal{U}_{p,q} \right\} \leq K 2^{\frac{n(q+1)}{2}} \frac{n^n}{n!}.$$

Therefore, for $s \geq 0$, we have

$$|F_n|^2_{-(p+s)} = \sum_{j_1,\cdots,j_n=0}^{\infty} |\langle F_n, \mathcal{H}_{\beta,j_1} \otimes \cdots \otimes \mathcal{H}_{\beta,j_n}\rangle|^2 \lambda_{\beta,j_1}^{-2(p+s)} \cdots \lambda_{\beta,j_n}^{-2(p+s)}$$

$$= \sum_{j_1,\cdots,j_n=0}^{\infty} |\langle F_n, \lambda_{\beta,j_1}^{-p} \mathcal{H}_{\beta,j_1} \otimes \cdots \otimes \lambda_{\beta,j_n}^{-p} \mathcal{H}_{\beta,j_n}\rangle|^2 \lambda_{\beta,j_1}^{-2s} \cdots \lambda_{\beta,j_n}^{-2s}$$

$$\leq K^2 2^{n(q+1)} \frac{n^{2n}}{n!^2} \sum_{j_1,\cdots,j_n=0}^{\infty} \lambda_{\beta,j_1}^{-2s} \cdots \lambda_{\beta,j_n}^{-2s}$$

$$\leq K^2 \left(e\, 2^{\frac{q+1}{2}}\right)^{2n} \|A_\beta^{-s}\|_{HS}^{2n}. \tag{18}$$

We put $\Phi(w) = \sum_{n=0}^{\infty} \frac{2^n (\beta+1)_n}{(2\beta)_n} \langle : w^{\otimes n} :_\beta, F_n\rangle$. For $s, q' \geq 0$, inequality (18) yields

$$\|\Phi\|^2_{\beta,-(p+s),-q'} = \sum_{n=0}^{+\infty} 2^{-nq'} M_{\beta,n} \frac{2^{2n}(\beta+1)_n^2}{(2\beta)_n^2} |F_n|^2_{-(p+s)}$$

$$\leq K^2 \sum_{n=0}^{+\infty} \frac{(\beta+1)_n}{(\beta)_n (2\beta)_n} \left(e^2 2^{q-q'+1} \|A_\beta^{-s}\|_{HS}^2\right)^n.$$

Choose $q' \in \mathbb{N}$ such that $e^2 2^{q-q'+1} \|A_\beta^{-s}\|_{HS}^2 < 1$, then the last series converges. This shows that $\Phi \in (\mathcal{N}_\beta)^*$. Furthermore, it is clear that $S_\beta \Phi(\xi) = \sum_{n=0}^{+\infty} \langle F_n, \xi^{\otimes n}\rangle = F(\xi)$. Thus, we have found $\Phi \in (\mathcal{N}_\beta)^*$ such that $S_\beta \Phi = F$. This proves that S_β is surjective and bicontinuous. □

To characterize the space $(\mathcal{N}_\beta)$ we start by the definition of the S_β-transform of test functions. First of all, note that for any $z \in \mathcal{N}'_\beta$, we can define the β-exponential $e_\beta(z,\cdot)$ in $(\mathcal{N}_\beta)^*$. Indeed, for $p \in \mathbb{N}$ such that $z \in \mathcal{N}_{\beta,-p,c}$, similarly as in (14), we consider the decomposition

$$e_\beta(z,w) := \sum_{n=0}^{+\infty} \frac{2^n}{n!} (\beta)_n \langle : w^{\otimes n} :_\beta, z^{\otimes n}\rangle.$$

Then, for $q \in \mathbb{N}$, we have

$$\|e_\beta(z,\cdot)\|^2_{\beta,-p,-q} = \sum_{n=0}^{+\infty} 2^{-nq} M_{\beta,n}(\beta)^2_n \frac{4^n}{(n!)^2} |z^{\otimes n}|^2_{-p} = G_\beta(2^{-q}|z|^2_{-p}) < \infty$$

if and only if $2^{-q}|z|^2_{-p} < 1$. Hence, for all $z \in \mathcal{N}'_\beta$ with $2^{-q}|z|^2_{-p} < 1$, we can define the S_β-transform of $\varphi \in (\mathcal{N}_\beta)$ as

$$S_\beta \varphi(z) := \langle\!\langle e_\beta(z,\cdot), \varphi \rangle\!\rangle. \tag{19}$$

Thus, for $\varphi \in (\mathcal{N}_{\beta,p})_q$, (19) defines $S_\beta \varphi$ as a function on the following open neighborhood of zero

$$\mathcal{U}_{-p,-q} = \left\{ z \in \mathcal{N}_{\beta,-p,c};\ 2^{-q}|z|^2_{-p} < 1 \right\}.$$

Let $\mathcal{H}(\mathcal{N}'_{\beta,c})$ stands for the space of all entire functions on $\mathcal{N}'_{\beta,c}$. By definition, any $\varphi \in \mathcal{H}(\mathcal{N}'_{\beta,c})$ is an entire function on any $\mathcal{N}_{\beta,-p,c}$, $p \in \mathbb{N}$. For $p,q \in \mathbb{N}$, we define the Banach space

$$\mathrm{Hyp}^{-q}(\mathcal{N}_{\beta,-p,c}) = \{ f \in \mathcal{H}(\mathcal{N}_{\beta,-p,c});\ \|f\|_{\beta,-p,-q} < \infty \}$$

where

$$\|f\|_{\beta,-p,-q} := \sup_{z \in \mathcal{U}_{-p,-q}} \left\{ |f(z)| G_\beta^{-1/2}\left(2^{-q}|z|^2_{-p}\right) \right\}.$$

We denote by $\mathrm{Hyp}(\mathcal{N}'_{\beta,c})$ the subspace of $\mathcal{H}(\mathcal{N}'_{\beta,c})$ of all entire functions on $\mathcal{N}'_{\beta,c}$ with hypergeometric growth and minimal type, i.e.

$$\mathrm{Hyp}(\mathcal{N}'_{\beta,c}) = \bigcap_{p,q \geq 0} \mathrm{Hyp}^{-q}(\mathcal{N}_{\beta,-p,c}).$$

With the above definition, we have the following characterization theorem for test functions.

Theorem 4.2. *The S_β-transform defines a topological isomorphism of the space of test functions $(\mathcal{N}_\beta)$ onto the space $\mathrm{Hyp}(\mathcal{N}'_{\beta,c})$*

Proof. Let $\varphi \in (\mathcal{N}_\beta)$ be given. For $p,q \in \mathbb{N}$ and $z \in \mathcal{U}_{-p,-q}$, Cauchy's inequality yields

$$|S_\beta \varphi(z)| \leq \sum_{n=0}^{+\infty} \frac{(2\beta)_n}{2^n(\beta+1)_n} |\varphi^{(n)}|_p |z|^n_{-p}$$

$$\leq \left(\sum_{n=0}^{+\infty} 2^{nq} M_{\beta,n} |\varphi^{(n)}|^2_p \right)^{1/2} \left(\sum_{n=0}^{+\infty} \frac{2^{-nq}(2\beta)^2_n}{2^{2n}(\beta+1)^2_n M_{\beta,n}} |z|^{2n}_{-p} \right)^{1/2}$$

$$\leq \|\varphi\|_{\beta,p,q} \left[G_\beta\left(2^{-q}|z|^2_{-p}\right) \right]^{1/2}.$$

This proves that S_β is injective and continuous from $(\mathcal{N}_\beta)$ to $\mathrm{Hyp}\,(\mathcal{N}'_{\beta,c})$.
Conversely, let $u \in \mathrm{Hyp}\,(\mathcal{N}'_{\beta,c})$ be given. There exist $p, q \in \mathbb{N}$ such that

$$u(z) = \sum_{n=0}^{+\infty} \langle u_n, z^{\otimes n} \rangle = \sum_{n=0}^{+\infty} D_z^n u(0)$$

converges uniformly and bounded on $\mathcal{U}_{-p,-q}$. For $z \in \mathcal{N}'_{\beta,c}$ with $|z|_{-p} = 1$ and $\lambda \in \mathbb{C}$ such that $|\lambda| < 2^{q/2}$, as u is bounded by a constant c on $\mathcal{U}_{-p,-q}$, Cauchy's integral formula yields

$$\left|\langle u_n, z^{\otimes n}\rangle\right| = \left|\frac{1}{2i\pi}\int_{|\lambda|=2^{\frac{q-1}{2}}} \frac{u(\lambda z)}{\lambda^{n+1}}\,d\lambda\right| \leq c\, 2^{\frac{-n(q-1)}{2}}.$$

The polarization identity gives

$$\sup\left\{\left|\langle u_n, z_1 \widehat{\otimes} \cdots \widehat{\otimes} z_n\rangle\right|, z_1, \cdots, z_n \in \mathcal{U}_{-p,-q}\right\} \leq c\, 2^{\frac{-n(q-1)}{2}} \frac{n^n}{n!}. \qquad (20)$$

For given $p' \in \mathbb{N}$, we choose $p \in \mathbb{N}$ such that the embedding operator $i_{p',p} : \mathcal{N}_{\beta,-p'} \longrightarrow \mathcal{N}_{\beta,-p}$ is Hilbert-Schmidt. Let $(e_j)_{j\in\mathbb{N}}$ be a complete orthonormal system in $\mathcal{N}_{\beta,-p'}$. Then (20) gives

$$|u_n|_{p'}^2 = \sum_{j_1,\cdots,j_n=0}^{\infty} \left|\langle u_n, e_{j_1}\widehat{\otimes}\cdots\widehat{\otimes} e_{j_n}\rangle\right|^2$$

$$\leq c^2\, 2^{-n(q-1)} \frac{n^{2n}}{(n!)^2} \sum_{j_1,\cdots,j_n=0}^{\infty} |i_{p',p}\,e_{j_1}|_{-p}^2 \cdots |i_{p',p}\,e_{j_n}|_{-p}^2$$

$$\leq c^2\left(e\, 2^{-\frac{q-1}{2}}\|i_{p',p}\|_{HS}\right)^{2n}. \qquad (21)$$

Put $\varphi(w) = \sum_{n=0}^{\infty} \frac{2^n(\beta+1)_n}{(2\beta)_n}\langle :w^{\otimes n}:_\beta, u_n\rangle$. Then, the formula (21) yields

$$\|\varphi\|_{\beta,p',q'}^2 = \sum_{n=0}^{+\infty} 2^{nq'} M_{\beta,n} \frac{2^{2n}(\beta+1)_n^2}{(2\beta)_n^2} |u_n|_{p'}^2$$

$$\leq c^2 \sum_{n=0}^{+\infty} \frac{(\beta+1)_n}{(\beta)_n(2\beta)_n}\left(e^2 2^{q'-q+1}\|i_{p',p}\|_{HS}^2\right)^n.$$

If we choose $q' \in \mathbb{N}$ such that $e^2 2^{q'-q+1}\|i_{p',p}\|_{HS}^2 < 1$, the last series converges and then $\varphi \in (\mathcal{N}_\beta)$. Furthermore, we get $S_\beta\varphi(z) = \sum_{n=0}^{+\infty}\langle u_n, z^{\otimes n}\rangle =$

$u(z)$. Thus, we have found $\varphi \in (\mathcal{N}_\beta)$ such that $S_\beta \varphi = u$. This proves that S_β is surjective and bicontinuous. □

References

1. S. Albeverio, Yu.L. Daletsky, Yu.G. Kondratiev and L. Streit, *Non-Gaussian infinite dimensional analysis*, J. Funct. Anal. 138 (1996), 311–350.
2. N. Asai, I. Kubo and H.-H. Kuo, *Multiplicative Renormalization and Generating Functions I.*, Taiwanese Journal of Mathematics, 7 (2003), 89–101.
3. N. Asai, I. Kubo and H.-H. Kuo, *Multiplicative Renormalization and Generating Functions II.*, Taiwanese Journal of Mathematics, Vol. 8, No. 4 (2004), 583–628.
4. A. Barhoumi, H. Ouerdiane and A. Riahi, *Infinite dimensional Gegenbauer functionals*, Non-commutative Harmonic Analysis with Applications to Probability, Banach Center Publ., Vol. 78, pp. 1–11. Polish Acad. Sci., Warsaw 2007.
5. T.S. Chihara, *An Introduction to Orthogonal Polynomials*, Gordon and Breach, New York, 1978.
6. G. Di Nunno, B. Øksendal and F. Proske, *White noise analysis for Lévy processes*, J. Funct. Anal. 206, No. 1 (2004), 109–148.
7. R. Gannoun, R. Hchaichi, H. Ouerdiane and A. Rezgui, *Un théorème de dualité entre espaces de fonctions holomorphes à croissance exponentielle* J. Funct. Anal. Vol. 171, No.1 (2000), 1–14.
8. G. Gasper and M. Rahman, *Basic Hypergeometric Series*, Encyclopedia of Mathematics and its Applications, Vol. 35, Cambridge 1999.
9. T. Hida, *Analysis of Brownian Functionals*, Carleton Mathematical Lecture Notes 13, 1975.
10. T. Hida and N. Ikeda, *Analysis on Hilbert space with reproducing kernel arising from multiple Wiener integral*, Proceeding of the 5th Berkeley Symposium on Mathematics, Statistics and Probability, Vol. II, Part. 1 (1967), 117–143.
11. T. Hida, H.-H. Kuo, J. Potthoff and L. Streit, *White Noise: An infinite Dimensional Calculus*, Kluwer Acad. Publ., Dordrecht, 1993.
12. Y. Itô, *Generalized Poisson Functionals*, Probab. Theory Related Fields 77 (1988), 1–28.
13. Yu.G. Kondratiev, L. Streit, W. Westerkamp and J.A. Yan, *Generalized functions in infinite dimensional analysis*, Hiroshima Math. J. 28 (1998), 213–260.
14. H.-H. Kuo, *White noise distrubition theory*, CRC press, Boca Raton 1996.
15. Y.-J. Lee and H.-H. Shih, *Analysis of generalized Lévy white noise functionals*, J. Funct. Anal. 211 (2004), 1–70.
16. E.W. Lytvynov, *Orthogonal decompositions for Lévy processes with an application to the gamma, Pacsal, and Meixner processes*, Infin. Dimens. Anal. Quantum Probab. Relat. Top. Vol. 6, No. 1 (2003), 73–102.
17. E.W. Lytvynov, *Polynomials of Meixner's type in infinite dimensions-Jacobi fields and orthogonality measurs*, J. Funct. Anal. 200 (2003), 118–149.

18. N. Obata, *White noise calculus and Fock space*, Lecture Notes in Math. Vol. 1577, Springer-Verlag, 1994.
19. W. Schoutens, *Stochastic processes and orthogonal polynomials*, Lecture Notes in Statist., Vol. 146, Springer Berlin 1999.
20. J.L. Silva, *Studies in non-Gaussian Analysis*, Ph.D. Dissertation, University of Maderira, 1998.

A PROBLEM OF POWERS AND THE PRODUCT OF SPATIAL PRODUCT SYSTEMS*

B.V. R. BHAT

Statistics and Mathematics Unit, Indian Statistical Institute Bangalore, R. V. College Post, Bangalore 560059, India, E-mail: bhat@isibang.ac.in, Homepage: http://www.isibang.ac.in/Smubang/BHAT/

V. LIEBSCHER

Institut für Mathematik und Informatik, Ernst-Moritz-Arndt-Universität Greifswald, 17487 Greifswald, Germany, E-mail: volkmar.liebscher@uni-greifswald.de, Homepage: http://www.math-inf.uni-greifswald.de/biomathematik/liebscher/

M. SKEIDE[†]

Dipartimento S.E.G.e S., Università degli Studi del Molise, Via de Sanctis, 86100 Campobasso, Italy, E-mail: skeide@math.tu-cottbus.de, Homepage: http://www.math.tu-cottbus.de/INSTITUT/lswas/_skeide.html

In the 2002 AMS summer conference on "Advances in Quantum Dynamics" in Mount Holyoke Robert Powers proposed a sum operation for spatial E_0-semigroups. Still during the conference Skeide showed that the Arveson system of that sum is the product of spatial Arveson systems. This product may but need not coincide with the tensor product of Arveson systems. The Powers sum of two spatial E_0-semigroups is, therefore, up to cocycle conjugacy Skeide's product of spatial noises.

Keywords: Quantum dynamics, quantum probability, Hilbert modules, product systems, E_0-semigroups. 2000 AMS-Subject classification: 46L53; 46L55; 46L08; 60J25; 12H20.

---

*This work is supported by a PPP-project by DAAD and DST and by a Research in Pairs project at MfO.
[†]MS is supported by research funds of the University of Molise and the Italian MIUR (PRIN 2005) and by research funds of the Dipartimento S.E.G.e S. of University of Molise.

1. Introduction

Let $\vartheta^i = (\vartheta^i_t)_{t\in\mathbb{R}_+}$ ($i = 1, 2$) be two E_0-semigroups on $\mathcal{B}(H)$ with associated Arveson systems $\mathfrak{H}^{i\otimes} = (\mathfrak{H}^i_t)_{t\in\mathbb{R}_+}$ (Arveson[1]). Furthermore, let $\Omega^i = (\Omega^i_t)_{t\in\mathbb{R}_+} \subset \mathcal{B}(H)$ be two semigroups of intertwining isometries for ϑ^i (units). Then

$$T_t \begin{pmatrix} a_{11} & a_{12} \\ a_{21} & a_{22} \end{pmatrix} = \begin{pmatrix} \vartheta^1_t(a_{11}) & \Omega^1_t a_{12} \Omega^{2*}_t \\ \Omega^2_t a_{12} \Omega^{1*}_t & \vartheta^2_t(a_{22}) \end{pmatrix} \quad (*)$$

defines a CP-semigroup $T = (T_t)_{t\in\mathbb{R}_+}$ on $\mathcal{B}(H \oplus H)$. In the 2002 AMS summer conference on "Advances in Quantum Dynamics" Robert Powers asked for the Arveson system associated with T (Bhat,[4] Arveson[2]). During that conference (see Ref.[20]) Skeide showed that this product system is nothing but the *product of the spatial product systems* introduced in Skeide[23] (published first in 2001). Meanwhile, Powers has formalized the above *sum operation* in Ref.[18] and he has proved that the product may, but need not coincide with the tensor product of the involved Arveson systems, a fact suspected already in Ref.[11,20]

In these notes we extend Powers' construction to the case of spatial E_0-semigroups ϑ^i on $\mathcal{B}^a(E^i)$ where E^i are Hilbert $\mathcal{B}$-modules. We obtain the same result as in Ref.,[20] namely, the product system of the minimal dilation of the CP-semigroup on $\mathcal{B}^a(E^1 \oplus E^2)$ defined in analogy with $(*)$ is the product of spatial product systems from Ref.[23]

Like in Ref.,[20] it is crucial to understand the following point (on which we will spend some time in Section 3): In Bhat and Skeide[9] to every CP-semigroup on a C^*-algebra $\mathcal{B}$ a product system of $\mathcal{B}$-algebra has been constructed. However, the C^*-algebra in question here is $\mathcal{B}^a(E^1 \oplus E^2)$, where $\mathcal{B}^a(E)$ denotes the algebra of all adjointable operators on a Hilbert module E. So what has the product system of $\mathcal{B}^a(E^1 \oplus E^2)$-correspondences to do with the product systems of the E_0-semigroups ϑ^1 and ϑ^2, which are product systems of $\mathcal{B}$-correspondences? The answer to this question, like in Ref.,[20] will allow to construct the Arveson system of a CP-semigroup on $\mathcal{B}(H)$ *without* having to find first its minimal dilation. To understand even the Hilbert space case already requires, however, module techniques.

In Section 2 we repeat the necessary facts about product systems (in particular, spatial ones and their product) and how they are derived from CP-semigroups and E_0-semigroups. Section 3, introduces the new construction of the product system of $\mathcal{B}$-correspondences assocociated with a strict CP-semigroup on $\mathcal{B}^a(E)$ for some Hilbert $\mathcal{B}$-module E. In Section 4, finally, we put together all the results from the preceding sections to prove our claim.

2. Product systems, CP-semigroups, E_0–semigroups and dilations

Throughout these notes, by $\mathcal{B}$ we denote a unital C^*-algebra. There are no spatial product systems where $\mathcal{B}$ is nonunital. There is no reasonable notion of unit for product systems of correspondences over nonunital C^*-algebras, where $\mathcal{B}$ could not easily be substituted by a unital ideal of $\mathcal{B}$.

2.1. Product systems

Product systems of Hilbert modules (*product system* for short) occurred in different contexts in Bhat and Skeide,[9] Skeide,[19] Muhly and Solel[13] and other more recent publications. Let $\mathcal{B}$ be a unital C^*-algebra. A product system is a family $E^\odot = (E_t)_{t \in \mathbb{R}_+}$ of *correspondences* E_t over $\mathcal{B}$ (that is, a (right) Hilbert $\mathcal{B}$-module with a unital representation of $\mathcal{B}$) with an associative identification

$$E_s \odot E_t = E_{s+t},$$

where $E_0 = \mathcal{B}$ and for $s = 0$ or $t = 0$ we get the canonical identifications. By $\odot$ we denote the (internal) tensor product of correspondences.

If we want to emphasize that we do not put any technical condition, we say *algebraic* product system. There are concise definitions of *continuous*[21] and *measurable* (separable!)[10] product systems of C^*-correspondences, and *measurable* (separable pre-dual!)[14] product systems of W^*-correspondences. Skeide[24] will discuss *strongly continuous* product systems of von Neumann correspondences. We do not consider such constraints in these notes. We just mention for the worried reader the result from Ref.[21,23] that the product of continuous spatial product systems is continuous.

2.2. Units

A *unit* for a product system $E^\odot$ is family $\xi^\odot = (\xi_t)_{t \in \mathbb{R}_+}$ of elements $\xi_t \in E_t$ such that

$$\xi_s \odot \xi_t = \xi_{s+t}$$

and $\xi_0 = 1 \in \mathcal{B} = E_0$. A unit may be *unital* ($\langle \xi_t, \xi_t \rangle = 1 \forall t \in \mathbb{R}_+$), *contractive* ($\langle \xi_t, \xi_t \rangle \leq 1 \forall t \in \mathbb{R}_+$), or *central* ($b\xi_t = \xi_t b \forall t \in \mathbb{R}_+, b \in \mathcal{B}$).

We do not pose technical conditions on the unit. But, sufficiently continuous units can be used to pose technical conditions on the product system in a nice way; see Ref.[20]

2.3. *The product system of a CP-semigroup*

Let $T = (T_t)_{t \in \mathbb{R}_+}$ be a (not necessarily unital) CP-semigroup on a unital C^*-algebra $\mathcal{B}$. According to Bhat and Skeide[9] there exists a product system $E^\odot$ with a unit $\xi^\odot$ determined uniquely up to isomorphism (of the pair $(E^\odot, \xi^\odot)$) by the following properties:

(1) $\langle \xi_t, b\xi_t \rangle = T_t(b)$.
(2) $E^\odot$ is **generated** by $\xi^\odot$, that is, the smallest subsystem of $E^\odot$ containing $\xi^\odot$ is $E^\odot$.

In analogy with Paschke's[16] GNS-construction for CP-maps, we call $(E^\odot, \xi^\odot)$ the **GNS-system** of T and we call $\xi^\odot$ the **cyclic unit**. In fact, $\mathcal{E}_t = \overline{\text{span}}\,\mathcal{B}\xi_t\mathcal{B}$ is the **GNS-module** of T_t with cyclic vector ξ_t. For the comparison of the product system of Powers' CP-semigroup with a product of product systems it is important to note that

$$E_t = \overline{\text{span}}\{x_{t_n}^n \odot \ldots \odot x_{t_1}^1 : n \in \mathbb{N}, t_n + \ldots + t_1 = t, x_{t_k}^k \in \mathcal{E}_{t_k}\} \\ = \overline{\text{span}}\{b_n \xi_{t_n} \odot \ldots \odot b_1 \xi_{t_1} b_0 : n \in \mathbb{N}, t_n + \ldots + t_1 = t, b_k \in \mathcal{B}\}. \quad (1)$$

In fact, the product system E_t can be obtained as an inductive limit of the expressions $\mathcal{E}_{t_n} \odot \ldots \odot \mathcal{E}_{t_1}$ over refinement of the partitions $t_n + \ldots + t_1 = t$ of $[0, t]$.

2.4. *The product system of an E_0-semigroup on $\mathscr{B}^a(E)$*

Let E be a Hilbert $\mathcal{B}$-module with a **unit vector** ξ (that is, $\langle \xi, \xi \rangle = \mathbf{1}$) and let $\vartheta = (\vartheta_t)_{t \in \mathbb{R}_+}$ be an E_0-**semigroup** (that is, a semigroup of unital endomorphisms) on $\mathscr{B}^a(E)$. Let us denote by xy^* $(x, y \in E)$ the **rank-one operator**

$$xy^* : z \longmapsto x \langle y, z \rangle.$$

Then $p_t := \vartheta_t(\xi\xi^*)$ is a projection and the range $E_t := p_t E$ is a Hilbert $\mathcal{B}$-submodule of E. By defining the (unital!) left action $bx_t = \vartheta_t(\xi b \xi^*)x_t$ we turn E_t into a $\mathcal{B}$-correspondence. One easily checks that

$$x \odot y_t \longmapsto \vartheta_t(x\xi^*)y_t$$

defines an isometry $u_t \colon E \odot E_t \to E$. Clearly, if ϑ_t is **strict** (that is, precisely, if $\overline{\operatorname{span}}\, \vartheta_t(EE^*)E = E$), then u_t is a unitary. Identifying $E = E \odot E_t$ and using the semigroup property, we find

$$\vartheta_t(a) = a \odot \operatorname{id}_{E_t} \qquad (E \odot E_s) \odot E_t = E \odot (E_s \odot E_t). \qquad (2)$$

The restriction of u_t to $E_s \odot E_t$ is a bilinear unitary onto E_{s+t} and the preceding associativity reads now $(E_r \odot E_s) \odot E_t = E_r \odot (E_s \odot E_t)$. Obviously, $E_0 = \mathcal{B}$ and the identifications $E_t \odot E_0 = E_t = E_0 \odot E_t$ are the canonical ones. Thus, $E^\odot = (E_t)_{t \in \mathbb{R}_+}$ is a product system.

For E_0-semigroups on $\mathcal{B}(H)$ the preceding construction is due to Bhat,[4] the extension to Hilbert modules to Skeide.[19] We would like to mention that Bhat's construction does not give the Arveson system of an E_0-semigroup, but its *opposite* Arveson system (all orders in tensor products reversed). By Tsirelson[25] the two need not be isomorphic. For Hilbert C^*-modules Arveson's construction does not work. For von Neumann modules it works, but gives a product system of von Neumann correspondences over $\mathcal{B}'$, the commutant of $\mathcal{B}$; see Ref.[20,22]

Existence of a unit vector is not a too hard requirement, as long as $\mathcal{B}$ is unital. (If E has no unit vector, then a finite multiple E^n will have one; see Ref.[22] And product systems do not change under taking direct sums.) We would like to mention a further method to construct the product system of an E_0-semigroup, that works also for nonunital $\mathcal{B}$. It relies on the representations theory of $\mathcal{B}^a(E)$ in Muhly, Skeide and Solel.[15] See Ref.[22] for details.

2.5. *Dilation and minimal dilation*

Suppose $E^\odot$ is a product system with a unit $\xi^\odot$. Clearly, $T_t := \langle \xi_t, \bullet \xi_t \rangle$ defines a CP-semigroup $T = (T_t)_{t \in \mathbb{R}_+}$, which is unital, if and only if $\xi^\odot$ is unital. Obviously, $E^\odot$ is the product systems of T, if and only if it is generated by $\xi^\odot$.

If $\xi^\odot$ is unital, then we may embed E_t as $\xi_s \odot E_t$ into E_{s+t}. This gives rise to an inductive limit E and a factorization $E = E \odot E_t$, fulfilling the associativity condition in (2). It follows that $\vartheta_t(a) = a \odot \operatorname{id}_{E_t}$ defines an E_0-semigroup $\vartheta = (\vartheta_t)_{t \in \mathbb{R}_+}$ on $\mathcal{B}^a(E)$. The embedding $E_t \to E_{s+t}$ is, in general, only right linear so that, in general, E is only a right Hilbert module.

Under the inductive limit all $\xi_t \in E_t \subset E$ correspond to the same unit vector $\xi \in E$. Moreover, $\xi = \xi \odot \xi_t$, so that the vector expectation $\varphi := \langle \xi, \bullet \xi \rangle$ fulfills $\varphi \circ \vartheta_t(\xi b \xi^*) = T_t(b)$, that is, (E, ϑ, ξ) is a **weak dilation** of T in the sense of Ref.[8,9] Clearly, the product system of ϑ (constructed with the unit vector ξ) is $E^{\odot}$.

Suppose ϑ is a strict E_0-semigroup on some $\mathscr{B}^a(E)$ and that ξ is a unit vector in E. One may show (see Ref.[19]) that $T_t(b) := \varphi \circ \vartheta_t(\xi b \xi^*)$ defines a (necessarily unital) CP-semigroup (which it dilates), if and only if the projections $p_t := \vartheta_t(\xi \xi^*)$ increase. In this case, the product system $E^{\odot}$ of ϑ has a unit $\xi^{\odot} = (\xi_t)_{t \in \mathbb{R}_+}$ with $\xi_t := p_t \xi$, which fulfills $T_t = \langle \xi_t, \bullet \xi_t \rangle$. We say the weak dilation (E, ϑ, ξ) of T is **minimal**, if the **flow** $j_t(b) := \vartheta_t(\xi b \xi^*)$ generates E out of ξ. One may show that this is the case, if and only if the product system of ϑ coincides with the product system of T. The minimal (weak) dilation is determined up to suitable unitary equivalence.

Remark 2.1. We would like to emphasize that in order to construct the minimal dilation of a unital CP-semigroup T, we first constructed the product system of T and then constructed the dilating E_0-semigroup ϑ (giving back the product system of T). It is not necessary to pass through minimal dilation to obtain the product system of T, but rather the other way round.

2.6. *Spatial product systems*

Following Ref.,[23] we call a product system $E^{\odot}$ **spatial**, if it has a central unital **reference unit** $\omega^{\odot} = (\omega_t)_{t \in \mathbb{R}_+}$. The choice of the reference unit is part of the spatial structure, so we will write a pair $(E^{\odot}, \omega^{\odot})$. For instance, a **morphism** $w^{\odot} \colon E^{\odot} \to F^{\odot}$ between product systems $E^{\odot}$ and $F^{\odot}$ is a family $w^{\odot} = (w_t)_{t \in \mathbb{R}_+}$ of mappings $w_t \in \mathscr{B}^{a,bil}(E_t, F_t)$ (that is, bilinear adjointable mappings from E_t to F_t) fulfilling $w_s \odot w_t = w_{s+t}$ and $w_0 = \mathrm{id}_{\mathcal{B}}$. To be a **spatial** morphism of spatial product systems, $w^{\odot}$ must send the reference unit of $E^{\odot}$ to the reference unit of $F^{\odot}$.

Our definition matches that of Powers[17] in that an E_0-semigroup ϑ on $\mathscr{B}^a(E)$ admits a so-called intertwining semigroup of isometries, if and only if the product system of ϑ is spatial. It does not match the usual definition for Arveson systems, where an Arveson system is *spatial*, if it has a unit. The principle result of Barreto, Bhat, Liebscher and Skeide[3] asserts that a product system of von Neumann correspondences is spatial, if it has a (continuous) unit. But, for Hilbert modules this statement fails. In fact, we show in Ref.[6] that, unlike for Arveson systems, a subsystem of a product system of Fock modules need not be spatial (in particular, it need not be Fock).

There are many interesting questions about spatial product systems, open even in the Hilbert space case. Does the spatial structure of the spatial product system depend on the choice of the reference unit? The equivalent question is, whether every spatial product system is *amenable*[5] in the sense that the product system automorphisms act transitively on the set of units. Tsirelson[26] claims they are not. But, still there is a gap that has not yet been filled. In contrast to this, the question raised by Powers,[18] whether the product defined in the next section depends on the reference units, or not, we can answer in the negative sense; see Ref.[7]

2.7. *The product of spatial product systems*

The basic motivation of Ref.[23] was to define an *index* of a product system and to find a *product* of product systems under which the index is *additive*. Both problems could not be solved in full generality, but precisely for the category of spatial product systems.

The mentioned result Ref.[6] is one of the reasons why it is hopeless to define an index for nonspatial product systems. However, once accepted the necessity to restrict to spatial product systems (anyway, the index of a nonspatial Arveson systems is somewhat an artificial definition), everything works as we know it from Arveson systems, provided we indicate the good product operation.

In the theory of Arveson systems, there is the tensor product (of arbitrary Arveson systems). However, for modules this does not work. (You may write down the tensor product of correspondences, but, in general, it is not possible to define a product system structure.)

The **product** of two spatial product systems $E^{i\odot}$ ($i = 1, 2$) with reference units $\omega^{i\odot}$ is the spatial product system $(E^1 \odot E^2)^\odot$ with reference unit $\omega^\odot$ which is characterized uniquely up to spatial isomorphism by the following properties:

(1) There are spatial isomorphisms $w^{i\odot}$ from $E^{i\odot}$ onto subsystems of $(E^1 \odot E^2)^\odot$.
(2) $(E^1 \odot E^2)^\odot$ is generated by these two subsystems.
(3) $\langle w_t^1(x_t^1), w_t^2(y_t^2)\rangle = \langle x_t^1, \omega_t^1\rangle\langle \omega_t^2, y_t^2\rangle$.

Existence of the product follows by an inductive limit; see Ref.[23] By Condition 1 we may and, usually, will identify the factors as subsystem of the product. Condition 3 means, roughly speaking, that the reference units of the two factors are identified, while components from different factors which are orthogonal to the respective reference unit are orthogonal in the product. Condition 2 means that

$$(E^1 \odot E^2)_t = \overline{\text{span}}\{x_{t_n}^n \odot \ldots \odot x_{t_1}^1 : \\ n \in \mathbb{N}, t_n + \ldots + t_1 = t, x_{t_k}^k \in E_{t_k}^i \ (i = 1, 2)\}.$$

It is important to note (crucial exercise!) that this may be rewritten in the form

$$(E^1 \odot E^2)_t = \overline{\text{span}}\{x_{t_n}^n \odot \ldots \odot x_{t_1}^1 : \\ n \in \mathbb{N}, t_n + \ldots + t_1 = t, x_{t_k}^k \in \mathcal{E}_{t_k}^i \ (i = 1, 2)\}, \quad (3)$$

where we put $\mathcal{E}_t := \mathcal{B}\omega_t \oplus (E_t^1 \ominus \mathcal{B}\omega_t^1) \oplus (E_t^2 \ominus \mathcal{B}\omega_t^2)$ (the direct sum of E^1 and E^2 with "identification of the reference vectors" and denoting the new reference vector by ω_t). Written in that way, it is easy to see that the subspaces are actually increasing over the partitions $t_n + \ldots + t_1 = t$ of $[0, t]$. This gives an idea how to obtain the product as an inductive limit; see Ref.[23]

3. The product system of $\mathcal{B}$-correspondences of a CP-semigroup on $\mathcal{B}^a(E)$

In Section 2.3 we have said what the product system of CP-semigroup on $\mathcal{B}$ is. It is a product system of $\mathcal{B}$-correspondences. On the other hand, if $\mathcal{B}(H)$-people speak about the product system of a unital CP-semigroup on $\mathcal{B}(H)$, they mean an Arveson system, that is, a product system of Hilbert spaces. Following Bhat[4] and Arveson,[2] the Arveson system of a unital CP-semigroup is the Arveson system of its minimal dilating E_0-semigroup. (To be specific, we mean the product system constructed as in Section 2.4 following Ref.,[4] not the product system constructed in Ref.,[1] which is anti-isomorphic to the former.) A precise understanding of the relation between the two product systems, one of $\mathcal{B}(H)$-modules, the other of Hilbert spaces, will allow to avoid the construction of the minimal dilation. But we will discuss it immediately for CP-semigroups on $\mathcal{B}^a(E)$.

Suppose we have a Hilbert $\mathcal{B}^a(E)$-module F. Then we may define the Hilbert $\mathcal{B}$-module $F \odot E$. Every $y \in F$ gives rise to a mapping $y \odot \text{id} \in \mathcal{B}^a(E, F \odot E)$ defined by $(y \odot \text{id}_E)x = y \odot x$ with adjoint $y^* \odot \text{id}_E : y' \odot x \mapsto \langle y, y' \rangle x$. These mappings fulfill $(y \odot \text{id}_E)^*(y' \odot \text{id}_E) = \langle y, y' \rangle$ and $ya \odot \text{id}_E = (y \odot \text{id}_E)a$ for every $a \in \mathcal{B}^a(E)$. Via $a \mapsto y \odot \text{id}_E$ we may identify F as a subset of $\mathcal{B}^a(E, F \odot E)$. This subset is strictly dense but, in general, it need not coincide. In fact, we have always $F \supset \mathcal{K}(E, F \odot E)$ where the **compact operators** between Hilbert $\mathcal{B}$-modules E_1 and E_2 are defined as $\mathcal{K}(E_1, E_2) := \overline{\text{span}}\{x_1 x_2^* : x_i \in E_i\}$, and $F = \mathcal{K}(E, F \odot E)$ whenever the right multiplication is strict (in the same sense as left multiplication, namely, $\overline{\operatorname{span}} FEE^* = F$).

Remark 3.1. The space $\mathscr{B}^a(E, F \odot E)$ may be thought of as the strict completion of F, and it is possible to define a strict tensor product of $\mathscr{B}^a(E)$-correspondences. We do not need this here, and refer the interested reader to Ref.[22]

Now suppose that F is a $\mathscr{B}^a(E)$-correspondence with strict left action. If E has a unit vector, then, doing as in Section 2.4, we see that F factors into $E \odot F_E$ (where the F_E is a suitable multiplicity correspondence from $\mathcal{B}$ to $\mathscr{B}^a(E)$) and $a \in \mathscr{B}^a(E)$ acts on $F = E \odot F_E$ as $a \odot \operatorname{id}_{F_E}$. For several reasons we do not follow Section 2.4, but refer to the representation theory of $\mathscr{B}^a(E)$ from Ref.[15] This representation theory tells us that F_E may be chosen as $E^* \odot F$, where E^* is the **dual** $\mathcal{B}$-$\mathscr{B}^a(E)$-correspondence of E with operations $\langle x^*, x'^* \rangle := xx'^*$ and $bx^*a := (a^*xb^*)^*$. Then, clearly,

$$F = \overline{\operatorname{span}} \mathscr{K}(E)F = \mathscr{K}(E) \odot F = (E \odot E^*) \odot F = E \odot (E^* \odot F) = E \odot F_E$$

explains both how the isomorphism is to be defined and what the action of a is. Putting this together with the preceding construction, we obtain

$$\mathscr{B}^a(E, E \odot E_F) \supset F \supset \mathscr{K}(E, E \odot E_F) = E \odot E_F \odot E^*,$$

where we defined the $\mathcal{B}$-correspondence $E_F := E^* \odot F \odot E$.

Remark 3.2. We do not necessarily have equality $F = \mathscr{K}(E, E \odot E_F)$. But if we have (so that F is a *full* Hilbert $\mathscr{K}(E)$-module), then the operation of *tensor conjugation* with E^* may be viewed as an operation of Morita equivalence for correspondences in the sense of Muhly and Solel.[12] In what follows, the generalization to Morita equivalence of product systems[22] is in the background. An elaborate version for the strict tensor product (see Remark 3.1) can be found in Ref.[22]

We observe that the assignment (the functor, actually) $F \mapsto E_F := E^* \odot F \odot E$ respects tensor products. Indeed, if F_1 and F_2 are $\mathscr{B}^a(E)$-correspondences with strict left actions, then

$$\begin{aligned} E_{F_1} \odot E_{F_2} &= (E^* \odot F_{F_1} \odot E) \odot (E^* \odot F_{F_2} \odot E) \\ &= E^* \odot F_{F_1} \odot (E \odot E^* \odot F_{F_2}) \odot E \\ &= E^* \odot F_{F_1} \odot F_{F_2} \odot E = E_{F_1 \odot F_2}. \end{aligned} \quad (6)$$

It is, clearly, associative. It respects inclusions and, therefore, inductive limits. If E is full, then $E^* \odot \mathscr{B}^a(E) \odot E = \mathcal{B}$. We summarize:

Proposition 3.1.[22] *Suppose that E is full (for instance, E has a unit vector). Suppose that $F^\odot = (F_t)_{t \in \mathbb{R}_+}$ is a product system of $\mathscr{B}^a(E)$-correspondences such that the left actions of all F_t are strict.*

Then the family $E^\odot = (E_t)_{t \in \mathbb{R}_+}$ of $\mathcal{B}$-correspondences $E_t := E^* \odot F_t \odot E$ with product system structure defined by (6) is a product system.

Moreover, if the F_t are inductive limits over families $\mathscr{F}_\mathfrak{t}$, then the E_t are inductive limits over the corresponding $\mathscr{E}_\mathfrak{t} := E^* \odot \mathscr{F}_\mathfrak{t} \odot E$.

Theorem 3.1. *Let $F^\odot$ be the GNS-system of a strict unital CP-semigroup T on $\mathscr{B}^a(E)$, and denote by (F, θ, ζ) the minimal dilation of T. Then $E^\odot$ (from Proposition 3.1) is the product system of the strict E_0-semigroup ϑ induced on $\mathscr{B}^a(F \odot E) \cong \mathscr{B}^a(F) \odot \mathrm{id}_E = \mathscr{B}^a(F)$ by θ.*

The triple $(F \odot E, \vartheta, p = \vartheta_0(\zeta \zeta^*))$ is the unique minimal dilation of T to the operators on a Hilbert $\mathcal{B}$-module in the sense that

$$p(F \odot E) = 1_{\mathscr{B}^a(E)} \odot E = E$$

and

$$p\vartheta_t(a)p = T_t(a).$$

Proof. We proceed precisely as in the proof of Theorem 5.12 in Ref.[22] We know (see Section 2.5) that the product system of the minimal θ is $F^\odot$. Though, we have a unit vector ζ in F, it is more suggestive to think of the correspondences F_t to be obtained as $F_t = F^* \odot {}_t F$ where ${}_t F$ is F viewed as $\mathscr{B}^a(E)$-correspondences via θ_t; see Ref.,[22] Section 2, for details. In the same way, the product system of ϑ is $(F \odot E)^* \odot {}_t(F \odot E)$. We find

$$(F \odot E)^* \odot {}_t(F \odot E) = (E^* \odot F^*) \odot ({}_t F \odot E)$$
$$= E^* \odot (F^* \odot {}_t F) \odot E = E^* \odot F_t \odot E = E_t.$$

(Note: The first step where ${}_t$ goes from outside the brackets into, is just the definition of ϑ_t.) This shows the first statement.

For the second statement, we observe that $x \mapsto \zeta \odot x$ provides an isometric embedding of E into $F \odot E$ and that p is the projection on the range $\zeta \odot E$ of this embedding. Clearly,

$$p\vartheta_t(a)p = (\zeta\zeta^* \odot \mathrm{id}_E)(\theta_t(a) \odot \mathrm{id}_E)(\zeta\zeta^* \odot \mathrm{id}_E)$$
$$= \left(\zeta\langle \zeta, \theta_t(a)\zeta\rangle\zeta^*\right) \odot \mathrm{id}_E = (\zeta T_t(a)\zeta^*) \odot \mathrm{id}_E = T_t(a),$$

when $\mathscr{B}^a(E)$ is identified with the corner $(\zeta\mathscr{B}^a(E)\zeta^*) \odot \mathrm{id}_E$ in $\mathscr{B}^a(F \odot E)$. $\square$

Remark 3.3. If T is a normal unital CP-semigroup on $\mathscr{B}(H)$ (normal CP-maps on $\mathscr{B}(H)$ are strict), then $E^\odot$ is nothing but the Arveson system of T (in the sense of Bhat's construction). Note that we did construct $E^\odot$ **without** constructing the minimal dilation first. In the theorem the minimal dilation occurred only, because we wanted to verify that our product system coincides with the one constructed via minimal dilation.

Remark 3.4. We hope that the whole discussion could help to clarify the discrepancy between the terminology and constructions in the case of CP-semigroups on $\mathscr{B}(H)$ and those for CP-semigroups on $\mathcal{B}$. The semigroups of this section lie in between, in that they are CP-semigroups on $\mathscr{B}^a(E)$, so not general $\mathcal{B}$ but also not just $\mathscr{B}(H)$. The operation that transforms the product system of $\mathscr{B}^a(E)$-correspondences into a product system of $\mathcal{B}$-correspondences is *cum grano salis* an operation of Morita equivalence. (In the von Neumann case and when E is full, it is Morita equivalence.) We obtain $\mathcal{B}$-correspondences because E is a Hilbert $\mathcal{B}$-module. For $\mathscr{B}(H)$ we obtain $\mathbb{C}$-correspondences (or Hilbert spaces), because H is a Hilbert $\mathbb{C}$-module.

4. Powers' CP-semigroup

We, finally, come to Powers' CP-semigroup and to the generalization to Hilbert modules of the result from Ref.[20] that its product system is the product of the involves spatial product systems.

Let ϑ^i ($i = 1, 2$) be two strict E_0-semigroup on $\mathscr{B}^a(E^i)$ (E^i two Hilbert $\mathcal{B}$-modules with unit vectors ω^i) with spatial product systems $E^{i\odot}$ (as in Section 2.4) and unital central reference units $\omega^{i\odot}$. Since ω^i_t commutes with $\mathcal{B}$, the mapping $b \mapsto \omega^i_t b$ is bilinear. Consequently, $\Omega^i_t := \mathrm{id}_{E^i} \odot \omega^i_t \colon x^i \mapsto x^i \odot \omega^i_t \in E^i \odot E^i_t = E^i$ defines a semigroup of isometries in $\mathscr{B}^a(E^i)$. (The isometries are *intertwining* in the sense that $\vartheta^i_t(a)\Omega^i_t = (a \odot \mathrm{id}_{E^i_t})(\mathrm{id}_{E^i} \odot \omega^i_t) = (\mathrm{id}_{E^i} \odot \omega^i_t)a = \Omega^i_t a$.) It follows that

$$T_t \begin{pmatrix} a_{11} & a_{12} \\ a_{21} & a_{22} \end{pmatrix} = \begin{pmatrix} \vartheta^1_t(a_{11}) & \Omega^1_t a_{12} \Omega^{2*}_t \\ \Omega^2_t a_{21} \Omega^{1*}_t & \vartheta^2_t(a_{22}) \end{pmatrix}$$

defines a unital semigroup on $\mathscr{B}^a\left(\begin{smallmatrix}E^1\\E^2\end{smallmatrix}\right)$. (We see later on that T_t is completely positive, by giving its GNS-module explicitly.) Using the identifications $E^i = E^i \odot E^i_t$ we find the more convenient form

$$T_t \begin{pmatrix} a_{11} & a_{12} \\ a_{21} & a_{22} \end{pmatrix} = \begin{pmatrix} a_{11} \odot \mathrm{id}_{E^1_t} & (\mathrm{id}_{E^1} \odot \omega^1_t) a_{12} (\mathrm{id}_{E^2} \odot \omega^{2*}_t) \\ (\mathrm{id}_{E^2} \odot \omega^2_t) a_{21} (\mathrm{id}_{E^1} \odot \omega^{1*}_t) & a_{22} \odot \mathrm{id}_{E^i_t} \end{pmatrix}$$

where T_t maps from $\mathscr{B}^a\left(\begin{smallmatrix}E^1\\E^2\end{smallmatrix}\right)$ to $\mathscr{B}^a\left(\begin{smallmatrix}E^1\odot E_t^1\\E^2\odot E_t^2\end{smallmatrix}\right) = \mathscr{B}^a\left(\begin{smallmatrix}E^1\\E^2\end{smallmatrix}\right)$.

Denote by $F^{\odot}$ the product system of T in the sense of Section 2.4, that is, the F_t are $\mathscr{B}^a\left(\begin{smallmatrix}E^1\\E^2\end{smallmatrix}\right)$-correspondences. By Proposition 3.1, setting $E_t := \left(\begin{smallmatrix}E^1\\E^2\end{smallmatrix}\right)^* \odot F_t \odot \left(\begin{smallmatrix}E^1\\E^2\end{smallmatrix}\right)$ we define a product system $E^{\odot}$ of $\mathcal{B}$-correspondences and

$$\mathscr{K}\left(\left(\begin{smallmatrix}E^1\\E^2\end{smallmatrix}\right), \left(\begin{smallmatrix}E^1\\E^2\end{smallmatrix}\right) \odot E_t\right) \subset F_t \subset \mathscr{B}^a\left(\left(\begin{smallmatrix}E^1\\E^2\end{smallmatrix}\right), \left(\begin{smallmatrix}E^1\\E^2\end{smallmatrix}\right) \odot E_t\right).$$

Theorem 4.1. $E^{\odot}$ is the product $(E^1 \odot E^2)^{\odot}$ of the spatial product systems $E^{1\odot}$ and $E^{2\odot}$.

Proof. Recall that, by (1), F_t is the inductive limit of expressions of the form

$$\mathscr{F}_{\mathbf{t}} := \mathscr{F}_{t_n} \odot \ldots \odot \mathscr{F}_{t_1}$$

over the partitions $\mathbf{t} = (t_n, \ldots, t_1)$ with $t_n + \ldots + t_1 = t$, where $\mathscr{F}_t$ is the GNS-module of T_t with cyclic vector ζ_t.

Put $\mathscr{E}_t = \left(\begin{smallmatrix}E^1\\E^2\end{smallmatrix}\right)^* \odot \mathscr{F}_t \odot \left(\begin{smallmatrix}E^1\\E^2\end{smallmatrix}\right)$. Then $\mathscr{F}_t \subset \mathscr{B}^a\left(\left(\begin{smallmatrix}E^1\\E^2\end{smallmatrix}\right), \left(\begin{smallmatrix}E^1\\E^2\end{smallmatrix}\right) \odot \mathscr{E}_t\right)$. We claim that $\mathscr{E}_t = \mathcal{B}\omega_t \oplus (E_t^1 \ominus \mathcal{B}\omega_t^1) \oplus (E_t^2 \ominus \mathcal{B}\omega_t^2)$ and that ζ_t is the operator given by

$$\zeta_t\begin{pmatrix}x^1\\x^2\end{pmatrix} = \begin{pmatrix}z^1\\0\end{pmatrix} \odot (\langle\omega_t^1, y_t^1\rangle, p_t^1 y_t^1, 0) + \begin{pmatrix}0\\z^2\end{pmatrix} \odot (\langle\omega_t^2, y_t^2\rangle, 0, p_t^2 y_t^2),$$

with $\begin{pmatrix}x^1\\x^2\end{pmatrix} = \begin{pmatrix}z^1 \odot y_t^1\\z^2 \odot y_t^2\end{pmatrix} \in \left(\begin{smallmatrix}E^1\\E^2\end{smallmatrix}\right) = \left(\begin{smallmatrix}E^1 \odot E_t^1\\E^2 \odot E_t^2\end{smallmatrix}\right)$ and $p_t^i := \mathrm{id}_{E_t^i} - \omega_t^i \omega_t^{i*}$. To show this, we must check two things. Firstly, we must check whether $\langle\zeta_t, a\zeta_t\rangle = T_t(a)$. This is straightforward and we leave it as an exercise. Secondly, we must check whether elements of the form on the right-hand side of

$$\begin{pmatrix}x^1\\x^2\end{pmatrix}^* \odot \zeta_t \odot \begin{pmatrix}z^1 \odot y_t^1\\z^2 \odot y_t^2\end{pmatrix} \longmapsto \left(\begin{pmatrix}x^1\\x^2\end{pmatrix}^* \odot \mathrm{id}_{\mathscr{E}_t}\right) \zeta_t \begin{pmatrix}z^1 \odot y_t^1\\z^2 \odot y_t^2\end{pmatrix}$$

are total in $\mathscr{E}_t$. For the right-hand side we find

$$\left\langle\begin{pmatrix}x^1\\x^2\end{pmatrix}, \begin{pmatrix}z_1\\0\end{pmatrix}\right\rangle(\langle\omega_t^1, y_t^1\rangle, p_t^1 y_t^1, 0) + \left\langle\begin{pmatrix}x^1\\x^2\end{pmatrix}, \begin{pmatrix}0\\z_2\end{pmatrix}\right\rangle(\langle\omega_t^2, y_t^2\rangle, 0, p_t^2 y_t^2)$$

$$= (\langle x_1, z_1\rangle\langle\omega_t^1, y_t^1\rangle + \langle x_2, z_2\rangle\langle\omega_t^2, y_t^2\rangle, \langle x_1, z_1\rangle p_t^1 y_t^1, \langle x_2, z_2\rangle p_t^2 y_t^2).$$

From this, totality follows.

By Proposition 3.1, we obtain E_t as inductive limit over the expressions

$$\mathscr{E}_{\mathbf{t}} := \mathscr{E}_{t_n} \odot \ldots \odot \mathscr{E}_{t_1},$$

which, by the preceding computation, precisely coincides with what is needed, according to (3), to obtain $E_t = (E^1 \odot E^2)_t$. $\square$

References

1. W. Arveson. *Continuous analogues of Fock space*. Number 409 in Mem. Amer. Math. Soc. American Mathematical Society, 1989.
2. W. Arveson. Minimal E_0–semigroups. In P. Fillmore and J. Mingo, editors, *Operator algebras and their applications*, number 13 in Fields Inst. Commun., pages 1–12. American Mathematical Society, 1997.
3. S.D. Barreto, B.V.R. Bhat, V. Liebscher, and M. Skeide. Type I product systems of Hilbert modules. *J. Funct. Anal.*, 212:121–181, 2004. (Preprint, Cottbus 2001).
4. B.V.R. Bhat. An index theory for quantum dynamical semigroups. *Trans. Amer. Math. Soc.*, 348:561–583, 1996.
5. B.V.R. Bhat. *Cocycles of CCR-flows*. Number 709 in Mem. Amer. Math. Soc. American Mathematical Society, 2001.
6. B.V.R. Bhat, V. Liebscher, and M. Skeide. Subsystems of Fock need not be Fock. in preparation, 2008.
7. B.V.R. Bhat, V. Liebscher, and M. Skeide. The product of Arveson systems does not depend on the reference units. in preparation, 2008.
8. B.V.R. Bhat and K.R. Parthasarathy. Kolmogorov's existence theorem for Markov processes in C^*–algebras. *Proc. Indian Acad. Sci. (Math. Sci.)*, 104:253–262, 1994.
9. B.V.R. Bhat and M. Skeide. Tensor product systems of Hilbert modules and dilations of completely positive semigroups. *Infin. Dimens. Anal. Quantum Probab. Relat. Top.*, 3:519–575, 2000. (Rome, Volterra-Preprint 1999/0370).
10. I. Hirshberg. C^*–Algebras of Hilbert module product systems. *J. Reine Angew. Math.*, 570:131–142, 2004.
11. V. Liebscher. Random sets and invariants for (type II) continuous tensor product systems of Hilbert spaces. Preprint, arXiv: math.PR/0306365, 2003. To appear in Mem. Amer. Math. Soc.
12. P.S. Muhly and B. Solel. On the Morita equivalence of tensor algebras. *Proc. London Math. Soc.*, 81:113–168, 2000.
13. P.S. Muhly and B. Solel. Quantum Markov processes (correspondences and dilations). *Int. J. Math.*, 51:863–906, 2002. (arXiv: math.OA/0203193).
14. P.S. Muhly and B. Solel. Quantum Markov semigroups (product systems and subordination). *Int. J. Math.*, 18:633–669, 2007. (arXiv: math.OA/0510653).
15. P.S. Muhly, M. Skeide, and B. Solel. Representations of $\mathscr{B}^a(E)$. *Infin. Dimens. Anal. Quantum Probab. Relat. Top.*, 9:47–66, 2006. (arXiv: math.OA/0410607).
16. W.L. Paschke. Inner product modules over B^*–algebras. *Trans. Amer. Math. Soc.*, 182:443–468, 1973.
17. R.T. Powers. A non-spatial continuous semigroup of $*$–endomorphisms of $\mathscr{B}(\mathfrak{H})$. *Publ. Res. Inst. Math. Sci.*, 23:1053–1069, 1987.
18. R.T. Powers. Addition of spatial E_0–semigroups. In *Operator algebras, quantization, and noncommutative geometry*, number 365 in Contemporary Mathematics, pages 281–298. American Mathematical Society, 2004.
19. M. Skeide. Dilations, product systems and weak dilations. *Math. Notes*, 71:914–923, 2002.

20. M. Skeide. Commutants of von Neumann modules, representations of $\mathscr{B}^a(E)$ and other topics related to product systems of Hilbert modules. In G.L. Price, B .M. Baker, P.E.T. Jorgensen, and P.S. Muhly, editors, *Advances in quantum dynamics*, number 335 in Contemporary Mathematics, pages 253–262. American Mathematical Society, 2003. (Preprint, Cottbus 2002, arXiv: math.OA/0308231).
21. M. Skeide. Dilation theory and continuous tensor product systems of Hilbert modules. In W. Freudenberg, editor, *Quantum Probability and Infinite Dimensional Analysis*, number XV in Quantum Probability and White Noise Analysis, pages 215–242. World Scientific, 2003. Preprint, Cottbus 2001.
22. M. Skeide. Unit vectors, Morita equivalence and endomorphisms. Preprint, arXiv: math.OA/0412231v5 (Version 5), 2004.
23. M. Skeide. The index of (white) noises and their product systems. *Infin. Dimens. Anal. Quantum Probab. Relat. Top.*, 9:617–655, 2006. (Rome, Volterra-Preprint 2001/0458, arXiv: math.OA/0601228).
24. M. Skeide. Dilations of product sytems and commutants of von Neumann modules. in preparation, 2008.
25. B. Tsirelson. From random sets to continuous tensor products: answers to three questions of W. Arveson. Preprint, arXiv: math.FA/0001070, 2000.
26. B. Tsirelson. On automorphisms of type II Arveson systems (probabilistic approach). Preprint, arXiv: math.OA/0411062, 2004.

FREE MARTINGALE POLYNOMIALS FOR STATIONARY JACOBI PROCESSES

N. DEMNI*

*LPMA, Paris VI University,
Paris, France
E-mail: demni@ccr.jussieu.fr

We generalize a previous result concerning free martingale polynomials for the stationary free Jacobi process of parameters $\lambda \in]0.1]$, $\theta = 1/2$. Hopelessly, apart from the case $\lambda = 1$, the polynomials we derive are no longer orthogonal with respect to the spectral measure. As a matter of fact, we use the multiplicative renormalization to write down the corresponding orthogonality measure.

Keywords: stationary free Jacobi process, multiplicative renormalization method, Tchebycheff polynomials.

1. Preliminaries

Let $(\mathscr{A}, \phi)$ a $W^\star$-non commutative probability space. Easily speaking, $\mathscr{A}$ is a unital von Neumann algebra and ϕ is a tracial faithful linear functional (state). In a previous work,[8] we defined, via matrix theory, and studied a two parameters-dependent self-adjoint free process, called free Jacobi process. Our focus will be on a particular case called the stationary Jacobi process since its spectral distribution does not depend on time. It is defined by

$$J_t := PUY_t QY_t^\star U^\star P$$

where

- $(Y_t)_{t\geq 0}$ is a free multiplicative Brownian motion[7].
- U is a Haar unitary operator in $(\mathscr{A}, \Phi)$.
- P is a projection with $\Phi(P) = \lambda\theta \leq 1$, $\theta \in]0,1]$.
- Q is a projection with $\Phi(Q) = \theta$.
- $QP = PQ = \begin{cases} P & \text{if } \lambda \leq 1 \\ Q & \text{if } \lambda > 1 \end{cases}$
- $\{U, U^\star\}$, $\{(Y_t)_{t\geq 0}, (Y_t^\star)_{t\geq 0}\}$ and $\{P, Q\}$ are free.[13]

Remark 1.1. Y^*, U^* are the adjoint operators of Y, U respectively when the latters are viewed as operators acting on a infinite Hilbert space.

Thus the process $(J_t)_{t\geq 0}$ takes values in *the compressed space* $(P\mathscr{A}P, (1/\phi(P))\phi)$. The spectral distribution has the following decomposition:

$$\mu_{\lambda,\theta}(dx) = \frac{1}{2\pi\lambda\theta} \frac{\sqrt{(x_+ - x)(x - x_-)}}{x(1-x)} 1_{[x_-, x_+]}(x)dx + a_0\delta_0(dx) + a_1\delta_1(dx),$$

where δ_y stands for the Dirac mass at y with corresponding weight a_y, $y \in \{0,1\}$ and

$$x_\pm = \left(\sqrt{\theta(1-\lambda\theta)} \pm \sqrt{\lambda\theta(1-\theta)}\right)^2.$$

Its Cauchy-Stieltjes transform writes, for $z \in \mathbb{C} \setminus [0,1]$,

$$G_{\mu_{\lambda,\theta}}(z) = \frac{(2 - (1/\lambda\theta))z + (1/\lambda - 1) + \sqrt{Az^2 - Bz + C}}{2z(z-1)}, \quad (1)$$

with $A = 1/(\lambda\theta)^2$, $B = 2((1/\lambda\theta)(1+1/\lambda) - 2/\lambda)$ et $C = (1-1/\lambda)^2$. It was shown in [8] that if $\lambda \in]0,1], 1/\theta \geq \lambda + 1$ then the process is injective in $P\mathscr{A}P$, that is $a_0 = a_1 = 0$. Moreover, $\mu_{1,1/2}(dx)$ fits the Beta distribution $B(1/2, 1/2)$:

$$\mu_{1,1/2}(dx) = \frac{1}{\pi\sqrt{x(1-x)}} 1_{[0,1]}(x)dx.$$

Recall that the Tchebycheff polynomials of the first kind are defined by

$$T_n(x) = \cos(n \arccos x), \quad n \geq 0, |x| \leq 1,$$

and that they are orthogonal with respect to the image of $\mu_{1,1/2}(dx)$ by the map $x \mapsto 2x - 1$. Their generating function is given by:

$$g(u,x) = \sum_{n\geq 0} T_n(x)u^n = \frac{1 - ux}{1 - 2ux + u^2}, \quad |u| < 1.$$

In [8], we proved that for $r > 0$

$$g(re^t, J_t) = ((1 + re^t)P - 2e^t J_t)((1 + re^t)^2 P - 4re^t J_t)^{-1}, \quad t < -\ln r$$

defines a free martingale with respect to the natural filtration of J, say $\mathscr{J}_t$. One can express this result using Tchebycheff polynomials of the second kind defined by

$$U_n(\cos\alpha) := \frac{\sin(n+1)\alpha}{\sin\alpha}, \quad \alpha \in \mathbb{R},$$

with generating function given by

$$\sum_{n \geq 0} U_n(x) u^n = \frac{1}{1 - 2ux + u^2}, \quad |x| \leq 1, \ |u| < 1.$$

and related to the T_n's by $2T_n = U_n - U_{n-2}, U_{-1} := 0$. It follows that $\{M_t^n := e^{nt}(U_n - U_{n-2})(2J_t - P), \ n \geq 1\}_{t \geq 0}$ is a family of free martingale polynomials. The aim of this work is to extend this claim to the range $\theta = 1/2, \lambda \in]0, 1]$. The motivation comes from the fact that from a matrix theory point of view, the choice $\theta = 1/2$ corresponds to the ultraspherical multivariate Beta distribution.[8] Moreover, to our best knowledge, there is only one result concerning martingale polynomials for the stationary (classical) Jacobi process, which is restricted to the one dimensional case.[12] Before stating the main result, let us write

$$x_- = \left(\frac{\sqrt{2-\lambda}}{2} - \frac{\sqrt{\lambda}}{2}\right)^2 \leq x \leq x_+ = \left(\frac{\sqrt{2-\lambda}}{2} + \frac{\sqrt{\lambda}}{2}\right)^2$$

therefore $-1 \leq \frac{2x-1}{\sqrt{\lambda(2-\lambda)}} \leq 1$, so that mapping $\mu_{\lambda,1/2}$ to a measure supported in $[-1, 1]$ (set $u = (2x-1)/\sqrt{\lambda(2-\lambda)}$), one gets

$$\nu_\lambda(du) = \frac{2(2-\lambda)}{\pi} \frac{\sqrt{1-u^2}}{1 - \lambda(2-\lambda)u^2} \mathbf{1}_{[-1,1]}(u) du$$

which already appeared in [9] and was considered in [10] with $a = 1/(2-\lambda)$.

2. Main result

Theorem 2.1. *Set*

$$a(\lambda) = \frac{(1-\lambda)}{\sqrt{\lambda(2-\lambda)}}, \quad x_{t,\lambda} = \frac{2J_t - P}{\sqrt{\lambda(2-\lambda)}}.$$

For each $n \geq 1$, the process defined by

$$M_t^n := [U_n(x_{t,\lambda}) - 2a(\lambda) U_{n-1}(x_{t,\lambda}) - U_{n-2}(x_{t,\lambda})] \left(\frac{e^t}{\lambda(2-\lambda)}\right)^n, \ t \geq 0,$$

is a ($\mathscr{I}_t$)-free martingale.

We will see that the polynomials written above are not orthogonal with respect to ν_λ except for $\lambda = 1$. We will use the multiplicative renormalization method to write down their orthogonality measure. They will be also seen from their Jacobi-Szegö parameters (or the Jacobi matrix of finite-type) as a rank-one deformation of the T_ns.

Proof.

First step: it consists in deriving a martingale function for all values of $\lambda \in]0,1], \theta \leq 1/2 \leq 1/(\lambda+1)$. Inspired by the above expression of $h(re^t, J_t)$, we will look for martingales of the form

$$R_t := K_t(P - Z_t J_t)^{-1} = K_t \sum_{n \geq 0} Z_t^n J_t^n := K_t H_t$$

where K, Z are differentiable functions of the variable t lying in some interval $[0, t_0[$ such that $0 < Z_t < 1$ for $t \in [0, t_0[$. The finite variation part of dR_t is given by

$$FV(dR_t) = K'_t H_t dt + K_t FV(dH_t).$$

Our main tool is the free stochastic calculus and more precisely the free stochastic differential equation already set for J_t^n, $n \geq 1$ ([8]):

$$dJ_t^n = dM_t + n(\theta P - J_t) J_t^{n-1} dt$$

$$+ \lambda \theta \sum_{l=1}^{n-1} l[m_{n-l}(P - J_t) J_t^{l-1} + (m_{n-l-1} - m_{n-l}) J_t^l] dt$$

where dM stands for the martingale part and m_n is the n-th moment of J_t in $P\mathscr{A}P$:

$$m_n := \tilde{\phi}(J_t^n) := \frac{1}{\phi(P)} \phi(J_t^n).$$

The finite variation part $FV(dJ_t^n)$ of J_t^n transforms to:

$$FV(dJ_t^n) = n(\theta P - J_t) J_t^{n-1} dt$$

$$+ \lambda \theta \left[\sum_{l=1}^{n-1} l \left[m_{n-l} J_t^{l-1} + \sum_{l=1}^{n-1} l(m_{n-l-1} - 2m_{n-l}) J_t^l \right] \right] dt$$

$$= n(\theta P - J_t) J_t^{n-1} dt$$

$$+ \lambda \theta \sum_{l=1}^{n-1} l m_{n-l} J_t^{l-1} + \sum_{l=1}^{n} (l-1)(m_{n-l} - 2m_{n-l+1}) J_t^{l-1} dt$$

$$= n(\theta P - J_t) J_t^{n-1} dt$$

$$+ \lambda \theta \sum_{l=1}^{n} [l m_{n-l} + (l-1)(m_{n-l} - 2m_{n-l+1})] J_t^{l-1} dt - n\lambda \theta J_t^{n-1} dt$$

$$= n\theta(1-\lambda) J_t^{n-1} dt - n J_t^n dt$$

$$+ \lambda \theta \sum_{l=1}^{n} [m_{n-l} + 2(l-1)(m_{n-l} - m_{n-l+1})] J_t^{l-1} dt.$$

Thus

$$FV(dH_t) = \sum_{n\geq 1} nZ'_t Z_t^{n-1} J_t^n dt + \sum_{n\geq 1} Z_t FV(J_t^n)$$

$$= \sum_{n\geq 1} nZ'_t Z_t^{n-1} J_t^n dt - \sum_{n\geq 0} nZ_t^n J_t^n dt + \theta(1-\lambda) \sum_{n\geq 1} nZ_t^n J_t^{n-1} dt$$

$$+ \lambda\theta \sum_{n\geq 1}\sum_{l=1}^n Z_t^n m_{n-l} J_t^{l-1} dt$$

$$+ 2\lambda\theta \sum_{n\geq 1}\sum_{l=1}^n [(l-1)Z_t^n(m_{n-l} - m_{n-l+1})] J_t^{l-1} dt$$

$$= \sum_{n\geq 1} n[Z'_t Z_t^{n-1} - Z_t^n] J_t^n dt + \theta(1-\lambda) \sum_{n\geq 0} (n+1) Z_t^{n+1} J_t^n dt$$

$$+ \lambda\theta \sum_{n\geq 0}\sum_{l\geq 0} Z_t^{n+l+1} m_n J_t^l dt$$

$$+ 2\lambda\theta \sum_{n\geq 0}\sum_{l\geq 0} lZ_t^{n+l+1}(m_n - m_{n+1}) J_t^l dt$$

$$= [Z'_t/Z_t - 1 + \theta(1-\lambda)Z_t] \sum_{n\geq 1} nZ_t^n J_t^n dt + \theta(1-\lambda) Z_t \sum_{n\geq 0} Z_t^n J_t^n dt$$

$$+ \lambda\theta \sum_{n\geq 0} Z_t^{n+1} m_n \sum_{l\geq 0} Z_t^l J_t^l dt$$

$$+ 2\lambda\theta \sum_{n\geq 0} [Z_t^{n+1}(m_n - m_{n+1})] \sum_{l\geq 0} lZ_t^l J_t^l dt.$$

Recall that the Cauchy-Stieltjes transform of a measure on the real line is defined by

$$G_\nu(z) = \int_\mathbb{R} \frac{1}{z-x} \nu(dx) = \sum_{n\geq 0} \frac{1}{z^{n+1}} \int_\mathbb{R} x^n \nu(dx)$$

for some values of z for which both the integral and the infinite sum make sense. Then, since $0 < Z < 1$ and $\mu_{\lambda,\theta}$ is supported in $[0,1]$, it is easy to see that

$$\sum_{n\geq 0} Z_t^{n+1}(m_n - m_{n+1}) = \left(1 - \frac{1}{Z_t}\right) G_{\mu_{\lambda,\theta}}\left(\frac{1}{Z_t}\right) + 1$$

with $G_{\mu_{\lambda,\theta}}$ given by (1). This gives

$$2\lambda\theta(1-z) G_{\mu_{\lambda,\theta}}(z) = \frac{(1-2\lambda\theta)z - \theta(1-\lambda) - \sqrt{z^2 - (\lambda\theta)^2 Bz + (\lambda\theta)^2 C}}{z},$$

so that

$$2\lambda\theta(1-Z_t^{-1})G_{\mu_{\lambda,\theta}}(Z_t^{-1})+2\lambda\theta = 1-\theta(1-\lambda)Z_t - \sqrt{1-(\lambda\theta)^2 BZ_t+(\lambda\theta)^2 CZ_t^2}.$$

We finally get:

$$FV(dH_t) = [Z_t'/Z_t - \sqrt{1-(\lambda\theta)^2 BZ_t+(\lambda\theta)^2 CZ_t^2}]\sum_{n\geq 1} nZ_t^n J_t^n dt$$

$$+ \left[\lambda\theta G_{\mu_{\lambda,\theta}}\left(\frac{1}{Z_t}\right) + \theta(1-\lambda)Z_t\right]\sum_{n\geq 0} Z_t^n J_t^n dt$$

In order to derive free martingales, we shall pick Z such that $Z_t' = Z_t\sqrt{1-(\lambda\theta)^2 BZ_t + (\lambda\theta)^2 CZ_t^2}$. This shows that Z is an increasing function and one can solve the above non linear differential equation as follows: use the variables change $u = Z_t$, $t < t_0$, then integrate to get:

$$\int_{[Z_0, Z_t]} \frac{du}{u\sqrt{1-2\theta(1+\lambda-2\lambda\theta)u+(\theta(1-\lambda))^2 u^2}} = t.$$

Let $c_1 = 2\theta(1+\lambda-2\lambda\theta)$, $c_2 = \theta^2(1-\lambda)^2$. Then, the function $u \mapsto 1 - c_1 u + c_2 u^2$ is decreasing for $u \in\,]0,1[$: in fact,

$$2c_2 u - c_1 < 2c_2 - c_1 = 2\theta^2(1-\lambda)^2 - 2\theta(1+\lambda-2\lambda\theta)$$

$$= 2\theta[\theta(1+\lambda^2) - (1+\lambda)] \leq 2\theta\left(\frac{1+\lambda^2}{1+\lambda} - (1+\lambda)\right)$$

$$= -\frac{4\lambda\theta}{1+\lambda} < 0$$

which yields $1 - c_1 u + c_2 u^2 > 1 - c_1 + c_2 = (1-\theta(1+\lambda))^2 \geq 0$.

Next, use the variable change $1 - vu = \sqrt{1 - c_1 u + c_2 u^2}$. This gives

$$u = \frac{2v - c_1}{v^2 - c_2}, \quad du = -2\frac{v^2 + c_2 - c_1 v}{(v^2 - c_2)^2}dv, \quad 1 - vu = -\frac{v^2 + c_2 - c_1 v}{v^2 - c_2}.$$

Moreover

$$u \mapsto v = \frac{1 - \sqrt{1 - c_1 u + c_2 u^2}}{u}, \quad 0 < u < 1$$

is an increasing function: in fact the numerator of its derivative writes

$$c_1 u - 2c_2 u^2 + 2(1 - c_1 u + c_2 u^2) - 2\sqrt{1 - c_1 u + c_2 u^2} = (2 - c_1 u) - 2\sqrt{1 - c_1 u + c_2 u^2}.$$

Since $2 - c_1 u > 2 - c_1 = 2(1-\theta(1+\lambda)) + 4\lambda\theta^2 > 0$, our claim follows from the fact that $c_1^2 - 4c_2 = 16\lambda\theta^2(1-\lambda\theta)(1-2\theta) \geq 0$.

Free Martingale Polynomials for Stationary Jacobi Processes   113

Finally, the integral transforms to

$$\int_{[v_0, v_t]} \frac{2dv}{2v - c_2} = \log \left| \frac{2v_t - c_1}{2v_0 - c_1} \right| = t$$

where $1 - Z_t v_t = \sqrt{1 - c_1 Z_t + c_2 Z_t^2}$, $1 - Z_0 v_0 = \sqrt{1 - c_1 Z_0 + c_2 Z_0^2}$.
Note also that $c_1^2 - 4c_2 \geq 0$ implies that for all $u \in [Z_0, Z_t] \subset]0, 1[$

$$v - \frac{c_1}{2} = \frac{1 - \sqrt{1 - c_1 u + c_2 u^2}}{u} - \frac{c_1}{2} = \frac{(1 - c_1 u/2) - \sqrt{1 - c_1 u + c_2 u^2}}{u}$$

$$= \frac{(1 - c_1 u/2)^2 - (1 - c_1 u + c_2 u^2)}{u((1 - c_1 u/2) + \sqrt{1 - c_1 u + c_2 u^2})} \geq 0$$

since $1 - c_1/2u \geq 1 - c_1/2 \geq 0$. Thus $v \geq c_1/2 \geq \sqrt{c_2}$.

$$v_t = [(2v_0 - c_1)e^t + c_1]/2 \Leftrightarrow \sqrt{1 - c_1 Z_t + c_2 Z_t^2} = 1 - \frac{(2v_0 - c_1)e^{\pm t} + c_1}{2} Z_t.$$

We finally get

$$Z_t = \frac{4(2v_0 - c_1)e^{\pm t}}{((2v_0 - c_1)e^t + c_1)^2 - 4c_2}, \quad t \leq t_0$$

where t_0 is the first time such that $Z_{t_0} = 1 \Leftrightarrow (2v_0 - c_1)e^{t_0} + c_1)^2 - 4c_2 = 4(2v_0 - c_1)e^{t_0}$. Set $r = r(\lambda, \theta) := (2v_0 - c_1)$ and $x_0 = e^{t_0} > 1$, then $r^2 x_0^2 + 2(c_1 - 2)r x_0 + c_1^2 - 4c_2 = 0$. The discriminant equals to $\Delta = 16r^2(1 + c_2 - c_1) = 16r^2(1 - \theta(1 + \lambda))^2$. Thus

$$x_0 = \frac{-(c_1 - 2) - 2(1 - \theta(1+\lambda))}{r} = \frac{2(1 - \theta(1+\lambda)) + 4\lambda\theta^2 - 2(1 - \theta(1+\lambda))}{r} = \frac{4\lambda\theta^2}{r} \geq 1.$$

The last inequality follows from the fact that

$$1 - \sqrt{c_2} u \geq 1 - \theta(1 + \lambda) \geq 0$$

and from

$$r - 4\lambda\theta^2 = 2v_0 - c_1 - 4\lambda\theta^2 = 2(v_0 - \theta(1 + \lambda)) = 2(v_0 - \sqrt{c_2}) \leq 0.$$

It gives $t_0 = -\ln(r/4\lambda\theta^2)$. Note also that the denominator is well defined for all $t \leq t_0$ since $c_1^2 \geq 4c_2$ and $2v_0 - c_1 \geq 0$.
For the ramaining terms, we shall choose K such that

$$K_t' + K_t \left[\lambda \theta G_{\mu_{\lambda,\theta}} \left(\frac{1}{Z_t} \right) + \theta(1 - \lambda) Z_t \right] = 0.$$

An easy computation shows that this equals to

$$K_t' + \frac{K_t}{2} \left[\theta(1 - \lambda) \frac{Z_t^2}{Z_t - 1} + (1 - 2\theta) \frac{Z_t}{Z_t - 1} - \frac{Z_t \sqrt{1 - c_1 Z_t + c_2 Z_t^2}}{Z_t - 1} \right] = 0.$$

Remembering the choice of the function Z, this writes
$$K'_t - \frac{K_t}{2}\left[\frac{Z'_t}{Z_t - 1} - (1 - 2\theta)\frac{Z_t}{Z_t - 1} - \theta(1-\lambda)\frac{Z_t^2}{Z_t - 1}\right] = 0$$
or equialently
$$K'_t - \frac{K_t}{2}\left[\frac{Z'_t}{Z_t - 1} - (1 - \theta - \lambda\theta)\frac{Z_t}{Z_t - 1} - \theta(1-\lambda)Z_t\right] = 0.$$
If $K_t \neq 0$, then
$$\log K_t = \frac{1}{2}\log(1 - Z_t) - \frac{1 - \theta - \lambda\theta}{2}\int \frac{Z_s}{Z_s - 1}ds - \frac{\theta(1-\lambda)}{2}\int Z_s ds + C.$$
If $\lambda \neq 1$, then the last term is given by
$$-\frac{\theta(1-\lambda)}{2}\int Z_s ds = \frac{\theta(1-\lambda)}{\sqrt{c_2}}\int \frac{(r/2\sqrt{c_2})e^t}{1 - \left(\frac{re^t + c_1}{2\sqrt{c_2}}\right)^2} = \arg\tanh\left(\frac{re^t + c_1}{2\sqrt{c_2}}\right)$$
where $\arg\tanh(u) = (1/2)\log((u+1)/(u-1)), |u| > 1$. The second term writes
$$\frac{Z_t}{Z_t - 1} = \frac{4re^t}{4c_2 + 4re^t - (re^t + c_1)^2}$$
$$= \frac{4re^t}{4c_2 - c_1^2 + (c_1 - 2)^2 - (re^t + c_1 - 2)^2}$$
$$= \frac{re^t}{c_2 + 1 - c_1 - \left(\frac{re^t + c_1 - 2}{2}\right)^2}$$
$$= \frac{1}{c_2 + 1 - c_1}\frac{re^t}{1 - \left(\frac{re^t + c_1 - 2}{2\sqrt{c_2 + 1 - c_1}}\right)^2}$$
$$= \frac{2}{\sqrt{c_2 + 1 - c_1}}\frac{(r/2\sqrt{c_2 + 1 - c_1})e^t}{1 - \left(\frac{re^t + c_1 - 2}{2\sqrt{c_2 + 1 - c_1}}\right)^2}$$
Observe that $2 - c_1 - re^t > 2 - c_1 - re^{t_0} = 2(1 - \theta(1+\lambda)) \geq 0$. Thus, if $\theta(1+\lambda) \neq 1$
$$\frac{1 - \theta(1+\lambda)}{2}\int \frac{Z_s}{Z_s - 1}ds = \arg\tanh\left(\frac{2 - c_1 - re^t}{2\sqrt{c_2 + 1 - c_1}}\right).$$
As a result, if $\lambda \neq 1$ ($\theta \leq 1/2 < 1/(\lambda + 1)$),
$$K_t = C(1 - Z_t)^{1/2}\left(\frac{re^t + c_1 + 2\sqrt{c_2}}{re^t + c_1 - 2\sqrt{c_2}}\right)^{1/2}\left(\frac{2 - c_1 - 2c_3 - re^t}{2 - c_1 + 2c_3 - re^t}\right)^{1/2}

where $c_3 := \sqrt{c_2 + 1 - c_1} = 1 - \theta(\lambda + 1)$. Note that for $\lambda = 1, \theta = 1/2$, $c_1 = 1, c_2 = 0, c_3 = 0$ and

$$K_t = C \frac{1 - re^t}{1 + re^t}, \quad t < t_0 = -\ln r.$$

Second step; the case $\theta = 1/2, \lambda \neq 1$. One has

$$c_1 = 1, \; c_2 = \frac{(1-\lambda)^2}{4}, \; c_3 = \sqrt{c_2} = \frac{1-\lambda}{2}, \; Z_t = \frac{4re^t}{(re^t + 1)^2 - (1-\lambda)^2}$$

$$c_1 + 2\sqrt{c_2} = 2(1 + c_3) - c_1 = 2 - \lambda, \; c_1 - 2\sqrt{c_2} = 2(1 - c_3) - c_1 = \lambda.$$

$$1 - Z_t = \frac{(re^t - 1)^2 - (1-\lambda)^2}{(re^t + 1)^2 - (1-\lambda)^2} = \frac{(re^t + \lambda - 2)(re^t - \lambda)}{(re^t + 2 - \lambda)(re^t + \lambda)}.$$

Thus, for $t < -\ln(r/\lambda)$,

$$K_t = C \frac{\lambda - re^t}{\lambda + re^t}$$

so that

$$R_t = C \frac{\lambda - re^t}{\lambda + re^t}(P - \frac{4re^t}{(re^t+1)^2 - (1-\lambda)^2} J_t)^{-1}$$

$$= C(\lambda - re^t)(2 - \lambda + re^t)(\lambda(2-\lambda)P + (re^t)^2 P - 2re^t(2J_t - P))^{-1}$$

$$= \frac{C(\lambda - re^t)(2-\lambda+re^t)}{\lambda(2-\lambda)}\left(P - \frac{2re^t}{\sqrt{\lambda(2-\lambda)}} \frac{(2J_t - P)}{\sqrt{\lambda(2-\lambda)}} + \frac{(re^t)^2}{\lambda(2-\lambda)}P\right)^{-1}$$

$$= C\left(1 - 2\frac{(1-\lambda)}{\sqrt{\lambda(2-\lambda)}}\frac{re^t}{\sqrt{\lambda(2-\lambda)}} - \frac{(re^t)^2}{\lambda(2-\lambda)}\right)$$

$$\times \left(P - \frac{2re^t}{\sqrt{\lambda(2-\lambda)}} \frac{(2J_t - P)}{\sqrt{\lambda(2-\lambda)}} + \frac{(re^t)^2}{\lambda(2-\lambda)}P\right)^{-1}$$

is a free martingale with respect to the natural filtration $\mathscr{J}_t$. Besides, since $\lambda \in]0,1]$, then $\lambda \leq \sqrt{\lambda(2-\lambda)}$, hence $(re^t)/(\sqrt{\lambda(2-\lambda)}) < 1$ for all $t < -\ln(r/\lambda)$. Now, let us consider the following generating function

$$g(u,x) = \frac{1 - 2au - u^2}{1 - 2xu + u^2}, \quad 0 < a, u < 1, \; |x| \leq 1,$$

or equivalently

$$g(u,x) = U_0(x) + (U_1(x) - 2a)u + \sum_{n \geq 2}[U_n(x) - 2aU_{n-1}(x) - U_{n-2}(x)]u^n$$

Setting $u_{t,\lambda} := re^t/(\sqrt{\lambda(2-\lambda)})$, $t < t_0$, then

$$R_t = C[P + (x_{t,\lambda} - 2a(\lambda)P)]u_{t,\lambda} + \sum_{n \geq 2}[U_n(x_{t,\lambda}) - 2a(\lambda)U_{n-1}(x_{t,\lambda}) - U_{n-2}(x_{t,\lambda})]u_{t,\lambda}^n.$$

Setting $U_{-1} = U_{-2} = 0$, it can be written as
$$R_t = C \sum_{n \geq 0} [U_n(x_{t,\lambda}) - 2a(\lambda)U_{n-1}(x_{t,\lambda}) - U_{n-2}(x_{t,\lambda})] u_{t,\lambda}^n.$$
□

Remark 2.1. In Ref. 11, authors provided orthogonal polynomials with respect to the measure with density given by:
$$f(x) = \frac{c\sqrt{1-x^2}}{\pi[b^2 + c^2 - 2b(1-c)x + (1-2c)x^2]} \mathbf{1}_{[-1,1]}(x),$$
where $|b| < 1 - c$, $0 < c \leq 1$. This fits the image of absolutely continuous part of $\mu_{\lambda,\theta}$ by the map
$$u = \frac{2x - s}{d} \in [-1, 1]$$
with $d = d(\lambda, \theta) = x_+ - x_- = 4\theta\sqrt{\lambda(1-\theta)(1-\lambda\theta)}$, $s = s(\lambda, \theta) = x_+ + x_- = 2\theta(1 + \lambda - 2\lambda\theta)$. One gets
$$\nu_{\lambda,\theta}(dx) = \frac{d^2}{2\pi\lambda\theta} \frac{\sqrt{1-x^2}}{s(2-s) + 2d(1-s)x - d^2x^2} dx$$
which provides the following relations
$$c = \frac{1}{2(1-\lambda\theta)}, \quad b = \sqrt{\frac{\lambda}{(1-\theta)(1-\lambda\theta)}}(2\theta - 1) \quad (2)$$

As a result, one can derive the correponding orthogonal polynomials for $\lambda \in]0,1], \theta \leq 1/(\lambda + 1)$ from the generating function:[11]
$$\phi(u, x) = \frac{1 - 2bu + (1-2c)u^2}{1 - 2ux + u^2}. \quad (3)$$
whence we deduce the Jacobi-szegö parameters $\alpha_0 = b, \alpha_n = 0$ for $n \geq 1$ and $\omega_1 = c/2, \omega_n = 1/4$ for $n \geq 1$. This gives us a realization of the spectral measure by means of the associated creation, annihilation and neutral operators defined on the appropriate one mode Interacting Fock space.[1]

3. Orthogonality measure for martingale polynomials

Consider the polynomials P_n^λ defined by
$$P_n^\lambda(x) = U_n(x) - 2a(\lambda)U_{n-1}(x) - U_{n-2}(x), \quad U_{-1} = U_{-2} := 0$$
with generating function
$$g(u, x) = \frac{1 - 2a(\lambda)u - u^2}{1 - 2xu + u^2}, \quad a(\lambda) = \frac{1-\lambda}{\lambda(2-\lambda)}, \quad 0 < u < 1.$$

The P_n^λ's appear in Ref. 2 as a limiting case of the q-Pollaczek polynomials. The coefficient of the highest monomial is equal to 2^n. Using the recurrence relation satisfied by $(U_n)_{n\geq 0}$,[6] one deduces that

$$2[x - a(\lambda)]P_0^\lambda(x) = P_1^\lambda(x)$$
$$2xP_1^\lambda(x) = P_2^\lambda(x) + 2P_0^\lambda(x)$$
$$2xP_n^\lambda(x) = P_{n+1}^\lambda(x) + P_{n-1}^\lambda(x), \quad n \geq 2.$$

Thus the Jacobi-Szegö parameters are given by $\alpha_0 = a(\lambda)$ and $\alpha_n = 0$ for all $n \geq 1$ and $\omega_1 = 1/2$, $\omega_n = 1/4$, $n \geq 2$ ($P_{-1}^\lambda = 0$).

One may use the multiplicative renormalization method[3–5,11] to write down the probability measure, ξ_λ, with respect to which the P_n^λs are orthogonal. Since $\alpha_0 \neq 0$, then ξ_λ is not symmetric. Moreover, with the same notations as in,[10] the moment generating function θ must be equal to

$$\theta(\rho(u)) := \theta\left(\frac{2u}{1+u^2}\right) = \frac{1+u^2}{1-2a(\lambda)-u^2}$$

so that

$$\theta(u) = \frac{1}{\sqrt{1-u^2} - a(\lambda)u}.$$

From the definition of θ, one deduces that

$$G_{\xi_\lambda}(u) := \int_{\mathbb{R}} \frac{1}{u-x}\xi_\lambda(dx) = \frac{1}{u}\theta\left(\frac{1}{u}\right) = \frac{\sqrt{u^2-1}+a(\lambda)}{u^2-(1+a^2(\lambda))}$$

for $|u| > 1$, $u \neq \pm\sqrt{1+a(\lambda)^2}$. Thus, ξ_λ possibly has two atoms $a_\pm$ at $\pm\sqrt{a^2(\lambda)+1}$ and an absolutely continuous part given by

$$a_\pm = -\lim_{y \to 0^+} y\Im G_{\xi_\lambda}(\pm\sqrt{a^2(\lambda)+1}+iy), \quad g(x) = -\frac{1}{\pi}\lim_{y \to 0^+} \Im G_{\xi_\lambda}(x+iy).$$

Using that the Cauchy-Stieltjes transform maps $\mathbb{C}^+$ to $\mathbb{C}^-$, one finally gets

$$\xi_\lambda(dx) = \frac{a(\lambda)}{\sqrt{a^2(\lambda)+1}}\delta_{\sqrt{a^2(\lambda)+1}}(dx) + \frac{1}{\pi}\frac{\sqrt{1-x^2}}{a^2(\lambda)+1-x^2}\mathbf{1}_{|x|<1}dx.$$

To see that this defines a probability measure for $\lambda \neq 1$, one proceeds as follows: write

$$\frac{1}{\pi}\int_{-1}^{1} \frac{\sqrt{1-x^2}}{a^2(\lambda)+1-x^2}dx = \frac{1}{\pi}\int_0^1 \frac{\sqrt{1-x}}{\sqrt{x}(a^2(\lambda)+1-x)}dx$$
$$= \frac{1}{2(a^2(\lambda)+1)}{}_2F_1\left(1,\frac{1}{2};2;\frac{1}{a^2(\lambda)+1}\right)$$

where $_2F_1$ denotes the Gauss hypergeometric function given by

$$_2F_1(e,b,c;z) = \frac{\Gamma(c)}{\Gamma(b)\Gamma(c-b)} \int_0^1 x^{b-1}(1-x)^{c-b-1}(1-zx)^{-e} dx,$$

with $\Re(b) \wedge \Re(c-b) > 0$ for $|u| < 1$. Then, use the identity

$$_2F_1(1,b,2;z) = \frac{1-(1-z)^{1-b}}{(1-b)z}$$

to get

$$\frac{1}{\pi}\int_{-1}^1 \frac{\sqrt{1-x^2}}{a^2(\lambda)+1-x^2} dx = 1 - \frac{a(\lambda)}{\sqrt{a^2(\lambda)+1}}.$$

Acknowledgments

The author wants to thank Professors K. Dykema and M. Anshelevich as well as the organization team for their financial support to attend the concentration week on probability and analysis at Texas A&M university, where the author started this work, and hospitality. A special thank to Professor P. Graczyk who invites the author to the 28-th conference on quantum probability and related topics, and to Professor R. Quezada Batalla for the financial support to attend the conference and talk about this work. The author is grateful to Professor L. Accardi for detailed explanations about one mode IFS and to Professor H. H. Kuo for useful remarks on the manuscript.

References

1. L. Accardi, M. Bozejko. Interacting Fock space and Gausssianization of probability measures. *Infin. Dimens. Anal. Quantum Probab. Relat. Top.1.* **4**. 1998, 663-670.
2. W. A. Al-Salam, T. S. Chihara. q-Pollaczek polynomials and a conjecture of Andrews and Askey. *SIAM J. Math. Anal.* **18** (1987), no. 1, 228–242.
3. N. Asai, I. Kubo, H. H. Kuo. Multiplicative renormalization and generating function I. *Taiwanese J. Math.* 7. **1**. 2003, 89-101.
4. N. Asai, I. Kubo, H. H. Kuo. Multiplicative renormalization and generating function II. *Taiwanese J. Math.* 8. **4**. 2004, 593-628.
5. N. Asai, I. Kubo, H. H. Kuo. Generating function method for orthogonal polynomials and Jacobi-Szegö parameters. *Probab. Math. Statist.* 23. **2**, 2003, 273-291. Acta Univ. Wratislav. No 2593.
6. G. E. Andrews, R. Askey, R. Roy. Special functions. *Cambridge University Press*. 1999.
7. P. Biane. Free Brownian motion, free stochastic calculus and random matrices. *Fie. Inst. Comm.* **12**, Amer. Math. Soc. Providence, RI, 1997, 1-19.

8. *N. Demni.* Free Jacobi process. To appear in *J. Theo. Proba.*
9. *H. Kesten.* Symmetric random walks on groups. *Trans. Amer. Math. Soc.* **92**, 1959, 336-354.
10. *I. Kubo, H. H. Kuo, S. Namli.* Interpolation of Chebyshev polynomials and interacting Fock spaces. *Infin. Dimens. Anal. Quantum Probab. Relat. Top.* *9.* **3**. 2006, 361-371.
11. *I. Kubo, H. H. Kuo, S. Namli.* The Characterization of a Class of Probability Measures by Multiplicative Renormalization. *Communications on Stochastic Analysis.* **1**, no. 3. 2007, 455–472.
12. *M. L. Silverstein.* Orthogonal polynomial martingales on spheres. *Sém. Probab. XX.* 1984/85, 419-422. Lecture notes in Math., **1204**, Springer, Berlin, 1986.
13. *R. Speicher.* Combinatorics of Free Probability Theory. *Lectures. I. H. P. Paris.* 1999.

ACCARDI COMPLEMENTARITY FOR $-1/2 < \mu < 0$ AND RELATED RESULTS

L. A. ECHAVARRÍA CEPEDA* AND S. B. SONTZ**

Centro de Investigación en Matemáticas, A.C. (CIMAT)
Jalisco S.N. Valenciana
C.P.36240 Guanajuato, Mexico.
*E-mail: lenin@cimat.mx
**E-mail: sontz@cimat.mx

C. J. PITA RUIZ VELASCO

Universidad Panamericana
Augusto Rodin 498, Insurgentes-Mixcoac
C.P. 03920 Mexico, D.F.
E-mail: cpita@up.edu.mx

We show that the momentum and position operators of μ-deformed quantum mechanics for $-1/2 < \mu < 0$ are not Accardi complementary. This proves an earlier conjecture of the last two authors as well as extending their analogous result for the case $\mu > 0$. We also prove some related formulas that were conjectured by the same authors.

Keywords: Accardi complementarity, μ-deformed quantum mechanics.

1. Introduction

In this article we present a new result in the same direction as the main result of the recent work by Pita-Sontz[10] as well as proving some formulas that were also conjectured there. This article should be considered as a sequel to Pita-Sontz[10]. For the reader's convenience, we collect in this section some of the basic material in Pita-Sontz[10].

First we present some relevant facts of the so-called *μ-deformed quantum mechanics*. For more details, refer to Angulo-Sontz[3,4], Marron[7], Pita-Sontz[8–10] and Rosenblum[11]. In this theory the mathematical objects of quantum mechanics (position and momentum operators, configuration space, phase space, etc.) are deformed by a parameter $\mu > -\frac{1}{2}$ (the undeformed theory corresponding to $\mu = 0$). We will be dealing with the complex Hilbert space $L^2(\mathbb{R}, m_\mu)$, where the measure m_μ is given by

$$dm_\mu(x) := \left(2^{\mu+\frac{1}{2}} \Gamma\left(\mu + \frac{1}{2}\right)\right)^{-1} |x|^{2\mu} dx$$

for $x \in \mathbb{R}$. Here dx is Lebesgue measure on $\mathbb{R}$ and Γ is the Euler gamma function. The normalization of this measure is chosen to give us a self-dual (μ-deformed) Fourier transform. See Rosenblum[11] for details. In this Hilbert space $L^2(\mathbb{R}, m_\mu)$ we have two unbounded self-adjoint operators: the μ-deformed position operator Q_μ and the μ-deformed momentum operator P_μ. These are defined for $x \in \mathbb{R}$ and certain elements $\psi \in L^2(\mathbb{R}, m_\mu)$ by

$$Q_\mu \psi(x) := x\psi(x),$$
$$P_\mu \psi(x) := \frac{1}{i}\left(\psi'(x) + \frac{\mu}{x}(\psi(x) - \psi(-x))\right).$$

We omit details about exact domains of definition. Interest in these operators originates in Wigner[12] where equivalent forms of them are used as examples of operators that do not satisfy the usual canonical commutation relation in spite of the fact that they do satisfy the equations of motion $i[H_\mu, Q_\mu] = P_\mu$ and $i[H_\mu, P_\mu] = -Q_\mu$ for the Hamiltonian $H_\mu := \frac{1}{2}(Q_\mu^2 + P_\mu^2)$. What does hold is the μ-deformed canonical commutation relation: $i[P_\mu, Q_\mu] = I + 2\mu J$, where I is the identity operator and J is the parity operator $J\psi(x) := \psi(-x)$.

Accardi[2] introduced a definition of complementary observables in quantum mechanics. We now generalize that definition to the current context. We use the usual identification of observables in quantum mechanics as self-adjoint operators acting in some Hilbert space.

Definition 1.1. We say that the (not necessarily bounded) self-adjoint operators S and T acting in $L^2(\mathbb{R}, m_\mu)$ are *Accardi complementary* if for any pair of bounded Borel subsets A and B of $\mathbb{R}$ we have that the operator $E^S(A)E^T(B)$ is trace class with trace given by

$$Tr\left(E^S(A)E^T(B)\right) = m_\mu(A)m_\mu(B).$$

Here E^S is the projection-valued measure on $\mathbb{R}$ associated with the self-adjoint operator S by the spectral theorem, and similarly for E^T.

So, $E^S(A)E^T(B)$ is clearly a bounded operator acting on $L^2(\mathbb{R}, m_\mu)$. But whether it is also trace class is another matter. And, given that it is trace class, it is a further matter to determine if the trace can be written as the product of measures, as indicated. Accardi's result[2] (which is also discussed in detail and proved in Cassinelli-Varadarajan[5]) is that $Q \equiv Q_0$

and $P \equiv P_0$ are Accardi complementary. Accardi also conjectured that this property of Q and P characterized this pair of operators acting on $L^2(\mathbb{R}, m_0)$. It turns out that this is not so. (See Ref. 5.)

The main result in the paper by Pita-Sontz[10] is the following theorem.

Theorem 1.1. *Let A and B be bounded Borel subsets of $\mathbb{R}$ with $0 \notin A^-$, the closure of A. Then $E^{Q_\mu}(A) E^{P_\mu}(B)$ is a trace class operator in $L^2(\mathbb{R}, m_\mu)$ for any $\mu > -\frac{1}{2}$ with*

$$0 \leq Tr\left(E^{Q_\mu}(A) E^{P_\mu}(B)\right) = \int_A dm_\mu(x) \int_B dm_\mu(k) \, |\exp_\mu(ikx)|^2 < \infty. \tag{1}$$

Moreover, if $\mu > 0$ and $m_\mu(A) m_\mu(B) \neq 0$ then we have that

$$Tr\left(E^{Q_\mu}(A) E^{P_\mu}(B)\right) < m_\mu(A) m_\mu(B). \tag{2}$$

In particular, the operators Q_μ and P_μ are not Accardi complementary if $\mu > 0$.

In Pita-Sontz[10] the conjecture is made that

$$Tr\left(E^{Q_\mu}(A) E^{P_\mu}(B)\right) > m_\mu(A) m_\mu(B) \tag{3}$$

for A, B bounded Borel sets of positive m_μ measure and $-1/2 < \mu < 0$. We shall prove this result under the extra technical hypothesis $0 \notin A^-$ and thereby establish that the operators Q_μ and P_μ are not Accardi complementary for $-1/2 < \mu < 0$.

The organization of this article is as follows. In the next section we present a theorem that will allow us to prove the main result in Section 3. Finally in Section 4, we prove several new identities of μ-deformed quantities, including all of those conjectured in Pita-Sontz[10].

2. Preliminary Results

We take $\mu > -\frac{1}{2}$ arbitrary unless otherwise stated. We denote by $\mathbb{N}$ the set of non-negative integers and by $\mathbb{Z}$ the set of all integers.

Definition 2.1. *The μ-deformed factorial function $\gamma_\mu : \mathbb{N} \to \mathbb{R}$ is defined by $\gamma_\mu(0) := 1$ and*

$$\gamma_\mu(n) := (n + 2\mu\theta(n)) \gamma_\mu(n-1),$$

where $n \geq 1$ and $\theta : \mathbb{N} \to \{0, 1\}$ is the characteristic function of the odd integers.

This definition can be found in Rosenblum[11]. In the case $\mu = 0$ we obtain the known object $\gamma_0(n) = n!$ (the factorial function). Next we define the μ-deformed exponential function, which also can be found in Rosenblum[11].

Definition 2.2. The μ-deformed exponential function $\exp_\mu : \mathbb{C} \to \mathbb{C}$ is defined for $z \in \mathbb{C}$ by

$$\exp_\mu(z) := \sum_{n=0}^{\infty} \frac{z^n}{\gamma_\mu(n)}.$$

It is easy to see that this series converges absolutely and uniformly on compact sets and so $\exp_\mu : \mathbb{C} \to \mathbb{C}$ is holomorphic (that is, it is an entire function). Observe also that, since $\gamma_0(n) = n!$, the undeformed exponential function $\exp_0$ is just the usual complex exponential function $\exp$.

Theorem 2.1. *Suppose that $x \in \mathbb{R} \setminus \{0\}$.*
(a) $\left|\exp_\mu(ix)\right| = 1$ if and only if $\mu = 0$.
(b) $\left|\exp_\mu(ix)\right| < 1$ if and only if $\mu > 0$.
(c) $\left|\exp_\mu(ix)\right| > 1$ if and only if $-\frac{1}{2} < \mu < 0$.

Remark 2.1. Clearly, $\exp_\mu(0) = 1$ for all $\mu > -1/2$. The implication $\Leftarrow$ of Part (b) was proved by another method in Pita-Sontz[10].

Proof. We let $J_\nu(z)$ denote the Bessel function of order ν with its standard domain of definition, namely the complex plane cut along the negative real axis: $\mathbb{C} \setminus (-\infty, 0]$. We will use formula (3.1.2) from Rosenblum[11]:

$$\exp_\mu(-ix) = \Gamma(\mu + \tfrac{1}{2}) 2^{\mu - \frac{1}{2}} \frac{J_{\mu - \frac{1}{2}}(x) - i J_{\mu + \frac{1}{2}}(x)}{x^{\mu - \frac{1}{2}}}. \tag{4}$$

We will only need this identity for real $x > 0$. Also, we will use the following two identities. First, we have for all $z \in \mathbb{C} \setminus (-\infty, 0]$ that

$$J_{\nu-1}(z) + J_{\nu+1}(z) = \frac{2\nu}{z} J_\nu(z), \tag{5}$$

which can be found as formula (9.1.27) in Abramowitz-Stegun[1] or as formula (5.3.6) in Lebedev[6]. Next, for all $z \in \mathbb{C} \setminus (-\infty, 0]$ we have that

$$\frac{d}{dz}\left\{z^{-\nu} J_\nu(z)\right\} = -z^{-\nu} J_{\nu+1}(z), \tag{6}$$

which comes from formula (9.1.30) in Abramowitz-Stegun[1] or formula (5.3.5) in Lebedev[6]. We take $x > 0$ in the following calculation. (Justifications of the steps are given afterwards.)

$$\frac{d}{dx}\{|\exp_\mu(-ix)|^2\}$$

$$= \frac{d}{dx}\left\{\Gamma(\mu+\tfrac{1}{2})^2 2^{2\mu-1}\left[\left(x^{\frac{1}{2}-\mu}J_{\mu-\frac{1}{2}}(x)\right)^2 + \left(x^{\frac{1}{2}-\mu}J_{\mu+\frac{1}{2}}(x)\right)^2\right]\right\}$$

$$= \Gamma(\mu+\tfrac{1}{2})^2 2^{2\mu-1}\left[2\left(x^{\frac{1}{2}-\mu}J_{\mu-\frac{1}{2}}(x)\right)\left(-x^{\frac{1}{2}-\mu}J_{\mu+\frac{1}{2}}(x)\right)\right.$$

$$\left. + 2\left(x^{\frac{1}{2}-\mu}J_{\mu+\frac{1}{2}}(x)\right)\left(-x^{\frac{1}{2}-\mu}J_{\mu+\frac{3}{2}}(x) + x^{-\mu-\frac{1}{2}}J_{\mu+\frac{1}{2}}(x)\right)\right]$$

$$= 2^{2\mu}\Gamma(\mu+\tfrac{1}{2})^2 x^{1-2\mu}J_{\mu+\frac{1}{2}}(x)\left[-J_{\mu-\frac{1}{2}}(x) - J_{\mu+\frac{3}{2}}(x) + \frac{1}{x}J_{\mu+\frac{1}{2}}(x)\right]$$

$$= 2^{2\mu}\Gamma(\mu+\tfrac{1}{2})^2 x^{1-2\mu}J_{\mu+\frac{1}{2}}(x)\left[-\frac{2(\mu+\tfrac{1}{2})}{x}J_{\mu+\frac{1}{2}}(x) + \frac{1}{x}J_{\mu+\frac{1}{2}}(x)\right]$$

$$= (-\mu)2^{2\mu+1}\Gamma(\mu+\tfrac{1}{2})^2 x^{-2\mu}\left(J_{\mu+\frac{1}{2}}(x)\right)^2$$

The first equality follows from equation (4) and the fact that $J_\nu(x)$ is real for $x > 0$. For the second equality we used equation (6) twice together with the identity

$$x^{\frac{1}{2}-\mu}J_{\mu+\frac{1}{2}}(x) = x\left(x^{-\frac{1}{2}-\mu}J_{\mu+\frac{1}{2}}(x)\right).$$

The third and fifth equalities follow from simple algebra, while the fourth is an application of equation (5).

So, for $x > 0$ the derivative of $|\exp_\mu(-ix)|^2$ has the same sign as $-\mu$ or is zero. Since $\phi(x) := |\exp_\mu(-ix)|^2 = \exp_\mu(-ix)\exp_\mu(ix)$ is an even function of $x \in \mathbb{R}$, it follows that its derivative $\phi'(x)$ is an odd function of $x \in \mathbb{R}$. So, for $x < 0$ the derivative $\phi'(x)$ has the same sign as μ or is zero. Of course, this agrees with the classical result when $\mu = 0$, namely that the derivative of

$$|\exp_0(-ix)|^2 = |\exp(-ix)|^2 = 1$$

is identically zero. We now consider the case when $\mu \neq 0$. Then $\phi(x) = \exp_\mu(-ix)\exp_\mu(ix)$ is clearly real analytic (in the variable $x \in \mathbb{R}$) and not constant. And this implies that the critical points of $\phi(x)$ are isolated. But $x = 0$ is a critical point of $\phi(x)$, since $\phi'(x)$ is odd and continuous, implying that $\phi'(0) = 0$. And the corresponding critical value is $\phi(0) = |\exp_\mu(0)|^2 = 1$.

Now the above analysis of the sign of the derivative of $|\exp_\mu(-ix)|^2$ in the intervals $(-\infty, 0)$ and $(0, \infty)$ shows that the critical value 1 at $x = 0$

is an absolute minimum if $-1/2 < \mu < 0$ while it is an absolute maximum if $\mu > 0$. And thus we have shown all three parts of the statement of the theorem. □

3. Main Result

We are now ready to state and prove our main result.

Theorem 3.1. *Let A and B be bounded Borel subsets of $\mathbb{R}$ with $0 \notin A^-$, the closure of A. If $-1/2 < \mu < 0$ and $m_\mu(A)m_\mu(B) \neq 0$ then we have that*

$$Tr\left(E^{Q_\mu}(A)E^{P_\mu}(B)\right) > m_\mu(A)m_\mu(B). \qquad (7)$$

In particular, the operators Q_μ and P_μ are not Accardi complementary for $-1/2 < \mu < 0$.

Proof. The formula (1) of Theorem 1.1 holds. So we use the lower bound of part (c) of Theorem 2.1 to estimate the integral in formula (1) from below. This gives the result. □

4. Some Identities

In this section we always will take $\mu > -\frac{1}{2}$. Recall that the μ-deformed factorial function $\gamma_\mu(n)$ has been defined in Definition 2.1.

Definition 4.1. *The μ-deformed binomial coefficient is defined for all $n \in \mathbb{N}$ and $k \in \mathbb{Z}$ by*

$$\binom{n}{k}_\mu := \frac{\gamma_\mu(n)}{\gamma_\mu(n-k)\gamma_\mu(k)}$$

if $0 \leq k \leq n$ and $\binom{n}{k}_\mu := 0$ for other integer values of k. For $n \in \mathbb{N}$ and $x, y \in \mathbb{C}$ the n-th μ-deformed binomial polynomial (or μ-deformed binomial polynomial of degree n) is defined by

$$p_{n,\mu}(x,y) := \sum_{k=0}^{n} \binom{n}{k}_\mu x^k y^{n-k}.$$

These definitions can be found in Rosenblum.[11] In the case $\mu = 0$ we obtain the known objects $\binom{n}{k}_0 = \binom{n}{k}$ (the binomial coefficient) and $p_{n,0}(x,y) = \sum_{k=0}^{n} \binom{n}{k} x^k y^{n-k}$ (the n-th binomial polynomial $(x+y)^n$). Note that $p_{0,\mu}(x,y) = 1$ and $p_{1,\mu}(x,y) = x+y$. So the μ-deformed binomial polynomials of degree 0 and 1 are the same as the undeformed binomial polynomials of the same degree. However, $p_{n,\mu}(x,y)$ does depend on μ for $n \geq 2$.

Clearly we have that $\gamma_\mu(n) > 0$ for all $n \in \mathbb{N}$ and thus $\binom{n}{k}_\mu \geq 0$ for all $n \in \mathbb{N}$ and $k \in \mathbb{Z}$. Observe also that for all $n \in \mathbb{N}$ and $\mu > -\frac{1}{2}$ we have that

$$\binom{n}{0}_\mu = \binom{n}{n}_\mu = 1$$

and

$$\binom{n}{k}_\mu = \binom{n}{n-k}_\mu.$$

The Pascal Triangle property $\binom{n}{k-1} + \binom{n}{k} = \binom{n+1}{k}$ for the binomial coefficients has the following form in the μ-deformed setting.

Theorem 4.1. *For $n \in \mathbb{N}$ and $k \in \mathbb{Z}$ we have that*

$$\binom{2n}{k-1}_\mu + \binom{2n}{k}_\mu = \binom{2n+1}{k}_\mu \tag{8}$$

and

$$\binom{2n+1}{k-1}_\mu + \binom{2n+1}{k}_\mu = \left(1 + \frac{2\mu\theta(k)}{n+1}\right)\binom{2n+2}{k}_\mu \tag{9}$$

Proof. Observe that formula (8) is trivial if $k \leq 0$ or $k \geq 2n+1$. So let us take $0 < k < 2n+1$. Since $\theta(2n+1-k) + \theta(k) = 1$, we have that

$$\binom{2n}{k-1}_\mu + \binom{2n}{k}_\mu$$
$$= \frac{\gamma_\mu(2n)}{\gamma_\mu(k-1)\gamma_\mu(2n-k+1)} + \frac{\gamma_\mu(2n)}{\gamma_\mu(k)\gamma_\mu(2n-k)}$$
$$= \frac{k + 2\mu\theta(k)}{2n+1+2\mu} \frac{\gamma_\mu(2n+1)}{\gamma_\mu(k)\gamma_\mu(2n-k+1)}$$
$$+ \frac{2n+1-k+2\mu\theta(2n+1-k)}{2n+1+2\mu} \frac{\gamma_\mu(2n+1)}{\gamma_\mu(k)\gamma_\mu(2n-k+1)}$$
$$= \frac{2n+1+2\mu(\theta(2n+1-k)+\theta(k))}{2n+1+2\mu}\binom{2n+1}{k}_\mu$$
$$= \binom{2n+1}{k}_\mu,$$

which proves (8). Similarly, formula (9) is trivial if $k \leq 0$ or $k \geq 2n+2$. So let us take $0 < k < 2n+2$. Since $\theta(2n+2-k) = \theta(k)$, we have that

$$\binom{2n+1}{k-1}_\mu + \binom{2n+1}{k}_\mu$$
$$= \frac{\gamma_\mu(2n+1)}{\gamma_\mu(k-1)\gamma_\mu(2n+2-k)} + \frac{\gamma_\mu(2n+1)}{\gamma_\mu(k)\gamma_\mu(2n+1-k)}$$
$$= \frac{k+2\mu\theta(k)}{2n+2} \frac{\gamma_\mu(2n+2)}{\gamma_\mu(k)\gamma_\mu(2n+2-k)}$$
$$+ \frac{2n+2-k+2\mu\theta(2n+2-k)}{2n+2} \frac{\gamma_\mu(2n+2)}{\gamma_\mu(k)\gamma_\mu(2n+2-k)}$$
$$= \left(1 + \frac{\mu(\theta(k) + \theta(2n+2-k))}{n+1}\right) \binom{2n+2}{k}_\mu$$
$$= \left(1 + \frac{2\mu\theta(k)}{n+1}\right) \binom{2n+2}{k}_\mu,$$

which proves (9). □

In the undeformed case we have $(x+y)(x+y)^n = (x+y)^{n+1}$. But, when working with μ-deformed binomial polynomials $p_{n,\mu}(x,y)$ for $\mu \neq 0$, the corresponding result is described in the following proposition.

Theorem 4.2. Let $x, y \in \mathbb{C}$ and $n \in \mathbb{N}$. Then we have that

$$p_{1,\mu}(x,y)\, p_{2n,\mu}(x,y) = p_{2n+1,\mu}(x,y) \tag{10}$$

and

$$p_{1,\mu}(x,y)p_{2n+1,\mu}(x,y) = p_{2n+2,\mu}(x,y) + \frac{2\mu}{n+1}\sum_{k=0}^{n}\binom{2n+2}{2k+1}_\mu x^{2k+1}y^{2n+1-2k}. \tag{11}$$

Remark 4.1. Formula (10) appears in Rosenblum[11] (Corollary 4.4), where μ is assumed to be a positive parameter.

Proof. By using (8) we have that

$$p_{1,\mu}(x,y)\, p_{2n,\mu}(x,y) = (x+y) \sum_{k=0}^{2n} \binom{2n}{k}_\mu x^k y^{2n-k}$$

$$= \sum_{k=1}^{2n+1} \binom{2n}{k-1}_\mu x^k y^{2n-k+1} + \sum_{k=0}^{2n} \binom{2n}{k}_\mu x^k y^{2n-k+1}$$

$$= \sum_{k=0}^{2n+1} \left(\binom{2n}{k-1}_\mu + \binom{2n}{k}_\mu \right) x^k y^{2n-k+1}$$

$$= \sum_{k=0}^{2n+1} \binom{2n+1}{k}_\mu x^k y^{2n-k+1}$$

$$= p_{2n+1,\mu}(x,y),$$

which proves (10). Now, by using (9) we have that

$$p_{1,\mu}(x,y)\, p_{2n+1,\mu}(x,y)$$

$$= (x+y) \sum_{k=0}^{2n+1} \binom{2n+1}{k}_\mu x^k y^{2n+1-k}$$

$$= \sum_{k=1}^{2n+2} \binom{2n+1}{k-1}_\mu x^k y^{2n+2-k} + \sum_{k=0}^{2n+1} \binom{2n+1}{k}_\mu x^k y^{2n+2-k}$$

$$= \sum_{k=0}^{2n+2} \left(\binom{2n+1}{k-1}_\mu + \binom{2n+1}{k}_\mu \right) x^k y^{2n+2-k}$$

$$= \sum_{k=0}^{2n+2} \left(1 + \frac{2\mu \theta(k)}{n+1} \right) \binom{2n+2}{k}_\mu x^k y^{2n+2-k}$$

$$= \sum_{k=0}^{2n+2} \binom{2n+2}{k}_\mu x^k y^{2n+2-k} + \frac{2\mu}{n+1} \sum_{k=0}^{2n+2} \theta(k) \binom{2n+2}{k}_\mu x^k y^{2n+2-k}$$

$$= p_{2n+2,\mu}(x,y) + \frac{2\mu}{n+1} \sum_{k=0}^{n} \binom{2n+2}{2k+1}_\mu x^{2k+1} y^{2n+1-2k},$$

which proves (11). □

Theorem 4.3.

(a) For $n \in \mathbb{N}$ we have that

$$p_{2n+1,\mu}(1,-1) = 0. \tag{12}$$

(b) $p_{0,\mu}(1,-1) = 1$.

(c) For $n \geq 1$ we have that

$$p_{2n,\mu}(1,-1) = \frac{2\mu}{n} \sum_{k=0}^{n-1} \binom{2n}{2k+1}_{\mu} \qquad (13)$$

(d) For $n \in \mathbb{N}$ we have that

$$p_{2n,\mu}(1,-1) = \frac{2^{2n}\mu}{n+\mu} \prod_{k=1}^{n} \frac{k+\mu}{k+2\mu}. \qquad (14)$$

(e) For $n \geq 1$ we have that

$$p_{4n,\mu}(1,-1) = \mu \frac{2^{2n} \prod_{k=n+1}^{2n-1}(\mu+k)}{\prod_{k=1}^{n}(\mu+k-1/2)} \qquad (15)$$

(f) For $n \geq 1$ we have that

$$p_{4n-2,\mu}(1,-1) = \mu \frac{2^{2n-1} \prod_{k=n+1}^{2n-1}(\mu+k-1)}{\prod_{k=1}^{n}(\mu+k-1/2)} \qquad (16)$$

Remark 4.2. Formulas (13), (15) and (16) were conjectured in Pita-Sontz[10]. Note that (14) is new and that it turns out, as we will show, to be a compact way of writing both (15) and (16).

Proof. (a) Though (12) is a direct consequence of (10) with $x = 1$ and $y = -1$ (since $p_{2n+1,\mu}(1,-1) = p_{1,\mu}(1,-1)p_{2n,\mu}(1,-1) = (1-1)p_{2n,\mu}(1,-1) = 0$), we would like to mention that one can prove (12) proceeding directly from the definition and using the symmetry property $\binom{n}{k}_{\mu} = \binom{n}{n-k}_{\mu}$ mentioned above:

$$p_{2n+1,\mu}(1,-1) = \sum_{k=0}^{2n+1} \binom{2n+1}{k}_\mu (-1)^k$$

$$= \sum_{k=0}^{n} \binom{2n+1}{k}_\mu (-1)^k + \sum_{k=n+1}^{2n+1} \binom{2n+1}{k}_\mu (-1)^k$$

$$= \sum_{k=0}^{n} \binom{2n+1}{k}_\mu (-1)^k + \sum_{j=n}^{0} \binom{2n+1}{2n+1-j}_\mu (-1)^{2n+1-j}$$

$$= \sum_{k=0}^{n} \binom{2n+1}{k}_\mu (-1)^k - \sum_{j=0}^{n} \binom{2n+1}{j}_\mu (-1)^j$$

$$= 0.$$

(b) This is immediate.

(c) First we observe that using (11) with $x=1$, $y=-1$ and n replaced by $n-1$, we obtain

$$p_{2n,\mu}(1,-1) + \frac{2\mu}{n}\sum_{k=0}^{n-1}\binom{2n}{2k+1}_\mu (-1) = p_{1,\mu}(-1,1)\, p_{2n-1,\mu}(1,-1) = 0,$$

and therefore

$$p_{2n,\mu}(1,-1) = \frac{2\mu}{n}\sum_{k=0}^{n-1}\binom{2n}{2k+1}_\mu \tag{17}$$

for $n \geq 1$. This proves Part (c).

(d) The case $n=0$ as well as the case $\mu=0$ are each trivial. (In the case when both $n=0$ and $\mu=0$ we use the convention that $\mu/(n+\mu)=1$. We also use the standard convention that a product over an empty index set is 1.) So hereafter we take $n \geq 1$ and $\mu \neq 0$.

Using the previous formula we obtain for $n \geq 1$ that

$$p_{2n,\mu}(1,-1) = \frac{2\mu}{n}\sum_{k=0}^{n-1}\binom{2n}{2k+1}_\mu$$

$$= \frac{\mu}{n}\left(\sum_{k=0}^{2n}\binom{2n}{k}_\mu - \sum_{k=0}^{2n}\binom{2n}{k}_\mu (-1)^k\right)$$

$$= \frac{\mu}{n}\left(p_{2n,\mu}(1,1) - p_{2n,\mu}(1,-1)\right),$$

and thus

$$p_{2n,\mu}(1,-1) = \frac{\mu}{n+\mu}p_{2n,\mu}(1,1). \tag{18}$$

From (10) with $x = y = 1$ we obtain

$$p_{2n+1,\mu}(1,1) = p_{1,\mu}(1,1)p_{2n,\mu}(1,1) = (1+1)p_{2n,\mu}(1,1) = 2p_{2n,\mu}(1,1). \tag{19}$$

Similarly from (11) with $x = y = 1$ we get

$$p_{1,\mu}(1,1)p_{2n-1,\mu}(1,1) = p_{2n,\mu}(1,1) + \frac{2\mu}{n}\sum_{k=0}^{n-1}\binom{2n}{2k+1}_\mu,$$

which by using $p_{1,\mu}(1,1) = 2$ and (17) becomes

$$2p_{2n-1,\mu}(1,1) = p_{2n,\mu}(1,1) + p_{2n,\mu}(1,-1)$$

This last expression together with (18) gives us

$$\begin{aligned}2p_{2n-1,\mu}(1,1) &= p_{2n,\mu}(1,1) + p_{2n,\mu}(1,-1) \\ &= p_{2n,\mu}(1,1) + \frac{\mu}{n+\mu}p_{2n,\mu}(1,1) \\ &= \frac{n+2\mu}{n+\mu}p_{2n,\mu}(1,1).\end{aligned}$$

So we have

$$p_{2n,\mu}(1,1) = \frac{2(n+\mu)}{n+2\mu}p_{2n-1,\mu}(1,1). \tag{20}$$

We claim that for $n \in \mathbb{N}$ we have that

$$p_{2n,\mu}(1,1) = 2^{2n}\prod_{k=1}^{n}\frac{k+\mu}{k+2\mu} \tag{21}$$

This is trivial for $n = 0$, while for $n = 1$ we have

$$\begin{aligned}p_{2,\mu}(1,1) &= \sum_{k=0}^{2}\binom{2}{k}_\mu = 2 + \binom{2}{1}_\mu = 2 + \frac{2}{1+2\mu} \\ &= 4\frac{1+\mu}{1+2\mu} = 2^{2(1)}\prod_{k=1}^{1}\frac{k+\mu}{k+2\mu}.\end{aligned}$$

Arguing by induction, we now assume that (21) is valid for a given $n \in \mathbb{N}$. Then by also using (19) and (20) we have

$$\begin{aligned}
p_{2n+2,\mu}(1,1) &= \frac{2(n+1+\mu)}{n+1+2\mu} p_{2n+1,\mu}(1,1) \\
&= \frac{2(n+1+\mu)}{n+1+2\mu} 2 p_{2n,\mu}(1,1) \\
&= \frac{2^2(n+1+\mu)}{n+1+2\mu} 2^{2n} \prod_{k=1}^{n} \frac{k+\mu}{k+2\mu} \\
&= 2^{2n+2} \prod_{k=1}^{n+1} \frac{k+\mu}{k+2\mu},
\end{aligned}$$

which proves (21) for $n+1$ and so proves our claim. Finally, from (18) and (21) we have that

$$\begin{aligned}
p_{2n,\mu}(1,-1) &= \frac{\mu}{n+\mu} p_{2n,\mu}(1,1) \\
&= \frac{2^{2n}\mu}{n+\mu} \prod_{k=1}^{n} \frac{k+\mu}{k+2\mu},
\end{aligned}$$

which proves (14) and so concludes the proof of Part (d).

(e) Using (14) for $2n$ in place of n we have that

$$\begin{aligned}
p_{4n}(1,-1) &= \frac{2^{4n}\mu}{2n+\mu} \prod_{k=1}^{2n} \frac{k+\mu}{k+2\mu} \\
&= \frac{2^{2n}\mu}{2n+\mu} \prod_{k=1}^{2n} \frac{2k+2\mu}{k+2\mu} \\
&= \frac{2^{2n}\mu}{2n+\mu} \cdot \frac{(2+2\mu)(4+2\mu)\cdots(4n-2+2\mu)(4n+2\mu)}{(1+2\mu)(2+2\mu)\cdots(2n-1+2\mu)(2n+2\mu)} \\
&= \frac{2^{2n}\mu}{2n+\mu} \cdot \frac{(2n+2+2\mu)(2n+4+2\mu)\cdots(4n-2+2\mu)(4n+2\mu)}{(1+2\mu)(3+2\mu)\cdots(2n-3+2\mu)(2n-1+2\mu)} \\
&= \frac{2^{2n}\mu}{2n+\mu} \cdot \frac{2^n}{2^n} \cdot \frac{(\mu+n+1)(\mu+n+2)\cdots(\mu+2n-1)(\mu+2n)}{(\mu+1/2)(\mu+3/2)\cdots(\mu+n-3/2)(\mu+n-1/2)} \\
&= 2^{2n}\mu \cdot \frac{\prod_{k=n+1}^{2n-1}(\mu+k)}{\prod_{k=1}^{n}(\mu+k-1/2)}
\end{aligned}$$

and this shows (15).

(f) Using (14) for $2n-1$ in place of n we have that

$$p_{4n-2}(1,-1) = \frac{2^{4n-2}\mu}{2n-1+\mu} \prod_{k=1}^{2n-1} \frac{k+\mu}{k+2\mu}$$

$$= \frac{2^{2n-1}\mu}{2n-1+\mu} \prod_{k=1}^{2n-1} \frac{2k+2\mu}{k+2\mu}$$

$$= \frac{2^{2n-1}\mu}{2n-1+\mu} \cdot \frac{(2+2\mu)(4+2\mu)\cdots(4n-4+2\mu)(4n-2+2\mu)}{(1+2\mu)(2+2\mu)\cdots(2n-2+2\mu)(2n-1+2\mu)}$$

$$= \frac{2^{2n-1}\mu}{2n-1+\mu} \cdot \frac{(2n+2\mu)(2n+2+2\mu)\cdots(4n-4+2\mu)(4n-2+2\mu)}{(1+2\mu)(3+2\mu)\cdots(2n-3+2\mu)(2n-1+2\mu)}$$

$$= \frac{2^{2n-1}\mu}{2n-1+\mu} \cdot \frac{2^n}{2^n} \cdot \frac{(\mu+n)(\mu+n+1)\cdots(\mu+2n-2)(\mu+2n-1)}{(\mu+1/2)(\mu+3/2)\cdots(\mu+n-3/2)(\mu+n-1/2)}$$

$$= 2^{2n-1}\mu \cdot \frac{\prod_{k=n+1}^{2n-1}(\mu+k-1)}{\prod_{k=1}^{n}(\mu+k-1/2)}$$

and this shows (16). □

Acknowledgments

This article was written while two of the authors (L.A.E.C. and S.B.S.) were on an academic visit at the University of Virginia. They would like to thank everyone who made this possible, but most especially Lawrence Thomas who served as their host. The research of L.A.E.C. was partially supported by CONACYT (Mexico) project 49187 and CONACYT (Mexico) student grant 165360. The research of S.B.S. and C.J.P.R.V. was partially supported by CONACYT (Mexico) project 49187.

References

1. M. Abramowitz and I.A. Stegun, *Handbook of Mathematical Functions*, Dover, New York, 1965. (9th printing, 1972.)
2. Accardi, L., *Some Trends and Problems in Quantum Probability*, In: Quantum Probability and Applications to the Quantum Theory of Irreversible Processes, Accardi, L., Frigerio, A and Gorini, V. (Eds.), Lecture Notes in Mathematics, Vol. 1055, Springer-Verlag, Berlin, 1984 pp. 1–19.
3. C. Angulo Aguila and S.B. Sontz, Reverse Inequalities in μ-deformed Segal-Bargmann analysis, J. Math. Phys. 47, 042103 (2006) (21 pages).
4. C. Angulo Aguila and S.B. Sontz, Direct and reverse log-Sobolev inequalities in μ-deformed Segal-Bargmann analysis, preprint, submitted for publication, 2007.
5. Cassinelli, G. and Varadarajan, V., *On Accardi's Notion of Complementary Observables*, Inf. Dim. Anal. Quantum Probab. Rel. Top. 5 (2002) 135–144.

6. N.N. Lebedev, *Special Functions and Their Applications*, Dover, New York, 1972.
7. C.S. Marron, Semigroups and the Bose-like oscillator, Ph.D. Dissertation, The University of Virginia, 1994.
8. C. Pita and S.B. Sontz, On Hirschman and log-Sobolev inequalities in μ-deformed Segal-Bargmann analysis, J. Phys. A: Math. Gen. **39** (2006) 8631–8662.
9. C. Pita and S.B. Sontz, On Shannon entropies in μ-deformed Segal-Bargmann analysis, J. Math. Phys. **47**, 032101 (2006) (31 pages).
10. C. Pita and S.B. Sontz, Accardi Complementarity in μ-deformed Quantum Mechanics, J. Geom. Symm. Phys., **6**, 101–108 (2006) in: *Proceedings of the "XXIV Workshop in Geometric Methods in Physics"*, (Białowieża, Poland, 26 June to 2 July, 2005).
11. M. Rosenblum, Generalized Hermite polynomials and the Bose-like oscillator calculus, in: *Operator Theory Advances and Applications*, Vol. 73, "Nonselfadjoint Operators and Related Topics", (A. Feintuch and I. Gohberg, eds.), Birkhäuser, 369–396 (1994).
12. E.P. Wigner, Do the equations of motion determine the quantum mechanical commutation relations?, Phys. Rev. **77** (1950) 711–712.

QUANTUM MODELS OF BRAIN ACTIVITIES I
RECOGNITION OF SIGNALS

K.-H. FICHTNER

Friedrich-Schiller-University Jena, Faculty of Mathematics and Information
Jena, 07737, Germany
fichtner@minet.uni-jena.de
www.uni-jena.de

L. FICHTNER

Friedrich-Schiller-University Jena, Faculty of Social and Behavioral Sciences
Jena, 07737, Germany
E-mail: Lars.Fichtner@uni-jena.de

One of the main activities of the brain is the recognition of signals. Based on the bosonic Fock space we consider a quantum model of recognition reflecting common expiriences of modern brain research.

Keywords: Bosonic Fock space, beam splitting, recognition, brain models

1. Introduction

Specialists in modern brain research are convinced that signals in the brain should be coded by populations of excited neurons. Considering models based on classical probability theory states of signals should be identified with probability distributions of certain random point fields located inside the volume of the brain. We have constructed a more general description of signals in terms of quantum point systems in order to explain some fundamental facts concerning the procedure of recognition of signals. As it was pointed out in some papers by Singer ([1]) and other specialists of modern brain research the procedure of recognition can be described as follows:
There is a set of complex signals stored in the memory. Choosing one of these signals may be interpreted as generating a hypothesis concerning an "expected view of the world". Then the brain compares a signal arising from our senses with the signal chosen from the memory. That procedure changes the state of both of the signals in such a manner that after the

procedure the signals coincide in a certain sense.

Furthermore, measurements of that procedure like EEG or MEG are based on the fact that ([2]) recognition of signals causes a certain loss of excited neurons, i. e., the neurons change their state from "to be excited" to "nonexcited" (firing of the neurons).

Now, our quantum model of the recognition process reflects both, that change of the signals and the loss of excited neurons.

Furthermore, due to the fact that quantum theory is non-local our model reflects the fact that activity in one part of the brain causes immediately certain changes of the other parts of the brain. For that reason our quantum model of the brain may answer the main question or problem of modern brain research, namely, how to explain that activities in one region of the brain have immediate consequences concerning the other regions (binding problem).

2. Some Postulates Concerning the Recognition of Signals

Wolf Singer is a well-known specialist in modern brian research. In his book ([1]) he summarizes some common experiences mainly found out in the last decade.

From this point of view a mathematical model of the procedure of recognition of signals should reflect the following postulates:

(P1) *The brain acts discrete in time.*

(P2) *Signals are represented by populations of excited neurons.*

(P3) *Signals can be decomposed into parts in compliance with the fact that there are different regions of the brain being responsible for different tasks,* e. g. the visual centre of the brain consists of more than 40 regions. Some regions are responsible for the recognition of different colors. Other parts deal only with certain geometrical objects etc.

(P4) *The brain acts parallel corresponding to the different regions.*

(P5) *The brain permanently created complex signals representing a hypothesis concerning an "expected view of the world".*

(P6) *Recognition of a signal produced by our senses is a random event which can occur as a consequence of the interaction of that signal and a signal created by the brain.*

(P7) *Recognition causes a loss of excited neurons in some regions of the brain,* e. g., the neurons change their state from excited to nonexcited (firing of neurons).

That is what medical doctors can measure with certain instruments like EEG and MEG.

(P8) *Recognition changes the state of the signal coming from our senses according to the "expected view of the world" represented by the signal chosen from the memory. You will be aware of that changed signal.*

Psychologists consider as a proven fact that nobody is able to recognize the real world. What one will be aware of is in some sense a mixture of the real world and what one expects to see partially depending on the experiences or on the knowledge. Especially in case of some diseases of the brain there is a large difference between the real world and what the patient believes to see - for instance: feeling some pains in a leg or arm the patient lost long ago; seeing white mouses is often a problem of alkoholics; schizophrenics hear some voices etc.

(P9) *Changes in some region of the brain have immediate consequences concerning the other regions.*

Now our brain contains about 100 billions of neurons. Thus a lot of signals could be described in terms of a classical probabilistic model according to postulate (P2). But there are many well-known facts being in contradiction to classical models. For instance (P9) is in contradiction to classical models. Further, there is no special region where the memory is localized, i. e., one cannot distinguish neurons supporting signals and neurons supporting the memory - they permanently change their purpose without any loss of information.

For further details we refer to ([3]). There H. Stapp especially argues for a quantum model of brain activities because *classical physics cannot explain consciousness because it cannot explain how the whole can be more than the parts.*

We will consider a quantum model of the process of recognition. Components of our model are

- a Hilbert space H representing signals
- a Hilbert space $\overline{H}$ representing the memory
- an isometry $D : \overline{H} \to \overline{H} \otimes H$ describing the creation of signals
- an unitary operator V on $H \otimes H$ describing the interaction of two signals

In the following we will specialize the components of our model according to the postulates mentioned above.

3. The Space of Signals

Starting point is suitable compact subset G of R^d representing the physical volume of the brain. If G is equiped with Lebesques measure μ then as usual the Hilbert space $L^2(G)$ one uses in order to describe a quantum particle inside of G. Now, a neuron is excited if a certain amount of energy (electric potential) is stored in that neuron. For that reason we identify:

neuron is excited $\Leftrightarrow$ a quantum particle is localized inside of the neuron

i. e. in the sense of our model the quantum particle represents that amount of energy which makes the difference between an excited and nonexcited neuron.

In this sense the bosonic Fock space $\Gamma(L^2(G))$ represents *populations of excited neurons* inside the volume G of the brain.

Now remember the postulate (P2) that signals are represented by populations of excited neurons. For that reason we will use that Fock space in order to describe signals, i. e. we put $H := \Gamma\left(L^2(G)\right)$

There is a well-known property of the bosonic Fock space:
Let $G_1, \ldots, G_n$ be a decomposition of G into measurable subsets. Then one can identify

$$\Gamma(L^2(G)) = \bigotimes_{r=1}^{n} \Gamma(L^2(G_r)) \tag{1}$$

For that reason the use of the bosonic Fock space ensures the postulate (P3), that signals can be decomposed into parts in compliance with the fact that there are different regions of the brain being responsible for different tasks.

Now let us consider the unitary operator V on $H \otimes H$ describing the interaction of two signals. We fix again a decomposition of G into measurable subsets $G_1, \ldots, G_n$ in compliance with the fact that there are different parts of the brain being responsible for different tasks.

Because of that decomposition (1) of the Fock space we can identify

$$H \otimes H = \bigotimes_{r=1}^{n} \left(\Gamma(L^2(G_r)) \otimes \Gamma(L^2(G_r)) \right) \tag{2}$$

Remember postulate (P4) that the brain acts parallel corresponding to the different regions. For that reason the unitary operator V should be of the type

$$V = \bigotimes_{r=1}^{n} V_{G_r} \qquad (3)$$

where V_{G_r} is an unitary operator on $\Gamma(L^2(G_r)) \bigotimes (L^2(G_r))$ $(r = 1, \ldots, n)$.

Now experiments show that the regions can increase or decrease ([4]). Further they can change their location on the surface of the brain ([5]).
On the other hand the physical structure of the brain doesn't change, i. e. there are always the same neurons, nonspecialized concerning the task they are involved.
For that reason (3) should hold for all decompositions $G_1, \ldots, G_n$ of G. That is a very strong condition. One checks ([6]) that there are only the following candidates

$$V = V^b, \text{ where } b = [b_1, b_2, b_3, b_4] : G \to C^4 \text{ with}$$
$$|b_1|^2 + |b_2|^2 \equiv 1 \equiv |b_3|^2 + |b_4|^2 \, , \, b_1\bar{b}_3 + b_2\bar{b}_4 \equiv 0$$

such that for exponential vectors hold

$$V^b \exp(f) \bigotimes \exp(g) = \exp(b_1 f + b_2 g) \bigotimes \exp(b_3 f + b_4 g) \quad (f, g \in L^2(G))$$

Now in ([1]) it was stated that the neurons are "nonspecialized". That implies that the functions b_k should be constant.
Finally, models of learning and dreaming should reflect further postulates leading to the conditions that $V(\psi \bigotimes \exp(0))$ should be always symmetrically and it should hold $VV = I_{H \otimes H}$. These conditions imply

$$b_4 = -\frac{1}{\sqrt{2}}, b_k = \frac{1}{\sqrt{2}} \quad (k = 1, 2, 3) \text{ or } b_4 = \frac{1}{\sqrt{2}}, b_k = -\frac{1}{\sqrt{2}} \quad (k = 1, 2, 3).$$

Without loss of generality we choose the first possibility.
Summarizing we can conclude that the interaction of two signals should be described by the symmetric **beam splitter** V on $H \bigotimes H$ characterized by

$$V \exp(f) \bigotimes \exp(g) = \exp\left(\frac{1}{\sqrt{2}}(f+g)\right) \bigotimes \exp\left(\frac{1}{\sqrt{2}}(f-g)\right) \qquad (4)$$
$$(f, g \in L^2(G))$$

Remark: The operator V is unitary and selfadjoint. Further V is a so-called exchange-operator, i. e., V preserves the number of excited neurons supporting both of the signals in each part of G. For that reason the use of the unitary operator V reflects that signals permanently exchange their supports without any loss of information (^{1}). Let me remark that we have investigated the class of all exchange-operators in (^{7}).

4. The Memory

Now let us consider the (longterm) memory.
We have used the Hilbert space $H = \Gamma\left(L^2(G)\right)$ in order to describe signals (coded by sets of excited neurons). The memory contains the information concerning sets of signals. For that reason it seems to be reasonable to describe the memory using the Fock space corresponding to the space of signals $\overline{H} := \Gamma(H) = \Gamma\left(\Gamma(L^2(G))\right)$. That Hilbert space is to large.
We have to take into account that

- it makes no sense to store the "empty" signal, i. e., one should remove the vacuum part
- it is well-known (^{1}) that there are finite sets of certain elementary signals corresponding to the different regions of the brain, and the only signals which can be stored in the memory are superpositions of such elementary signals.

For that reason we have constructed a certain subspace $H^{reg} \subset H$ describing so-called regular signals as follows:
We fix again a decomposition of G into measurable subsets $G_1, \ldots, G_n$ in compliance with the fact that there are different parts of the brain being responsible for different tasks.
We identify the set of elementary signals of type $r(=1,\ldots,n)$ with a certain finite ONS $|1,r\rangle, \ldots, |m_r, r\rangle$ in $\Gamma\left(L^2(G_r)\right)$ being coherent vectors where the vacuum part was removed (^{8}). Then we consider the subspace H_r of $\Gamma\left(L^2(G_r)\right)$ spanned by that ONS. Finally we put

$$H^{reg} := \bigotimes_{r=1}^{n} H_r \subset \Gamma\left(L^2(G)\right) .$$

Only regular signals can be stored in the memory. For that reason we describe the memory using the Hilbert space

$$\overline{H} := \Gamma(H^{reg}) \subset \Gamma\left(\Gamma(L^2(G))\right) .$$

Let me remark that the ONS we have used consists of certain coherent states where the vacuum part was removed. That corresponds to a certain procedure of accumulation of signals (^{9}).

Now the isometry $D : \overline{H} \to \overline{H} \otimes H$ describing the creation of signals according to postulate (P5) can be characterized using the coherent vectors from $\overline{H}$ as follows:

$$D\exp(\psi) = (N+1)^{-\frac{1}{2}} \exp(\psi) \otimes \psi \quad (\psi \in H^{reg})$$

where N denotes the number operator on $\Gamma(H)$.

5. Recognition of Signals

In order to describe the basic idea of recognition let us consider the special case that both the input signal ψ^{in} and the signal ψ^{mem} chosen from the memory correspond to coherent vectors:

$$\psi^{in} = |\exp(g)\rangle \quad \text{--input-signal}$$
$$\psi^{mem} = |\exp(f)\rangle \quad \text{--signal chosen from the memory}$$

Then the interaction of the signals gives again a pair of coherent states

$$V\left(\psi^{mem} \otimes \psi^{in}\right) = \left|\exp\left(\frac{1}{\sqrt{2}}(f+g)\right)\right\rangle \otimes \left|\exp\left(\frac{1}{\sqrt{2}}(f-g)\right)\right\rangle$$

If we have $f = g$ then $\left|\exp\left(\frac{1}{\sqrt{2}}(f-g)\right)\right\rangle = \exp(0)$ represents the vacuum state. Even in the case $f \neq g$ the vacuum can occur with probability $e^{\frac{1}{2}\|f-g\|^2}$. In that case the next step gives the pair of states

$$V\left|\exp\left(\frac{1}{\sqrt{2}}(f+g)\right)\right\rangle \otimes \exp(0)$$
$$= \left|\exp\left(\frac{1}{2}(f+g)\right)\right\rangle \otimes \left|\exp\left(\frac{1}{2}(f+g)\right)\right\rangle$$

i. e. the pair of created signal and signal arising from the senses coincide after that two steps of exchange if the above mentioned event occurs.
For that reason we will interpret this event as "full recognition" (postulate (P6)). If nothing happens the second step of exchange reconstructs the original pair of states

$$VV\left(\psi^{mem} \otimes \psi^{in}\right) = \psi^{mem} \otimes \psi^{in}$$

and the procedure can start again.
Observe that the event of recognition causes a change of the signals. Furthermore, the repetitions of the procedure after the first recognition will not cause further changes of the pair of states. That reflects postulate (P8).

Finally the recognition causes a certain loss of excites neurons, with reflects postulate (P7). In the case of our example the expectation of that loss is

$$\|f\|^2 + \|g\|^2 - \frac{1}{2}\|f+g\|^2 = \frac{1}{2}\|f-g\|^2 .$$

Furthermore the model reflects another well-known experience, namely that in case of signals being "unexpected" from the point of view of the memory, the probability of recognition is small but the measured activity, i. e. the loss of excited neurons is large.

Normally full recognition doesn't occur - only some kind of partial recognition.

We fix again a decomposition of G into measurable subsets $G_1, \ldots, G_n$. Let $\hat{H}_r$ be the subspace of $\Gamma\left(L^2(G_r)\right) \otimes \Gamma\left(L^2(G_r)\right)$ corresponding to the projection

$$Pr_{\hat{H}_r} := V\left(I_{L^2(M(G_r))} \otimes |\exp(0)\rangle \langle \exp(0)|\right) V \quad (r = 1, \ldots, n)$$

We define linear channels K_r^1 and K_r^0 from the set of positiv trace class operators ρ on $\Gamma(L^2(G_r)) \otimes \Gamma(L^2(G_r))$ into this set putting

$$K_r^1(\rho) := Pr_{\hat{H}_r} \rho Pr_{\hat{H}_r}; \quad K_r^0(\rho) := \rho \quad (r = 1, \ldots, n) .$$

Then we define linear channels from the set of positiv trace class operators on $\Gamma\left(L^2(G)\right) \otimes \Gamma\left(L^2(G)\right)$ into this set putting

$$K_{\varepsilon_1, \ldots, \varepsilon_n} := \bigotimes_{r=1}^{n} K_r^{\varepsilon_r} \quad (\varepsilon_1, \ldots, \varepsilon_n \in \{0, 1\}) .$$

If $tr K_{\varepsilon_1, \ldots, \varepsilon_n}(\rho) \neq 0$ then we put

$$\hat{K}_{\varepsilon_1, \ldots, \varepsilon_n}(\rho) := \frac{1}{tr K_{\varepsilon_1, \ldots, \varepsilon_n}(\rho)} K_{\varepsilon_1, \ldots, \varepsilon_n}(\rho) .$$

Now let us consider again the special case of coherent states

$|\exp(g)\rangle$ – input-signal;
$|\exp(f)\rangle$ – signal chosen from the memory

i. e. we consider a pure state ρ corresponding to

$$|\exp(f)\rangle \otimes |\exp(g)\rangle = \bigotimes_{r=1}^{n} |\exp(f\chi_{G_r})\rangle \otimes |\exp(g\chi_{G_r})\rangle$$

where χ_{G_r} denotes the indicator function of the set $G_r \subseteq G$. Then we get

$$\hat{K}_{\varepsilon_1, \ldots, \varepsilon_n}(\rho) = \bigotimes_{r=1}^{n} \rho_r$$

where the ρ_r are pure states on $\Gamma\left(L^2(G_r)\right) \otimes \Gamma\left(L^2(G_r)\right)$ corresponding to $|\exp(f)\rangle \otimes |\exp(g)\rangle$ if $\varepsilon_r = 0$. If we have "partial" recognition in the r-th region, i. e. $\varepsilon_r = 1$ then ρ_r is the pure state corresponding to

$$\left|\exp\left(\frac{1}{\sqrt{2}}(f+g)\chi_{G_r}\right)\right\rangle \otimes \left|\exp\left(\frac{1}{\sqrt{2}}(f+g)\chi_{G_r}\right)\right\rangle$$

i. e., the state ρ will be transformed into the state $\hat{K}_{\varepsilon_1,\ldots,\varepsilon_n}(\rho)$ if we have partial recognition concerning the r-th regions of the brain with $\varepsilon_r = 1$. In our special case of the state ρ the probability of that event is

$$\prod_{r=1}^{n}\left(e^{-\frac{\varepsilon_r}{2}\|(f-g)\chi_{G_r}\|^2}(1-(1-\varepsilon_r))\,e^{-\frac{1-\varepsilon_r}{2}\|(f-g)\chi_{G_r}\|^2}\right)$$

Now, recognition is a process with discrete time (see postulate (P1)). The procedure described above represents one step of that process. Step by step we get a sequence of random variables

$$\bar{\varepsilon}^m = \lfloor \varepsilon_1^m, \ldots, \varepsilon_n^m \rfloor \in \{0,1\}^n \quad (m \geq 1)$$

Where $\varepsilon_r^m = 1$ means that there was partial recognition in the r-th region at time m. The sequence $(\bar{\varepsilon}^m)_{m\geq 1}$ is increasing, i. e. if we have $\varepsilon_r^m = 1$ then it will be $\varepsilon_r^{m+k} = 1$ for each ($k \geq 1$). That means the recognition of the signal will be improved step by step up to the maximal level. Following (^{5}) one step of this process lasts $10^{-13} - 10^{-10}$ sec, and the whole procedure lasts $10^{-3} - 10^{-1}$ sec.

6. Measurements of the EEG-Typ

Let u be a function on G representing the electric potential of one excited neuron measured by an electrode placed on the surface of the scalp. If φ denotes a finite subset of G representing the positions of a configuration of excited neurons then electric potential of that configuration measured by the electrode is given by

$$U(\varphi) := \left(\sum_{x \in \varphi} u(x)\right)$$

Now, the bosonic Fock space H can be identified with a certain L^2-space of functions of finite point configurations, i. e. functions of finite subsets φ of G (^{10}), and the Fock space $\overline{H}$ can be identified with a certain L^2-space of functions of finite sets Φ of finite subsets of G. For that reason the elements of $\overline{H} \otimes H \otimes H$ are functions $\Psi(\Phi, \varphi_1, \varphi_2)$ where φ_1 and φ_2 are finite subsets of G and Φ denotes a finite set of finite subsets of G. One may interprete

φ_1 — "support" of the created signal ,
φ_1 — "support" of the arriving signal ,
$\bigcup_{\varphi \in \Phi} \varphi$ — "support" of all signals in the memory

Then
$$Z(\Phi, \varphi_1, \varphi_2) := \varphi_1 \cup \varphi_2 \cup \bigcup_{\varphi \in \Phi} \varphi$$

represents the positions of all excited neurons in the brain. Consequently the operator of multiplication corresponding to the function $U(Z(\varphi_1, \varphi_2, \Phi))$ represents the measurement of the electric potential of the brain corresponding to one electrode.

Now, in the case of an EEG-device one uses a sequence of potentials $(u_k)_{k=1}^r$ being of the typ $u_k(x) := u(x - y_k) \quad (x \in G)$.

Hereby, y_k represents the position of the k-th electrode placed on the surface of the scalp. The corresponding operators of multiplication commute ([8]), i. e., one is able to perform the corresponding measurements simultaneously.

References

1. W. Singer, *Der Beobachter im Gehirn* (Suhrkamp, 2002).
2. J. W. Phillips, R. M. Leahy and J. C. Mosher, *IEEE Engineering in Medicine and Biology* **16**, 34 (1997).
3. H. P. Stapp, *Mind, Matter and Quantum Mechanics*, 2nd edn. (Springer, 2003).
4. A. Nakamura and et.al., *NeuroImage* **7**, 377 (1998).
5. A. Sterr and et.al., *The journal of neuroscience* **18**, p. 4417 (1998).
6. L. Fichtner and M. Gaebler, Characterizations of beam splitters, Revised to 2007, (2007).
7. W. Fichtner, K.-H.; Freudenberg and V. Liebscher, *Random operators and stochastic equations* **12**, 331 (2004).
8. K.-H. Fichtner and L. Fichtner, *Quantum Markov Chains and the Process of Recognition*, Jenaer Schriften zur Mathematik und Informatik Math/Inf/02/07, FSU Jena, Faculty of Mathematics and Informatics (07737 Jena, Germany, 2007).
9. K.-H. Fichtner and L. Fichtner, *Bosons and a Quantum Model of the Brain*, Jenaer Schriften zur Mathematik und Informatik Math/Inf/08/05, FSU Jena, Faculty of Mathematics and Informatics (07737 Jena, Germany, 2005).
10. K.-H. Fichtner and W. Freudenberg, *Communications in mathematical Physics* **137**, 315 (1991).

KALLMAN DECOMPOSITIONS OF ENDOMORPHISMS OF VON NEUMANN ALGEBRAS

R. FLORICEL

Department of Mathematics and Statistics, University of Regina, Regina, SK, Canada, S4S 0A2 E-mail: floricel@math.uregina.ca

We show that any unital normal *-endomorphism of a von Neumann algebra admits a Kallman type decomposition, i.e., it can be decomposed uniquely as a central direct sum of a family of k-inner endomorphisms and a properly outer endomorphism. This decomposition is stable under conjugacy and cocycle conjugacy.

Keywords: von Neumann algebras; endomorphisms; conjugacy

Introduction

A classical result of R. Kallman, Theorem 1.11 in Ref.,[6] asserts that any automorphism of a von Neumann algebra can be decomposed as the direct sum of an inner and a freely acting automorphism. This decomposition result plays a central role in the theory of von Neumann algebras, being used extensively in their classification.[2]

Our main purpose in this paper is to show that a conceptually similar decomposition holds in a more general setting, i.e., for (not necessarily injective) unital normal *-endomorphisms of von Neumann algebras. This Kallman-type decomposition shows that the classification of endomorphisms of von Neumann algebras can be reduced to the classification of two particular classes of endomorphisms: the classes of k-inner and properly outer endomorphisms. One should note that a slightly similar reduction holds for E_0-semigroups,[4] but the present theory carries much richer structure.

This paper is organized as follows. In Section 2 we introduce the classes of k-inner, properly outer and freely acting endomorphisms, and discuss some of their properties, as well as some examples. In Section 3 we prove our main theorem: any unital normal *-endomorphism of a von Neumann algebra decomposes uniquely as the direct sum of a family of k-inner endomorphisms and a properly outer endomorphism. We also discuss the structure of ergodic endomorphisms. In Section 4 we show that this decomposition is stable under conjugacy and cocycle conjugacy.

We close our introduction with a few remarks on notation, most of which is standard. It this paper, we consider only separable Hilbert spaces. Let $M \subset \mathcal{B}(\mathcal{H})$ be a von Neumann algebra with center $\mathcal{Z}(M)$ and lattice of projections $\mathcal{P}(M)$, and let $\mathrm{End}(M)$ be the semigroup of all unital normal *-endomorphisms of M. For $\rho, \sigma \in \mathrm{End}(M)$, we consider the space of intertwiners between ρ and σ,

$$\mathrm{Hom}_M(\rho, \sigma) = \{u \in M \mid \sigma(x)u = u\rho(x),\ x \in M\}.$$

The fixed-point algebra of $\rho \in \mathrm{End}(M)$ is defined as $M^\rho = \{x \in M \mid \rho(x) = x\}$. If M^ρ reduces to scalars, then we say that ρ is an ergodic endomorphism of M.

1. Preliminaries on k-inner and properly outer endomorphisms

We begin this section by recalling briefly some definitions and properties about inner endomorphisms of von Neumann algebras. For a detailed treatment of this matter, we refer the reader to Ref.[8,11]

Let M be a von Neumann algebra. By a Hilbert space in a M (Ref.[11]) we understand a norm-closed linear subspace H of M that satisfies the following conditions:

(1) $u^*u \in \mathbb{C} \cdot 1$, for every $u \in H$;
(2) $xH \neq \{0\}$, for every $x \in M$, $x \neq 0$.

Any Hilbert space H in M has a natural inner product $\langle u, v \rangle = v^*u \cdot 1$, $u, v \in H$, and this inner product makes H into a genuine (complex) Hilbert space. An orthonormal basis for H is then given by a family of isometries $\{u_i\}_{i=\overline{1,k}}$ of M satisfying the Cuntz relations:[3]

$$u_i^* u_j = \delta_{i,j} \cdot 1 \quad \text{and} \quad \sum_{i=1}^{k} u_i u_i^* = 1, \tag{1}$$

where, if $k = \infty$, then the last sum is understood as convergence with respect to the strong topology of M.

As noted by J. Roberts, Lemma 2.2 in Ref.,[11] the Hilbert spaces in M implement a canonical class of endomorphisms of M, called inner endomorphisms. More precisely, for any Hilbert space H in M, there exists an endomorphism $\rho_H \in \text{End}(M)$, uniquely determined by the condition $H \subseteq \text{Hom}_M(Id, \rho_H)$. In fact, if $\{u_i\}_{i=\overline{1,k}}$ is an orthonormal basis for H, then the endomorphism ρ_H has the form

$$\rho_H(x) = \sum_{i=1}^{k} u_i x u_i^*, \ x \in M.$$

1.1. k-inner endomorphisms

Concerning the concept of inner endomorphism, one can naturally ask whether an inner endomorphism of M can be implemented by Hilbert spaces in M of different dimensions. If M is a factor, then the answer of this question is known to be negative: any inner endomorphism ρ of a factor M is implemented by a unique Hilbert space in M, and this Hilbert space is exactly $\text{Hom}_M(Id, \rho)$, Proposition 2.1 in Ref.[8] If M is an arbitrary von Neumann algebra, then the implementing Hilbert space in M of an inner endomorphism of M is unique up to isomorphism, as explained in the following lemma:

Lemma 1.1. *Let M be a von Neumann algebra, $\rho_1, \rho_2 \in \text{End}(M)$ be inner endomorphisms of M, and let $k, l \in \mathbb{N} \cup \{\infty\}$. Assume that ρ_1 is implemented by a k-dimensional Hilbert space in M with orthonormal basis $\{u_i\}_{i=\overline{1,k}}$, and that ρ_2 is implemented by a l-dimensional Hilbert space in M with orthonormal basis $\{v_j\}_{j=\overline{1,l}}$. Then the following conditions are equivalent:*

(i) $\rho_1 = \rho_2$;
(ii) $k = l$, and there exists a unitary matrix $[a_{ij}]_{i,\,j=\overline{1,k}}$ with coefficients in the center of M such that $v_j = \sum_{i=1}^{k} u_i a_{ij}$, $j = 1, 2, \ldots, k$.

Proof. First of all, we note that if $\rho \in \text{End}(M)$ is an inner endomorphism implemented by a k-dimensional Hilbert space in M, $k \in \mathbb{N} \cup \{\infty\}$, then the set $\text{Hom}_M(Id, \rho)$ is a free $\mathcal{Z}(M)$-bimodule of rank k. This can be easily seen by noticing that if $\{u_j\}_{j=\overline{1,k}}$ is an orthonormal basis for the Hilbert space in M that implements ρ, then $\text{Hom}_M(Id, \rho) = \left\{\sum_{j=1}^{k} u_j m_j : m_j \in \mathcal{Z}(M)\right\}$, and the latter set is a free $\mathcal{Z}(M)$-bimodule of rank k. We now proceed with the proof of our lemma.
(i) $\Rightarrow$ (ii). Since the rank of a free module over a commutative ring is unique, the above remark implies that $k = l$. Moreover, for any $x \in M$ we have

$$xu_i^* v_j = u_i^* \rho_1(x) v_j = u_i^* \rho_2(x) v_j = u_i^* v_j x, \text{ for every } i, j.$$

Therefore $u_i^* v_j \in \mathcal{Z}(M)$, for all $i, j = 1, \ldots, k$. We then define
$$a = [u_i^* v_j]_{i,j} \in M_k(\mathcal{Z}(M)),$$
and we see that a is a unitary matrix. Indeed, $(a^*a)_{i,j} = \sum_{s=1}^{k}(u_s^* v_i)^* u_s^* v_j = \sum_{s=1}^{k} v_i^* u_s u_s^* v_j = v_i^* v_j = \delta_{i,j} \cdot 1$, and similarly, $aa^* = 1$. Moreover $v_j = \left(\sum_{i=1}^{k} u_i u_i^*\right) v_j = \sum_{i=1}^{k} u_i a_{ij}$, for every $j = 1, 2, \ldots, k$.
(ii)$\Rightarrow$ (i) Straightforward computation. □

In order to distinguish among inner endomorphisms implemented by Hilbert spaces of different dimensions, we introduce the following definition:

Definition 1.1. Let $k \in \mathbb{N} \cup \{\infty\}$ and $\rho \in \mathrm{End}(M)$. We say that ρ is a k-inner endomorphism (or an inner endomorphism of dimension k) if ρ is implemented by a k-dimensional Hilbert space in M. Equivalently, ρ is k-inner if and only if there exists a set of isometries $\{v_i\}_{i \in \overline{1,k}} \subset \mathrm{Hom}_M(Id, \rho)$ that satisfies relations (1). Such a set will be called an implementing set for ρ. The set of all k-inner endomorphisms of M will be denoted by $\mathrm{End}_k(M)$.

Concerning the existence of k-inner endomorphisms, we note that if M is a properly infinite von Neumann algebra, then $\mathrm{End}_k(M)$ is non-empty, for every $k \in \mathbb{N} \cup \{\infty\}$. Moreover, if M is a type I_∞ factor, then every $\rho \in \mathrm{End}(M)$ is a k-inner endomorphism of M (Ref.[1]). Here k is the Powers index of ρ, i.e. that k such that $\mathrm{Hom}_M(\rho, \rho)$ is isomorphic to a factor of type I_k.[10]

If M is a finite von Neumann algebra, then $\mathrm{End}_k(M)$ is empty for $k \geq 2$. Obviously, the 1-inner endomorphisms are exactly the inner automorphisms of M.

1.2. *Properly outer endomorphisms and freely acting endomorphisms*

The properly outer endomorphisms are defined by complementarity. Loosely speaking, a properly outer endomorphism is an endomorphism which does not have inner parts.

Definition 1.2. Let M be a von Neumann algebra and $\rho \in \mathrm{End}(M)$. We say that ρ is a properly outer endomorphism of M, if for any central projection $p \in M^\rho$, we have
$$\rho \upharpoonright_{Mp} \notin \bigcup_{k=1}^{\infty} \mathrm{End}_k(Mp).$$

By appropriating the notion of freely acting automorphisms,[6] we define the corresponding class of endomorphisms:

Definition 1.3. An endomorphisms ρ of a von Neumann algebra M is said to be freely acting, if $\text{Hom}_M(Id, \rho) = \{0\}$.

It is easily seen that the freely acting endomorphisms must be properly outer. In the case of automorphisms, these two concepts are equivalent.[6] However, at least in the factor case, one can construct examples of properly outer endomorphisms that are not freely acting.

Example 1.1. In Proposition 3.2, Ref.,[5] M. Izumi has shown implicitly the existence of an endomorphism ρ of a type III factor M, satisfying the following Lee-Yang fusion rule:

$$[\rho^2] = [Id] \oplus [\rho]. \qquad (2)$$

Here $[\rho]$ denotes the sector associated to ρ.[9] Equation (2) implies the existence of two isometries u_1, u_2 of M that satisfy the Cuntz relations (1), such that

$$u_1 \in \text{Hom}_M(Id, \rho^2) \text{ and } u_2 \in \text{Hom}_M(\rho, \rho^2).$$

We also note that the Hilbert space $\text{Hom}_M(Id, \sigma^2)$ is 1-dimensional. It then follows that the endomorphism ρ^2 is properly outer, but not freely acting.

Observation 1.1. Let M be an infinite factor and $\rho \in \text{End}(M)$ be a properly outer endomorphism which is not freely acting. Conceptually, ρ has no "unital inner parts". However, it has a "non-unital inner part", as well as a "non-unital freely acting part". Indeed, let $\{u_i\}_{i \in I}$ be an orthonormal basis for $\text{Hom}_M(Id, \rho)$, and consider the projection $p = \sum_{i \in I} u_i u_i^* \in \text{Hom}_M(\rho, \rho)$. Then the "non-unital inner part" of ρ is the non-unital endomorphism $\rho_1 : M \to M_p \subset M$, $\rho_1(x) = \sum_{i \in I} u_i x u_i^*$, $x \in M$. One can easily see that the mapping $\rho_2 : M \to M_{(1-p)}$, $\rho_2(x) = \rho(x) - \rho_1(x)$, $x \in M$, is a *-homomorphism, and that ρ_2 is "freely acting" in the following sense: if $a \in M_{(1-p)}$ satisfies $\rho_2(x)a = ax$ for all $x \in M$, then $a = 0$. Indeed, if a is as above, then $a \in \text{Hom}_M(Id, \rho)$, and since $\langle a, u_j \rangle_M = u_j^* a = u_j^*(1-p)a = u_j^* \left(1 - \sum_{i \in I} u_i u_i^*\right) a = 0$, for every $j \in I$, we obtain $a = 0$. Thus ρ can be decomposed as the sum of an "inner" *-homomorphism and a "freely acting" *-homomorphism.

We end this section by discussing some examples of freely acting endomorphisms.

Example 1.2. Let M be a factor and $\rho \in \text{End}(M)$, $\rho(M) \neq M$. If M is a type II_1 factor, or if ρ is irreducible, i.e., $\text{Hom}_M(\rho, \rho) = \mathbb{C} \cdot 1_M$, then ρ is a freely acting endomorphism. Indeed, if $u \in \text{Hom}_M(Id, \rho)$, then $u^*u \in \mathcal{Z}(M) = \mathbb{C} \cdot 1$, and $uu^* \in \text{Hom}_M(\rho, \rho)$, and the conclusion follows.

Example 1.3. Let Γ be a discrete group, and let $\text{VN}(\Gamma) = \{\lambda_g \mid g \in \Gamma\}''$ be the group von Neumann algebra generated by the left regular representation λ of Γ. If $\rho : \Gamma \to \Gamma$ is an injective unital endomorphism of the group Γ, then ρ induces an endomorphism $\widetilde{\rho} \in \text{End}(\text{VN}(\Gamma))$ that acts on generators as $\widetilde{\rho}(\lambda_g) = \lambda_{\rho(g)}$, $g \in \Gamma$. By using an adaptation of Kallman's original argument in Theorem 2.2, Ref.,[6] one can check that $\widetilde{\rho}$ is freely acting if and only if the set $\{\rho(g)hg^{-1} \mid g \in \Gamma\}$ is infinite, for every $h \in \Gamma$. We note that if Γ is an ICC group, then $\widetilde{\rho}$ is automatically a freely acting endomorphism (see Example 1.2).

As an immediate exemplification, let Γ be the (non-ICC) group of all 2×2 matrices of the form $\begin{pmatrix} e & k \\ 0 & e' \end{pmatrix}$, where $e, e' = \pm 1$ and $k \in \mathbb{Z}$, and let ρ be the unital injective endomorphism of Γ given by

$$\rho(\begin{pmatrix} e & k \\ 0 & e' \end{pmatrix}) = \begin{pmatrix} e & 2k \\ 0 & e' \end{pmatrix}.$$

Then $\{\rho(g)hg^{-1} \mid g \in \Gamma\}$ is infinite, for every $h \in \Gamma$. Thus $\widetilde{\rho}$ is freely acting.

2. Decomposition of endomorphisms

We start with the following lemma, which has its roots in Ref.[6] Lemma 1.9, and Ref.[2] Proposition 1.5.1.

Lemma 2.1. *Let M be a von Neumann algebra, $\rho \in \text{End}(M)$ and $k \in \mathbb{N} \cup \{\infty\}$. If we define*

$$Q_k(\rho) = \{p \in \mathcal{P}(M) \cap \mathcal{Z}(M) \cap M^\rho \mid \rho \upharpoonright_{Mp} \in \text{End}_k(Mp)\},$$

then the set $Q_k(\rho)$ is closed under taking suprema.

Proof. Let $\{p_i\}_{i \in I} \subseteq Q_k(\rho)$, and define $p = \bigvee_{i \in I} p_i$. We want to show that

$$\rho \upharpoonright_{Mp} \in \text{End}_k(Mp).$$

By applying Zorn's Lemma, one can find a maximal (countable) family $\{q_j\}_{j \in J}$ of mutually orthogonal central projections of M, having the property that for any $j \in J$, there exists an $i \in I$ such that $q_j \leq p_i$. We then have that $\sum_{j \in J} q_j \leq p$, and we claim that the equality holds. Indeed, if $\sum_{j \in J} q_j \neq p$, then there exists $i \in I$ such that $p_i \nleq \sum_{j \in J} q_j$. Then it is easily seen that $(p - \sum_{j \in J} q_j) p_i$ is a nonzero central projection, and that $\left(p - \sum_{j \in J} q_j\right) p_i q_l = 0$, for every $l \in J$. This contradicts the maximality of the family $(q_j)_{j \in J}$. Thus $p = \sum_{j \in J} q_j$.

Since the direct sum of a family of k-inner endomorphisms is still a k-inner endomorphism, in order to complete this proof it is enough to show that

$$\rho \restriction_{Mq_j} \in \text{End}_k(Mq_j), \text{ for every } j \in J.$$

For this purpose, let $j \in J$ be fixed, and $i \in I$ be such that $q_j \leq p_i$. We choose an implementing set $\{u_l\}_{l=\overline{1,k}} \subset \text{Hom}_{Mp_i}(Id, \rho \restriction_{Mp_i})$ for $\rho \restriction_{Mp_i}$. Then $\rho(q_j) = \sum_{l=1}^{k} u_l q_j u_l^* = q_j \left(\sum_{l=1}^{k} u_l u_l^*\right) = q_j p_i = q_j$, so $\rho \restriction_{Mq_j} \in \text{End}(Mq_j)$. Moreover, the set $\{u_l q_j\}_{l=\overline{1,k}} \subset \text{Hom}_{Mq_j}(Id, \rho \restriction_{Mq_j})$ is an implementing set for the endomorphism $\rho \restriction_{Mq_j}$, so $\rho \restriction_{Mq_j} \in \text{End}_k(Mq_j)$. □

Definition 2.1. Let M be a von Neumann algebra. For any $\rho \in \text{End}(M)$ and $k \in \mathbb{N} \cup \{\infty\}$, we shall denote by $p_k(\rho)$ the projection $\bigvee Q_k(\rho)$. We also define the central projection

$$p(\rho) = \sum_{k \in I_\rho} p_k(\rho),$$

where $I_\rho = \{k \in \mathbb{N} \cup \{\infty\} \mid p_k(\rho) \neq 0\}$.

Lemma 2.2. *If $\rho \in \text{End}(M)$ and $k, l \in \mathbb{N} \cup \{\infty\}$, $k \neq l$, then $p_k(\rho)$ and $p_l(\rho)$ are orthogonal projections.*

Proof. Let $p \in Q_k(\rho)$ and $q \in Q_l(\rho)$. Suppose that $pq \neq 0$. Since $\rho \restriction_{Mp}$ is a k-inner endomorphism, and $\rho \restriction_{Mq}$ is a l-inner endomorphism, it follows that

$$\rho \restriction_{Mpq} \in \text{End}_k(Mpq) \cap \text{End}_l(Mpq).$$

This contradicts Lemma 1.1. Hence $pq = 0$, and the conclusion follows. □

We conclude that the projections $p_k(\rho)$ and $p(\rho)$ fully characterize the classes of k-inner and properly outer endomorphisms: an endomorphism ρ of a von Neumann algebra M is k-inner if and only if $p_k(\rho) = 1$. It is properly outer if and only if $p(\rho) = 0$. We are now ready to prove our main result.

Theorem 2.1. *Let M be a von Neumann algebra and $\rho \in \text{End}(M)$. Then there exists a set $I \subseteq \mathbb{N} \cup \{\infty\}$ such that M decomposes as a central direct* sum of von Neumann subalgebras $M = (\sum_{k \in I}^{\oplus} M_k) \bigoplus M_0$, and ρ decomposes as a direct sum of endomorphisms $\rho = (\sum_{k \in I}^{\oplus} \rho_k) \oplus \rho_0$ having the following properties:

(1) $\rho_k \in \mathrm{End}_k(M_k)$, for every $k \in I$;
(2) $\rho_0 \in \mathrm{End}(M_0)$ is a properly outer endomorphism of M_0.

This decomposition is unique.

Proof. First of all, we establish the existence of such a decomposition. For this purpose, we take $I = I_\rho$, where I_ρ is as in Definition 2.1. For any $k \in I$, we define

$$M_k = M p_k(\rho), \quad \rho_k = \rho \restriction_{M_k},$$

as well as

$$M_0 = M(1 - p(\rho)), \quad \rho_0 = \rho \restriction_{M_0}.$$

Then it is easily seen that this setting gives the required decomposition.

Next, we show that this decomposition is unique. For this purpose, let $J \subseteq \mathbb{N} \cup \{\infty\}$ be a set of indices, and let $\{q_l\}_{l \in J} \subset M^\rho$ be a family of mutually orthogonal central projections of M such that

(1') $\rho \restriction_{M q_l} \in \mathrm{End}_l(M q_l)$, for every $l \in J$;
(2') $\rho \restriction_{M(1-q)}$ is a properly outer endomorphism of $M(1-q)$,

where $q = \sum_{l \in J} q_l$. By construction, $J \subseteq I_\rho$, and we claim that $J = I_\rho$ and $p_k(\rho) = q_k$, for every $k \in I_\rho$.

First of all, we show that if $k \in I_\rho$, then $p_k(\rho) q_l = 0$, for every $l \in J$, $l \neq k$. Assume that $p_k(\rho) q_l \neq 0$. Since $M p_k(\rho) q_l = M p_k(\rho) \cap M q_l$, we obtain that

$$\rho \restriction_{M p_k(\rho) q_l} \in \mathrm{End}_k (M p_k(\rho) q_l) \cap \mathrm{End}_l (M p_k(\rho) q_l).$$

We then deduce from Lemma 1.1 that $k = l$.

Secondly, we show that for any $l \in J$, we have $p_l(\rho) = q_l$. Indeed, the construction of the projections $p_l(\rho)$ guarantees that $q_l \leq p_l(\rho)$, for every $l \in J$. Assume that there exists $l \in J$ such that $p_l(\rho) \neq q_l$. Then

$$p_l(\rho)(1 - q) = p_l(\rho) - q_l \neq 0,$$

and since $\rho \restriction_{M(1-q)}$ is a properly outer endomorphism, we obtain that

$$\rho \restriction_{M p_l(\rho)(1-q)} \notin \mathrm{End}_l(M p_l(\rho)(1-q)).$$

On the other hand, since $Mp_l(\rho)(1-q) \subset Mp_l(\rho)$, we have that

$$\rho \upharpoonright_{Mp_l(\rho)(1-q)} \in \mathrm{End}_l(Mp_l(\rho)(1-q)).$$

This is a contradiction. Thus $p_l(\rho) = q_l$, for every $l \in J$.

We are now ready to show that $I_\rho = J$. Assume that there exists $k \in I_\rho \setminus J$. It then follows that $p_k(\rho) \leq 1-q$, and since $\rho \upharpoonright_{M(1-q)}$ is properly outer, we obtain that

$$\rho \upharpoonright_{Mp_k(\rho)} \notin \mathrm{End}_k(Mp_k(\rho)).$$

This contradicts the definition of the projection $p_k(\rho)$. Therefore $I_\rho = J$, and the theorem is proved. □

An an immediate consequence, we obtain the following characterization of the ergodic endomorphisms of von Neumann algebras.

Corollary 2.1. *Let M be a von Neumann algebra, and $\rho \in \mathrm{End}(M)$ be an ergodic endomorphism. Then ρ is either a k-inner endomorphism ($k \geq 2$), or a properly outer endomorphism. Moreover, if ρ is a k-inner endomorphism, then M must be a factor.*

Proof. If ρ is an ergodic endomorphism, then either $p(\rho) = p_k(\rho) = 1$, for some $k \in \mathbb{N} \cup \{\infty\}$, $k \geq 2$, or $p(\rho) = 0$. The last statement follows from the fact that if ρ is a k-inner endomorphism, then $\mathcal{Z}(M)$ is a subset of M^ρ. □

Although there are no ergodic inner automorphisms acting on von Neumann algebras, one can construct a plethora of ergodic k-inner endomorphisms ($k \geq 2$). The construction is based on the following remark: if $\rho \in \mathrm{End}_k(M)$, ($k \geq 2$), and if $\{u_i\}_{i=\overline{1,k}}$ is an implementing set for ρ, then

$$M^\rho = \mathcal{O}'_k \cap M, \tag{1}$$

where $\mathcal{O}_k$ is the Cuntz algebra generated by $\{u_i\}_{i=\overline{1,k}}$.[3] The proof of this equality is straightforward, and we leave the details to the reader. We note that this result was also obtained by M. Laca in the case of type I_∞ factors, Proposition 3.1 in Ref.[7]

To construct ergodic k-inner endomorphisms, we start with the Cuntz algebra $\mathcal{O}_k$ with k-generators $\{u_i\}_{i=\overline{1,k}}$. Let ϕ be a factor state of $\mathcal{O}_k$, and let π_ϕ be the GNS representation of $\mathcal{O}_k$ with respect to ϕ. Then the canonical endomorphism σ of $\mathcal{O}_k$, defined by

$$\sigma(x) = \sum_{i=1}^{k} u_i x u_i^*, \quad x \in \mathcal{O}_k,$$

can be extended to a k-inner endomorphism of $\pi_\phi(\mathcal{O}_k)''$. Since $\pi_\phi(\mathcal{O}_k)''$ is a factor, it follows from (1) that this k-inner endomorphism is ergodic.

3. Reduction of Conjugacy and cocycle conjugacy

Two endomorphisms $\rho \in \text{End}(M)$ and $\sigma \in \text{End}(N)$ are said to be conjugate, if there exists a *-isomorphism θ of N onto M such that $\rho \circ \theta = \theta \circ \sigma$. They are called cocycle conjugate, if there exists a unitary u of N such that the endomorphisms $\text{Ad}(u) \circ \sigma$ and ρ are conjugate.

Theorem 2.1 allows us to reduce the classification up to conjugacy and cocycle conjugacy of arbitrary endomorphisms to the classification up to conjugacy, respectively cocycle conjugacy, of the classes of k-inner endomorphisms and properly outer endomorphisms.

Theorem 3.1. *Let M be a von Neumann algebra and ρ, $\sigma \in \text{End}(M)$ be two endomorphisms of M, decomposed as in Theorem 2.1:*

$$\rho = (\sum_{k \in I_\rho}^\oplus \rho_k) \oplus \rho_0, \quad \sigma = (\sum_{l \in I_\sigma}^\oplus \sigma_l) \oplus \sigma_0.$$

Then ρ and σ are conjugate if and only if the following conditions are satisfied:

(i) $I_\rho = I_\sigma$;
(ii) ρ_k and σ_k are conjugate endomorphisms, for every $k \in I_\rho$;
(iii) ρ_0 and σ_0 are conjugate endomorphisms.

Proof. ($\Longrightarrow$) Suppose that ρ and σ are conjugate, and let $\theta \in \text{Aut}(M)$ be such that $\rho \circ \theta = \theta \circ \sigma$. Then for any $l \in I_\sigma$, we have $\theta^{-1} \circ \rho \circ \theta \mid_{M p_l(\sigma)} = \sigma_l$. Since $\sigma_l(p_l(\sigma)) = p_l(\sigma)$, we deduce that $\theta(p_l(\sigma)) \in M^\rho$.

Let now $\{u_i\}_{i=\overline{1,l}} \subset \text{Hom}_{M p_l(\sigma)}(Id, \sigma_l)$ be an implementing set for the l-inner endomorphism σ_l. Then for any $x \in M p_l(\sigma)$ and $i, j = 1, 2, \ldots, l$, we have $\rho(\theta(x))\theta(u_i) = \theta(u_i)\theta(x)$, as well as

$$\theta(u_i)^* \theta(u_j) = \delta_{i,j} \cdot \theta(p_l(\sigma)) \quad \text{and} \quad \sum_{i=1}^k \theta(u_i)\theta(u_i)^* = \theta(p_l(\sigma)).$$

Therefore $\{\theta(u_i)\}_{i=\overline{1,l}}$ is an implementing set for the endomorphism $\rho \mid_{M\theta(p_l(\sigma))}$. Hence

$$\rho \upharpoonright_{M\theta(p_l(\sigma))} \in \text{End}_l(M\theta(p_l(\sigma))).$$

It then follows from Definition 2.1 that $l \in I_\rho$, and that $\theta(p_l(\sigma)) \leq p_l(\rho)$.

On the other hand, by repeating the above calculation for $\rho = \theta \circ \sigma \circ \theta^{-1}$, we obtain that $I_\rho = I_\sigma$, and that

$$\theta(p_k(\sigma)) = p_k(\rho), \text{ for every } k \in I_\rho.$$

In particular, $\theta \restriction_{Mp_k(\sigma)}$ is a *-isomorphism of $Mp_k(\sigma)$ onto $Mp_k(\rho)$ which implements the conjugacy between ρ_k and σ_k, for every k. By construction, $\theta \restriction_{M(1-p(\sigma))}$ is also a *-isomorphism of $M(1 - p(\sigma))$ onto $M(1 - p(\rho))$ which implements the conjugacy between ρ_0 and σ_0.

($\Longleftarrow$) We assume that the endomorphisms ρ and σ satisfy conditions (i), (ii), and (iii). For any $k \in I_\rho$, let $\theta_k : Mp_k(\sigma) \to Mp_k(\rho)$ be a *-isomorphism such that $\rho_k \circ \theta_k = \theta_k \circ \sigma_k$. Also, let $\theta_0 : M(1 - p(\sigma)) \to M(1 - p(\rho))$ be a *-isomorphism such that $\rho_0 \circ \theta_0 = \theta_0 \circ \sigma_0$. Thus, if we define

$$\theta = \left(\sum_{k \in I}^{\oplus} \theta_k \right) \oplus \theta_0,$$

then $\theta \in \text{Aut}(M)$ satisfies $\rho \circ \theta = \theta \circ \sigma$. $\square$

Corollary 3.1. *Let ρ and σ be as in Theorem 3.1. Then ρ and σ are cocycle conjugate if and only if the following conditions are satisfied:*

(i) $I_\rho = I_\sigma$;
(ii) the von Neumann algebras $Mp_k(\rho)$ and $Mp_k(\sigma)$ are isomorphic, for every $k \in I_\rho$;
(iii) ρ_0 and σ_0 are cocycle conjugate endomorphisms.

Proof. ($\Longrightarrow$) Let u be a unitary of M such that the endomorphisms ρ and $\text{Ad}(u) \circ \sigma$ are conjugate. If we denote by $u_l = up_l(\sigma)$, $l \in I_\sigma$, and by $u_0 = u(1 - p(\sigma))$, then the endomorphism $\text{Ad}(u) \circ \sigma$ decomposes as

$$\text{Ad}(u) \circ \sigma = \left(\sum_{l \in I_\sigma}^{\oplus} \text{Ad}(u_l) \circ \sigma_l \right) \oplus \left(\text{Ad}(u_0) \circ \sigma_0 \right),$$

where $\text{Ad}(u_l) \circ \sigma_l \in \text{End}_l(Mp_l(\sigma))$ for all $l \in J$, and $\text{Ad}(u_0) \circ \sigma_0$ is a properly outer endomorphism of $M(1 - p(\sigma))$. The required result follows then from Theorem 3.1.

($\Longleftarrow$) First of all, we show that the endomorphisms ρ_k and σ_k are cocycle conjugate, for every $k \in I_\rho (= I_\sigma)$. For this purpose, let $\{u_i\}_{i=\overline{1,k}} \subset \text{Hom}_{Mp_k(\rho)}(Id, \rho_k)$, respectively $\{v_i\}_{i=\overline{1,k}} \subset \text{Hom}\, Mp_k(\sigma)(Id, \sigma_k)$, be implementing sets for ρ_k, respectively σ_k, and let $\theta_k : Mp_k(\sigma) \to Mp_k(\rho)$ be a *-isomorphism. Let $w_k = \sum_{i=1}^{k} u_i \theta_k(v_i)^*$. Then w_k is a unitary of $M_{p_k(\rho)}$, and we have

$$\rho_k(x) = \mathrm{Ad}(w_k)\left(\sum_{i=1}^{k}\theta_k(v_i)x\theta_k(v_i)^*\right) = \mathrm{Ad}(w_k)\circ\theta_k\circ\sigma_k\circ\theta_k^{-1}(x),$$

for every $x \in M_{p_k(\rho)}$. Therefore ρ_k and σ_k are cocycle conjugate endomorphisms.

On the other hand, since ρ_0 and σ_0 are cocycle conjugate, there exists a *-isomorphism $\theta_0 : M(1-p(\sigma)) \to M(1-p(\rho))$, and a unitary w_0 of $M(1-p(\rho))$ such that $\rho_0 = \mathrm{Ad}(w_0)\circ\theta_0\circ\sigma_0\circ\theta_0^{-1}$. Therefore, if we define $w = \left(\sum_{k\in I_\rho}^{\oplus} w_k\right) \oplus w_0$ and $\theta = \left(\sum_{k\in I}^{\oplus}\theta_k\right)\oplus\theta_0$, then w is a unitary of M, $\theta \in \mathrm{Aut}(M)$ and $\rho = \mathrm{Ad}(w)\circ\theta\circ\sigma\circ\theta^{-1}$. □

References

1. W. Arveson, *Continuous analogues of Fock space*, Mem. Amer. Math. Soc., **80** (1989), no. 409.
2. A. Connes, *Une classification des facteurs de type* III, C. R. Acad. Sci. Paris, Ser. A-B, **275** (1972), A523–A525.
3. J. Cuntz, *Simple C^*-algebras generated by isometries*, Comm. Math. Phys. **57** (1977), no. 2, 173–185.
4. R. Floricel, *A decomposition of E_0-semigroups*, Advances in quantum dynamics (South Hadley, MA,) Contemp. Math. **335** (2002), Amer. Math. Soc., Providence, RI, (2003), 131–138.
5. M. Izumi, *Application of fusion rules to classification of subfactors*, Publ. Res. Inst. Math. Sci. **27** (1991), no. 6, 953–994.
6. R. Kallman, *A generalization of free action*, Duke Math. J. **36** (1969), 781–789.
7. M. Laca, *Endomorphisms of $B(H)$ and Cuntz algebras*, J. Operator Theory **30** (1993), no. 1, 85–108.
8. R. Longo, *Simple injective subfactors*, Adv. in Math. **63** (1987), no. 2, 152–171.
9. R. Longo, *Index of subfactors and statistics of quantum fields. I.* Comm. Math. Phys. **126** (1989), 217–247.
10. R. Powers, *An index theory for semigroups of *-endomorphisms of $\mathfrak{B}(\mathcal{H})$ and type II_1 factors*, Canad. J. Math. **40** (1988), no. 1, 86–114.
11. J. Roberts, *Cross products of von Neumann algebras by group duals*, Symposia Mathematica **XX** (1976), 335–363.

SCHRÖDINGER EQUATION, L^p-DUALITY AND THE GEOMETRY OF WIGNER-YANASE-DYSON INFORMATION

P. GIBILISCO

Dipartimento SEFEMEQ,
Facoltà di Economia, Università di Roma "Tor Vergata",
Via Columbia 2, 00133 Rome, Italy
E-mail: gibilisco@volterra.uniroma2.it
URL: http://www.economia.uniroma2.it/sefemeq/professori/gibilisco

D. IMPARATO

Dipartimento di Matematica,
Politecnico di Torino,
Corso Duca degli Abruzzi, 10129 Turin, Italy
E-mail: daniele.imparato@polito.it

T. ISOLA

Dipartimento di Matematica,
Università di Roma "Tor Vergata",
Via della Ricerca Scientifica, 00133 Rome, Italy
E-mail: isola@mat.uniroma2.it
URL: http://www.mat.uniroma2.it/~isola

We discuss the geometry of Wigner-Yanase-Dyson information via the so-called Amari-Nagaoka embeddings in L^p-spaces of quantum trajectories.

1. Introduction

The Wigner-Yanase-Dyson information was introduced in 1963.[28] Wigner and Yanase observed that "According to quantum mechanical theory, some observables can be measured much more easily than others: the observables which commute with the additive conserved quantities ... can be measured with microscopic apparatuses; those which do not commute with these quantities need for their measurements macroscopic systems. Hence the problem of defining a measure of our knowledge with respect to the latter quantities arises ...". After the discussion of the requirements such a measure should satisfy (convexity, ...) they proposed, tentatively, the following formula and called it *skew information*:

$$I_\rho(A) := -\frac{1}{2}\text{Tr}([\rho^{\frac{1}{2}}, A]^2).$$

More generally they defined (following a suggestion by Dyson)

$$I_\rho^\beta(A) := -\frac{1}{2}\text{Tr}([\rho^\beta, A] \cdot [\rho^{1-\beta}, A]), \qquad \beta \in [0,1].$$

The latter is known as WYD-information. The skew information should be considered as a measure of information contained in a state ρ with respect to a conserved observable A.

From that fundamental work WYD-information has found applications in a manifold of different fields. A possibly incomplete list should mention: i) strong subadditivity of entropy;[22,23] ii) homogeneity of the state space of factors (of type III$_1$);[6] hypothesis testing[3] iii) measures for quantum entanglement;[4,19] iv) uncertainty relations.[7,10–13,21,24,25,27]

Such a variety should be not surprising at the light of the result showing that WYD-information is just an example of monotone metric, namely it is a member of the vast family of quantum Fisher informations.[9] On the other hand one can prove that, among the family of all the quantum Fisher informations, the geometry of WYD-information is rather special.[8,16]

In this paper we want to discuss the particular features of WYD-information emphasizing the relation with the embedding of quantum dynamics in L^p-spaces.

2. Preliminary notions of matrix analysis

Let $M_n := M_n(\mathbb{C})$ (resp.$M_{n,sa} := M_n(\mathbb{C})_{sa}$) be the set of all $n \times n$ complex matrices (resp. all $n \times n$ self-adjoint matrices). We shall denote general matrices by $X, Y, ...$ while letters $A, B, ...$ (or H) will be used for self-adjoint matrices. Let D_n be the set of strictly positive elements of M_n while $D_n^1 \subset D_n$ is the set of density matrices namely

$$D_n^1 = \{\rho \in M_n | \text{Tr}\rho = 1, \rho > 0\}.$$

The tangent space to D_n^1 at ρ is given by $T_\rho D_n^1 \equiv \{A \in M_{n,sa} : \text{Tr}(A) = 0\}$, and can be decomposed as $T_\rho D_n^1 = (T_\rho D_n^1)^c \oplus (T_\rho D_n^1)^o$, where $(T_\rho D_n^1)^c := \{A \in T_\rho D_n^1 : [A, \rho] = 0\}$, and $(T_\rho D_n^1)^o$ is the orthogonal complement of $(T_\rho D_n^1)^c$, with respect to the Hilbert-Schmidt scalar product $\langle A, B \rangle := \langle A, B \rangle_{HS} := \text{Tr}(A^*B)$ (the Hilbert-Schmidt norm will be denoted by $\|\cdot\|$).

A typical element of $(T_\rho D_n)^o$ has the form $A = i[\rho, H]$, where H is self-adjoint.

In what follows we shall need the following result (pag. 124 in²).

Proposition 2.1. *Let $A \in M_{n,sa}$ be decomposed as $A = A^c + i[q, H]$ where $q \in D_n$, $[A^c, q] = 0$ and $H \in M_{n,sa}$. Suppose $\varphi \in C^1(0, +\infty)$. Then*

$$(D_q \varphi)(A) = \varphi'(q) A^c + i[\varphi(q), H].$$

3. Schrödinger equation and quantum dynamics

Let $\rho(t)$ be a curve in D_n^1 and let $H \in M_{n,sa}$ We say that $\rho(t)$ satisfy the Schrödinger equation w.r.t. H if $\frac{d}{dt} \rho(t) = i[\rho(t), H]$. This equation is also known in the literature as the Landau-von Neumann equation.

The solution of the above evolution equation (please note that H is time independent) is given by

$$\rho_H(t) := e^{-itH} \rho e^{itH}. \tag{1}$$

Therefore the commutator $i[\rho, H]$ appears as the tangent vector to the quantum trajectory (1) (at the initial point $\rho = \rho_H(0)$) generated by H. Suppose we are considering two different evolutions determined, through the Schrödinger equation, by H and K. If we want to quantify how "different" the trajectories $\rho_H(t), \rho_K(t)$ are, then it would be natural to measure the "area" spanned by the tangent vectors $i[\rho, H], i[\rho, K]$ (with respect to some scalar product[10]).

4. L^p-embedding for states and trajectories

The functions

$$\rho \to \frac{\rho^\beta}{\beta}, \quad \beta \in (0,1)$$

are known as Amari-Nagaoka embeddings.[1,14] They can be considered as an immersion of the state manifold into L^p-spheres.

Proposition 4.1. *Let $\rho(t)$ be a curve in D_n^1, let $H \in M_{n,sa}$ and let $\beta \in (0,1)$. The following differential equations are equivalent*

$$\frac{d}{dt} \rho(t) = i[\rho(t), H], \tag{1}$$

$$\frac{d}{dt} (\rho(t)^\beta) = i[\rho(t)^\beta, H]. \tag{2}$$

Proof. Let $\phi_\beta(\rho) := \rho^\beta$. By Proposition 2.1 we get

$$\frac{d}{dt}\left(\rho(t)^\beta\right) = D_\rho \phi_\beta \circ \frac{d}{dt}\rho(t) = D_\rho\phi_\beta(i[\rho(t), H])$$
$$= (i[\phi_\beta(\rho(t)), H]) = i[\rho(t)^\beta, H].$$

Therefore, Equation (1) implies Equation (2). Analogously, again using Proposition 2.1, Equation (2) implies Equation (1) because we have

$$\frac{d}{dt}(\rho(t)) = \frac{d}{dt}\left((\rho(t)^\beta)^{\frac{1}{\beta}}\right) = D_{(\rho(t)^\beta)}\phi_\beta^{-1} \circ \frac{d}{dt}\left(\rho(t)^\beta\right)$$
$$= D_{(\rho(t)^\beta)}\phi_\beta^{-1} \circ i[\rho(t)^\beta, H] = D_{(g(t))}\phi_\beta^{-1} \circ i[g(t), H]$$
$$= i[\phi_\beta^{-1}(g(t)), H] = i[\rho(t), H]. \qquad \square$$

5. WYD-information by pairing of dual trajectories

The Wigner-Yanase-Dyson information is defined as

$$I_\rho^\beta(H) := -\frac{1}{2}\mathrm{Tr}([\rho^\beta, H] \cdot [\rho^{1-\beta}, H]), \qquad \beta \in (0,1).$$

Let us explain the link between L^p-embeddings and WYD-information. Let V, W be vector spaces over $\mathbb{R}$ (or $\mathbb{C}$). One says that there is a duality pairing if there exists a separating bilinear form

$$\langle \cdot, \cdot \rangle : V \times W \to \mathbb{R}\,(\mathbb{C}).$$

In the case of L^p spaces the pairing is given by the L^2 scalar product. In our case this is just the HS-scalar product.

Note that using the function $\rho \to \rho^\beta$ we may look at dynamics as a curve on a $L^{\frac{1}{\beta}}$-sphere. The function $\rho \to \rho^{1-\beta}$ does the same on the dual space $\left(L^{\frac{1}{\beta}}\right)^* = L^{\frac{1}{1-\beta}}$.

Proposition 5.1. *If $\rho(t)$ satisfies the Schrödinger equation w.r.t. H then*

$$\langle \frac{d}{dt}\rho(t)^\beta, \frac{d}{dt}\rho(t)^{1-\beta}\rangle = 2 \cdot I_{\rho(t)}^\beta(H) \qquad \beta \in (0,1).$$

Proof. Apply Proposition 4.1 to obtain

$$\langle \frac{d}{dt}\left(\rho(t)^\beta\right), \frac{d}{dt}\left(\rho(t)^{1-\beta}\right)\rangle = \langle i[\rho(t)^\beta, H], i[\rho(t)^{1-\beta}, H]\rangle$$
$$= -\mathrm{Tr}([\rho(t)^\beta, H] \cdot [\rho(t)^{1-\beta}, H]). \qquad \square$$

In this way WYD-information appears as the "pairing" of the dual L^p-embeddings of the same quantum trajectory.

6. Quantum Fisher informations

In the commutative case a Markov morphism is a stochastic map $T : \mathbb{R}^n \to \mathbb{R}^k$. In the noncommutative case a Markov morphism is a completely positive and trace preserving operator $T : M_n \to M_k$. Let

$$\mathcal{P}_n := \{\rho \in \mathbb{R}^n | \rho_i > 0\} \qquad \mathcal{P}_n^1 := \{\rho \in \mathbb{R}^n | \sum \rho_i = 1, \rho_i > 0\}.$$

In the commutative case a monotone metric is a family of Riemannian metrics $g = \{g^n\}$ on $\{\mathcal{P}_n^1\}$, $n \in \mathbb{N}$, such that

$$g^m_{T(\rho)}(TX, TX) \leq g^n_\rho(X, X)$$

holds for every Markov morphism $T : \mathbb{R}^n \to \mathbb{R}^m$ and all $\rho \in \mathcal{P}_n^1$ and $X \in T_\rho \mathcal{P}_n^1$.

In perfect analogy, a monotone metric in the noncommutative case is a family of Riemannian metrics $g = \{g^n\}$ on $\{\mathcal{D}_n^1\}$, $n \in \mathbb{N}$, such that

$$g^m_{T(\rho)}(TX, TX) \leq g^n_\rho(X, X)$$

holds for every Markov morphism $T : M_n \to M_m$ and all $\rho \in \mathcal{D}_n^1$ and $X \in T_\rho \mathcal{D}_n^1$.

Let us recall that a function $f : (0, \infty) \to \mathbb{R}$ is called operator monotone if, for any $n \in \mathbb{N}$, any $A, B \in M_n$ such that $0 \leq A \leq B$, the inequalities $0 \leq f(A) \leq f(B)$ hold. An operator monotone function is said symmetric if $f(x) := xf(x^{-1})$. With such operator monotone functions f one associates the so-called Chentsov–Morotzova functions

$$c_f(x, y) := \frac{1}{yf(xy^{-1})} \qquad \text{for} \qquad x, y > 0.$$

Define $L_\rho(A) := \rho A$, and $R_\rho(A) := A\rho$. Since L_ρ and R_ρ commute we may define $c(L_\rho, R_\rho)$ (this is just the inverse of the operator mean associated to f by Kubo-Ando theory[10]). Now we can state the fundamental theorems about monotone metrics. In what follows uniqueness and classification are stated up to scalars (for reference see[26]).

Theorem 6.1. *(Chentsov 1982) There exists a unique monotone metric on $\mathcal{P}_n^1$ given by the Fisher information.*

Theorem 6.2. *(Petz 1996) There exists a bijective correspondence between monotone metrics on $\mathcal{D}_n^1$ and symmetric operator monotone functions. For $\rho \in \mathcal{D}_n^1$, this correspondence is given by the formula*

$$g_f(A, B) := g_{f,\rho}(A, B) := \mathrm{Tr}(A \cdot c_f(L_\rho, R_\rho)(B)).$$

Because of these two theorems, the terms "Monotone Metrics" and "Quantum Fisher Informations" are used with the same meaning.

Note that usually monotone metrics are normalized so that $[A, \rho] = 0$ implies $g_{f,\rho}(A, A) = \text{Tr}(\rho^{-1} A^2)$, that is equivalent to set $f(1) = 1$.

7. The WYD monotone metric

The following functions are symmetric, normalized and operator monotone (see[9,16]). Let

$$f_\beta(x) := \beta(1-\beta)\frac{(x-1)^2}{(x^\beta-1)(x^{1-\beta}-1)} \qquad \beta \in (0,1).$$

Proposition 7.1. *For the QFI associated to f_β one has*

$$g_{f_\beta}(i[\rho, H], i[\rho, K]) = -\frac{1}{\beta(1-\beta)}\text{Tr}([\rho^\beta, H] \cdot [\rho^{1-\beta}, K]) \qquad \beta \in (0,1).$$

One can find a proof in.[9,16] Because of the above Proposition, g_β is known as $WYD(\beta)$ monotone metric.

Of course what we have seen about L^p-embedding of quantum dynamics applies to this example of quantum Fisher information. Indeed we can summarize everything into the following final result.

Proposition 7.2.
Let H, K be selfadjoint matrices and ρ be a density matrix. Choose two curves $\rho(t), \sigma(t) \subset D_n^1$ such that
i) $\rho(t)$ satisfies the Schrödinger equation w.r.t. H;
ii) $\sigma(t)$ satisfies the Schrödinger equation w.r.t. K;
iii) $\rho = \rho(0) = \sigma(0)$.
One has

$$g_{f_\beta}(i[\rho, H], i[\rho, K]) = \langle \frac{d}{dt}\left(\frac{\rho(t)^\beta}{\beta}\right), \frac{d}{dt}\left(\frac{\sigma(t)^{1-\beta}}{1-\beta}\right)\rangle\big|_{t=0} \qquad \beta \in (0,1)$$

Proof. From Proposition 7.1, one gets

$$g_{f_\beta}(i[\rho, H], i[\rho, K]) = -\tfrac{1}{\beta(1-\beta)}\text{Tr}([\rho^\beta, H] \cdot [\rho^{1-\beta}, K])$$

$$= -\tfrac{1}{\beta(1-\beta)}\text{Tr}([\rho(t)^\beta, H] \cdot [\sigma(t)^{1-\beta}, K])\big|_{t=0}$$

$$= \langle \tfrac{d}{dt}\left(\tfrac{\rho(t)^\beta}{\beta}\right), \tfrac{d}{dt}\left(\tfrac{\sigma(t)^{1-\beta}}{1-\beta}\right)\rangle\big|_{t=0} \qquad \square$$

8. Conclusion

All the ingredients of the above construction make sense on a von Neumann algebra: WYD-information, quantum dynamics, L^p-spaces, Amari-Nagoka embeddings and so on.[14,20] Nevertheless we are not aware of any attempt to see geometry of WYD-information along the lines described in the present paper, in the infinite-dimensional context. We plan to address this problem in future work.

References

1. S. Amari and H. Nagaoka. *Methods of information geometry*, volume 191 of *Translations of Mathematical Monographs*. American Mathematical Society, Providence, RI, 2000. Translated from the 1993 Japanese original by Daishi Harada.
2. R. Bhatia, Matrix analysis, Springer-Verlag, New York, (1997).
3. J. Calsamiglia, R. Munoz-Tapia, L. Masanes, A. Acin and E. Bagan, Quantum Chernoff bound as a measure of distinguishability between density matrices: Application to qubit and gaussian states, *Phys. Rev. A*, 77, 032311, (2008).
4. Z. Chen. Wigner-Yanase skew information as tests for quantum entanglement, *Phys. Rev. A*, 71, 052302, (2005).
5. N. N. Čencov, *Statistical decision rules and optimal inference*. American Mathematical Society, Providence, R.I., (1982). Translation from the Russian edited by Lev J. Leifman.
6. A. Connes and E.Stormer. Homogeneity of the state space of factors of type III_1, *J. Funct. Anal.*, 28, 187–196, (1978).
7. F. Hansen. Metric adjusted skew information, arXiv:math-ph/0607049v3, (2006).
8. H. Hasegawa, Dual geometry of the Wigner-Yanase-Dyson information content, *Inf. Dim. Anal. Quant Prob. & Rel. Top.*, 6, 413-431, (2003).
9. H. Hasegawa and D.Petz, Noncommutative extension of the information geometry II. In *Quantum communications and measurement*, pages 109–118. Plenum, New York, (1997).
10. P. Gibilisco, D. Imparato, and T. Isola, Uncertainty principle and quantum Fisher information II. *J. Math. Phys.*, 48: 072109, (2007).
11. P. Gibilisco, D. Imparato and T. Isola, Inequality for quantum Fisher information. To appear on *Proc. Amer. Math. Soc.* arXiv:math-ph/0702058, (2007).
12. P. Gibilisco, D. Imparato, D. and T. Isola, A volume inequality for quantum Fisher information and the uncertainty principle. *J. Stat. Phys.*, 130(3): 545-559, (2008).
13. P. Gibilisco, D. Imparato and T. Isola, A Robertson-type uncertainty principle and quantum Fisher information. *Lin. Alg. Appl.*, 428(7), 1706-1724, (2008).
14. P.Gibilisco and T. Isola. Connections on Statistical manifolds of Density Op-

erators by Geometry of Noncommutative L^p-Spaces, *Inf. Dim. Anal. Quant Prob. & Rel. Top.*, **2** 169-178, (1999).
15. P.Gibilisco and T. Isola. Wigner-Yanase information on quantum state space: the geometric approach, *J. Math. Phys.*, **44**(9): 3752-3762, (2003).
16. P.Gibilisco and T. Isola. , On the characterization of paired monotone metrics, *Ann. Ins. Stat. Math.*, **56**(2): 369-381, (2004).
17. P.Gibilisco and T. Isola. On the monotonicity of scalar curvature in classical and quantum information geometry, *J. Math. Phys.*, **46**(2): 023501,14, (2005).
18. P.Gibilisco and G. Pistone. Connections on nonparametric statistical manifolds by Orlicz space geometry, *Inf. Dim. Anal. Quant Prob. & Rel. Top.*, **1** 325-347, (1998).
19. A. Klyachko, B. Oztop and A. S. Shumovsky. Measurable entanglement, *Appl. Phys. Lett.*, 88, 124102, (2006).
20. H. Kosaki. Interpolation theory and the Wigner-Yanase-Dyson-Lieb concavity. *Comm. Math. Phys.*, 87, no. 3, 315–329, (1982/83).
21. H. Kosaki. Matrix trace inequalities related to uncertainty principle, *Inter. Jour. Math.*, **6** 629-645, (2005).
22. E. Lieb, Convex trace functions and the Wigner-Yanase-Dyson conjecture. Advances in Math. 11, 267–288, (1973) .
23. E. Lieb and M.B. Ruskai, A fundamental property of the quantum mechanical entropy, *Phys. Rev. Lett.* 30, 434436, (1973).
24. S. Luo and Q. Zhang. On skew information, *IEEE Trans. Infor. Theory*, **50**(8), 1778-1782, (2004).
25. S. Luo and Z. Zhang. An informational characterization of Schrödinger's uncertainty relations, *J. Stat. Phys.*, **114**, 1557-1576, (2004).
26. D. Petz, Monotone metrics on matrix spaces. *Linear Algebra Appl.*, 244:81–96, (1996).
27. K. Yanagi, S. Furuichi and K. Kuriyama, A generalized skew information and uncertainty relation. *IEEE Trans. Inform. Theory*, 51(12):4401–4404, (2005).
28. E. P. Wigner and M. M. Yanase , Information content of distributions. *Proc. Nat. Acad. Sci. USA* 49: 910–918, (1963).

MULTIPLICATIVE RENORMALIZATION METHOD FOR ORTHOGONAL POLYNOMIALS

H.-H. KUO

Department of Mathematics
Louisiana State University
Baton Rouge, LA 70803, USA
E-mail: kuo@math.lsu.edu

We briefly explain the multiplicative renormalization method for the derivation of orthogonal polynomial generating functions of probability measures on the real line. Such an OP-generating function of μ can be used to calculate the associated μ-orthogonal polynomials and the Jacobi–Szegö parameters. Moreover, this method can be used to derive probability measures from certain functions $h(x)$. The case $h(x) = e^x$ is due to I. Kubo. We describe the recent results for the case $h(x) = (1-x)^{-1}$ obtained by I. Kubo, H.-H. Kuo, and S. Namli.

1. Orthogonal polynomials and Jacobi–Szegö parameters

Suppose μ is a probability measure on $\mathbb{R}$ with infinite support and finite moments of all orders. We can apply the Gram-Schmidt orthogonalization process to the sequence $\{x^n\}_{n=0}^{\infty}$ of monomials to obtain a μ-orthogonal sequence $\{P_n(x)\}_{n=0}^{\infty}$ such that $P_0(x) = 1$ and $P_n(x)$ is a polynomial of degree n with leading coefficient 1. These orthogonal polynomials satisfy the following recursion formula:

$$(x - \alpha_n)P_n(x) = P_{n+1}(x) + \omega_{n-1}P_{n-1}(x), \quad n \geq 0, \tag{1}$$

where $\omega_{-1} = P_{-1} = 0$ by convention. The numbers $\alpha_n, \omega_n, n \geq 0$, are known as the Jacobi–Szegö parameters of μ. We have $\omega_n > 0$, $n \geq 0$, since μ has infinite support. The family $\{P_n, \alpha_n, \omega_n\}_{n=0}^{\infty}$ plays an important role for the interacting Fock space associated with μ, see Accardi–Bożejko.[2]

Here is a natural question: *Given such a probability measure μ, how can we derive* $\{P_n, \alpha_n, \omega_n\}_{n=0}^{\infty}$?

Consider a simple example when μ is Poisson with parameter $\lambda > 0$. Being motivated by the concept of multiplicative renormalization in white noise theory (see, e.g., the book by Kuo[16]), we define the multiplicative renormalization $\psi(t,x)$ of the function $\varphi(t,x) = (1+t)^x$ by

$$\psi(t,x) := \frac{\varphi(t,x)}{E_\mu \varphi(t,\cdot)} = e^{-\lambda t}(1+t)^x, \quad (2)$$

which has the following power series expansion in t,

$$e^{-\lambda t}(1+t)^x = \left(\sum_{j=0}^{\infty} \frac{(-\lambda)^j}{j!} t^j\right)\left(\sum_{k=0}^{\infty} \frac{p_{x,k}}{k!} t^k\right) = \sum_{n=0}^{\infty} \frac{1}{n!} C_n(x) t^n, \quad (3)$$

where $p_{x,0} = 1$, $p_{x,k} = x(x-1)\cdots(x-k+1)$, $k \geq 1$, and

$$C_n(x) = \sum_{k=0}^{n} \binom{n}{k}(-\lambda)^k p_{x,n-k}, \quad n \geq 0. \quad (4)$$

These polynomials $C_n(x)$, $n \geq 0$, being μ-orthogonal, are known as the Charlier polynomials.

The above idea of multiplicative renormalization works miraculously for classical distributions. It is developed by Asai–Kubo–Kuo[6,7] to become a method for deriving a generating function (e.g., the one in Equation (3) for a Poisson measure) which can then be used to compute the sequence $\{P_n, \alpha_n, \omega_n\}_{n=0}^{\infty}$. For example, when μ is the Poisson measure μ, we have $P_n(x) = C_n(x)$ in Equation (4) and $\alpha_n = \lambda + n$, $\omega_n = \lambda(n+1)$, $n \geq 0$.

The multiplicative renormalization method turns out to be more than just a method for deriving $\{P_n, \alpha_n, \omega_n\}_{n=0}^{\infty}$ for certain classes of probability measures. It can also be used to discover new probability measures. In Section 3 we will explain some of such results from the recent papers by Kubo–Kuo–Namli.[14,15] For further results, see Namli.[18]

2. Multiplicative renormalization method

An *orthogonal polynomial generating function* (OP-generating function) for μ is a function which can be expanded as a power series in t,

$$\psi(t,x) = \sum_{n=0}^{\infty} c_n P_n(x) t^n, \quad (1)$$

where c_n's are nonzero real numbers and P_n's are the polynomials specified by Equation (1).

Suppose we have an OP-generating function $\psi(t,x)$ for μ. Then we can proceed to find $\{P_n, \alpha_n, \omega_n\}_{n=0}^{\infty}$ as follows:

1. Expand $\psi(t,x)$ as a power series in t as in Equation (1) to find c_n and P_n. (Recall that $P_n(x)$ is a polynomial of degree n with leading coefficient 1.)

2. If μ is symmetric, we have $\alpha_n = 0$ for all n. Then use the identity from Asai–Kubo–Kuo[7]

$$\int_{\mathbb{R}} \psi(t,x)^2 \, d\mu(x) = \sum_{n=0}^{\infty} c_n^2 \lambda_n t^{2n}$$

to find λ_n. And then find ω_n from the equality $\omega_n = \frac{\lambda_{n+1}}{\lambda_n}$.

3. In general, we can find ω_n as in Item 2. Then use another identity from Asai–Kubo–Kuo[7] to find α_n:

$$\int_{\mathbb{R}} x\psi(t,x)^2 \, d\mu(x) = \sum_{n=0}^{\infty} \left(c_n^2 \alpha_n \lambda_n t^{2n} + 2 c_n c_{n-1} \lambda_n t^{2n-1} \right), \qquad (2)$$

where $c_{-1} = 0$ by convention. In fact, this identity can be used to find both α_n and λ_n (hence ω_n) without going through Item 2.

Therefore, the crucial question is "*Given a probability measure μ, how can we find an OP-generating function $\psi(t,x)$ for μ?*"

We start with a "reasonably good" function $h(x)$ with $h^{(n)}(0) \neq 0$ for all $n \geq 0$, e.g., $h(x) = e^x$ or $h(x) = (1-x)^c$, c not a positive integer. Then define the following two functions:

$$\theta(t) = \int_{\mathbb{R}} h(tx) \, d\mu(x),$$

$$\widetilde{\theta}(t,s) = \int_{\mathbb{R}} h(tx) h(sx) \, d\mu(x). \qquad (3)$$

Theorem 2.1. (Asai–Kubo–Kuo[6,7]) *Suppose $\rho(t)$ is an analytic function in some neighborhood of 0 with $\rho(0) = 0$ and $\rho'(0) \neq 0$. Then the multiplicative renormalization*

$$\psi(t,x) := \frac{h(\rho(t) x)}{\theta(\rho(t))} \qquad (4)$$

is an OP-generating function for μ if and only if the function

$$\Theta_\rho(t,s) := \frac{\widetilde{\theta}(\rho(t), \rho(s))}{\theta(\rho(t)) \theta(\rho(s))} \qquad (5)$$

defined in some neighborhood of $(0,0)$ is a function of the product ts.

We will say that μ is *multiplicative renormalization method applicable* (MRM-applicable) for a function $h(x)$ if there exists an analytic function $\rho(t)$ in some neighborhood of 0 with $\rho(0) = 0$, $\rho'(0) \neq 0$, such that the function $\Theta_\rho(t,s)$ defined by Equation (5) is a function of ts.

If μ is MRM-applicable for a function $h(x)$, then by Theorem 2.1 the multiplicative renormalization $\psi(t,x)$ in Equation (4) is an OP-generating function for μ.

Here is the procedure of the multiplicative renormalization method:

$$\mu \longmapsto \{h(x), \theta(t), \widetilde{\theta}(t,s)\} \longmapsto \{\Theta_\rho(t,s), \rho(t)\} \longmapsto \psi(t,x)$$

Obviously, the key step in this procedure is to find a function $\rho(t)$ so that the resulting function $\Theta_\rho(t,s)$ is a function of the product ts. Then by Theorem 2.1 $\psi(t,x)$ is an OP-generating function for μ.

For example, consider the Poisson measure μ discussed in Section 1. We start with the function $h(x) = e^x$. It is easy to check that

$$\theta(t) = e^{\lambda(e^t - 1)},$$
$$\widetilde{\theta}(t,s) = e^{\lambda(e^{t+s} - 1)}.$$

Therefore, by Equation (5), we have

$$\Theta_\rho(t,s) = e^{\lambda(e^{\rho(t)} - 1)(e^{\rho(s)} - 1)}.$$

Hence in order for $\Theta_\rho(t,s)$ to be a function of ts we can take $e^{\rho(t)} - 1 = t$, which can be solved for the function $\rho(t)$,

$$\rho(t) = \ln(1+t).$$

Then we use Equation (4) to get an OP-generating function

$$\psi(t,x) = \frac{h(\rho(t)x)}{\theta(\rho(t))} = e^{-\lambda t}(1+t)^x.$$

Thus we have used the multiplicative renormalization method to derive the OP-generating function $\psi(t,x)$, which confirms the one given by Equation (2). Having found such a function $\psi(t,x)$, we can then compute the Charlier polynomials as explained in Section 1. Moreover, we can use Equation (2) to obtain the Jacobi–Szegö parameters

$$\alpha_n = \lambda + n, \quad \omega_n = \lambda(n+1), \quad n \geq 0.$$

Below in the chart are some further examples of using the multiplicative renormalization method to derive OP-generating functions.

μ	$h(x)$	$\theta(t)$	$\rho(t)$	$\psi(t,x)$
Gaussian $N(0,\sigma^2)$	e^x	$e^{\frac{1}{2}\sigma^2 t^2}$	t	$e^{tx-\frac{1}{2}\sigma^2 t^2}$
Poisson $\text{Poi}(\lambda)$	e^x	$e^{\lambda(e^t-1)}$	$\ln(1+t)$	$e^{-\lambda t}(1+t)^x$
gamma $\Gamma(\alpha)$	e^x	$\frac{1}{(1-t)^\alpha}$	$\frac{t}{1+t}$	$(1+t)^{-\alpha}e^{\frac{tx}{1+t}}$
uniform on $[-1,1]$	$\frac{1}{\sqrt{1-x}}$	$\frac{2}{\sqrt{1+t}+\sqrt{1-t}}$	$\frac{2t}{1+t^2}$	$\frac{1}{\sqrt{1-2tx+t^2}}$
arcsine on $[-1,1]$	$\frac{1}{1-x}$	$\frac{1}{\sqrt{1-t^2}}$	$\frac{2t}{1+t^2}$	$\frac{1-t^2}{1-2tx+t^2}$
semi-circle on $[-1,1]$	$\frac{1}{1-x}$	$\frac{2}{1+\sqrt{1-t^2}}$	$\frac{2t}{1+t^2}$	$\frac{1}{1-2tx+t^2}$
beta on $[-1,1]$ $\beta > -\frac{1}{2}, \beta \neq 0$	$\frac{1}{(1-x)^\beta}$	$\frac{2^\beta}{(1+\sqrt{1-t^2})^\beta}$	$\frac{2t}{1+t^2}$	$\frac{1}{(1-2tx+t^2)^\beta}$
Pascal $r > 0, 0 < q < 1$	e^x	$\frac{(1-q)^r}{(1-qe^t)^r}$	$\ln\frac{1+t}{1+qt}$	$(1+t)^x(1+qt)^{-x-r}$
stochastic area	e^x	$\sec t$	$\tan^{-1} t$	$\frac{e^{x\tan^{-1}t}}{\sqrt{1+t^2}}$

3. Characterization theorems

A natural question concerning the multiplicative renormalization method is the following: "*Are there other probability measures than those listed in the chart in Section 2?*"

The answer is obviously yes since we can take translations and dilations of these probability measures. For example, take the Poisson measure μ with parameter $\lambda > 0$ and define a probability measure ν by

$$\nu(B) = \mu\left(\frac{B-a}{b}\right), \quad B \in \mathcal{B}(\mathbb{R}),$$

where $a \in \mathbb{R}$ and $b \neq 0$. We can carry out calculations similar to those in Section 2 for μ to get the following functions for ν:

$$\theta(t) = e^{at+\lambda(e^{bt}-1)},$$
$$\rho(t) = \frac{1}{b}\ln(1+t),$$
$$\psi(t,x) = e^{-\lambda t}(1+t)^{\frac{1}{b}(x-a)}.$$

Then from the OP-generating function $\psi(t,x)$ we can derive the associated orthogonal polynomials and Jacobi–Szegö parameters:

$$P_n(x) = b^n C_n\left(\frac{x-a}{b}\right), \quad \alpha_n = a + \lambda b + bn, \quad \omega_n = \lambda b^2(n+1),$$

where $C_n(x)$'s are the Charlier polynomials defined by Equation (4).

The above example for ν leads to the problem: "*Given a fixed function $h(x)$, find all probability measures that are MRM-applicable for $h(x)$.*"

The case $h(x) = e^x$ has been solved by Kubo.[13] It turns out that these probability measures are exactly those in the Meixner class,[1,17] namely, those five probability measures with the function $h(x) = e^x$ in the chart in Section 2, subject to translations and dilations, in addition to the binomial distribution.

The next interesting case is when $h(x) = \frac{1}{1-x}$, which covers the arcsine and semi-circle distributions in view of the chart in Section 2. Are there other distributions except their translations and dilations? The answer is given in the recent papers by Kubo–Kuo–Namli,[14,15] which we will describe for the rest of this section.

Let μ be a probability measure on $\mathbb{R}$ with infinite support and finite moments of all orders. We will assume that the function

$$\theta(t) = \int_{\mathbb{R}} h(tx)\, d\mu(x) = \int_{\mathbb{R}} \frac{1}{1-tx}\, d\mu(x) \tag{1}$$

is analytic in some neighborhood of 0. Then we can show that

$$\widetilde{\theta}(t,s) = \frac{1}{t-s}\{t\theta(t) - s\theta(s)\}, \quad t \neq s, \tag{2}$$

and $\widetilde{\theta}(t,t)$ is defined to equal $\theta(t) + t\theta'(t)$.

Suppose $\rho(t)$ is an analytic function in some neighborhood of 0 with $\rho(0) = 0$, $\rho'(0) \neq 0$, such that

$$\Theta_\rho(t,s) = \frac{\widetilde{\theta}(\rho(t), \rho(s))}{\theta(\rho(t))\theta(\rho(s))} \tag{3}$$

is a function $J(ts)$ of the product ts with $J'(0) \neq 0$ and $J''(0) \neq 0$. Then it is shown by Kubo–Kuo–Namli[15] that $\rho(t)$ and $\theta(\rho(t))$ must be given by

$$\rho(t) = \frac{2t}{\alpha + 2\beta t + \gamma t^2}, \tag{4}$$

$$\theta(\rho(t)) = \frac{1}{1 - (b + at)\rho(t)}, \tag{5}$$

where $\alpha, \beta, \gamma, b, a \in \mathbb{R}$ and α, γ, a are nonzero numbers of the same sign. Conversely, suppose Equations (4) and (5) hold. Then we can put them and

Equation (2) into Equation (3) to show that

$$\Theta_\rho(t, s) = 1 + \frac{2ats}{\alpha - \gamma ts}.$$

Thus $\Theta_\rho(t, s)$ is indeed a function of the product ts. Hence by Theorem 2.1 the probability measure μ is MRM-applicable for $h(x) = \frac{1}{1-x}$. This leads to the OP-generating function $\psi(t, x)$ given in the next theorem.

Theorem 3.1. (Kubo–Kuo–Namli[15]) *Suppose the function $\theta(t)$ defined by Equation (1) is analytic in some neighborhood of 0 and let $\rho(t)$ be analytic in some neighborhood of 0 with $\rho(0) = 0$ and $\rho'(0) \neq 0$. Assume that the function $\Theta_\rho(t, s)$ in Equation (3) is a function of ts. Then μ is MRM-applicable for $h(x) = \frac{1}{1-x}$ and has an OP-generating function given by*

$$\psi(t, x) = \frac{\alpha + 2(\beta - b)t + (\gamma - 2a)t^2}{\alpha - 2t(x - \beta) + \gamma t^2}, \tag{6}$$

where α, β, γ, b, and a are specified by Equations (4) and (5).

Once the OP-generating function $\psi(t, x)$ in Equation (6) is derived, we can find the associated orthogonal polynomials and the Jacobi–Szegö parameters of the probability measure μ:

$$P_n(x) = \left(\frac{\sqrt{\alpha\gamma}}{2}\right)^n U_n\left(\frac{x - \beta}{\sqrt{\alpha\gamma}}\right) + (\beta - b)\left(\frac{\sqrt{\alpha\gamma}}{2}\right)^{n-1} U_{n-1}\left(\frac{x - \beta}{\sqrt{\alpha\gamma}}\right)$$

$$+ \frac{\alpha(\gamma - 2a)}{4}\left(\frac{\sqrt{\alpha\gamma}}{2}\right)^{n-2} U_{n-2}\left(\frac{x - \beta}{\sqrt{\alpha\gamma}}\right), \quad n \geq 0,$$

$$\alpha_n = \begin{cases} b, & \text{if } n = 0, \\ \beta, & \text{if } n \geq 1, \end{cases}$$

$$\omega_n = \begin{cases} \dfrac{a\alpha}{2}, & \text{if } n = 0, \\ \dfrac{\alpha\gamma}{4}, & \text{if } n \geq 1, \end{cases}$$

where $U_{-2} = U_{-1} = 0$ by convention and $U_n(x)$, $n \geq 0$, are the Chebyshev polynomials of the second kind,

$$U_n(x) = \frac{\sin[(n+1)\cos^{-1} x]}{\sin(\cos^{-1} x)} = \sum_{k=0}^{[n/2]} (-1)^k \binom{n-k}{k} 2^{n-2k} x^{n-2k}, \quad n \geq 0.$$

Next we ask the question: "*What is the class of probability measures that are MRM-applicable for $h(x) = \frac{1}{1-x}$?*" To answer this question, we need to find those probability measures whose corresponding functions $\rho(t)$ and $\theta(\rho(t))$ are given by Equation (4) and (5), respectively.

First consider the case $\alpha = \gamma = 1$ and $\beta = 0$. Then we have

$$\rho(t) = \frac{2t}{1+t^2}$$

and such a probability measure μ is determined by the numbers a and b in Equation (5). The next theorem gives an answer to the above question for the special case when μ has a density function.

Theorem 3.2. (Kubo–Kuo–Namli[15]) *A probability measure μ with density function $f(x)$ is MRM-applicable for $h(x) = \frac{1}{1-x}$ with $\rho(t) = \frac{2t}{1+t^2}$ if and only if $f(x)$ is given by*

$$f(x) = \begin{cases} \dfrac{a\sqrt{1-x^2}}{\pi[a^2 + b^2 - 2b(1-a)x + (1-2a)x^2]}, & \text{if } |x| < 1, \\ 0, & \text{otherwise,} \end{cases}$$

where $a > 0$ and $|b| \leq 1 - a$.

Next consider the general case when μ is specified by five numbers $\alpha, \beta, \gamma, b, a$ given by Equations (4) and (5). Here α, γ, and a must be of the same sign and can be taken to be positive in view of the above theorem. We need to find the unique probability measure μ satisfying the following equation

$$\int_{\mathbb{R}} \frac{1}{1 - \rho(t)x} \, d\mu(x) = \frac{1}{1 - (b + at)\rho(t)}, \quad \forall \text{ small } t, \tag{7}$$

where $\rho(t) = \frac{2t}{\alpha + 2\beta t + \gamma t^2}$. Let $t = \sqrt{\frac{\alpha}{\gamma}} \frac{z}{1+\sqrt{1-z^2}}$ and let

$$x = \sqrt{\alpha\gamma}\, y + \beta, \quad d\nu(y) = d\mu(\sqrt{\alpha\gamma}\, y + \beta), \quad A = \frac{a}{\gamma}, \quad B = \frac{b - \beta}{\sqrt{\alpha\gamma}}. \tag{8}$$

Then Equation (7) is equivalent to the new equation for ν,

$$\int_{\mathbb{R}} \frac{1}{1 - zy} \, d\nu(y) = \frac{1}{1 - A - Bz + A\sqrt{1-z^2}}, \quad \forall \text{ small } z. \tag{9}$$

By replacing $d\nu(y)$ with $d\nu(-y)$, if necessary, we may assume $B \geq 0$. For convenience, we divide the region $\{(A, B)\,;\, A > 0, B \geq 0\}$ into the disjoint union of the following regions

$$R_1 = \{(A,B); 0 < A \le 1, 0 \le B \le 1 - A\},$$
$$R_2 = \{(A,B); 0 < A \le 1, B > 1 - A\},$$
$$R_3 = \{(A,B); A > 1, 0 \le B \le A - 1\},$$
$$R_4 = \{(A,B); A > 1, B > A - 1\}.$$

Theorem 3.3. (Kubo–Kuo–Namli[15]) *For $A > 0$ and $B \ge 0$, the unique probability measure ν satisfying Equation (9) is given by*

$$d\nu(y) = W_0 \frac{\sqrt{1-y^2}}{\pi(1-py)(1-qy)} 1_{(-1,1)}(y)\, dy + W_1\, d\delta_{\frac{1}{p}}(y) + W_2\, d\delta_{\frac{1}{q}}(y),$$

where δ_r denotes the Dirac delta measure at r and p, q, W_0, W_1, W_2 are the numbers given by

$$p = \frac{B(1-A) + A\sqrt{B^2 + 2A - 1}}{A^2 + B^2},$$

$$q = \frac{B(1-A) - A\sqrt{B^2 + 2A - 1}}{A^2 + B^2},$$

$$W_0 = \frac{A}{A^2 + B^2},$$

$$W_1 = \begin{cases} 0, & \text{on } R_1, \\ \dfrac{B\sqrt{B^2 + 2A - 1} - A(1-A)}{A(B^2 + 2A - 1) + B(1-A)\sqrt{B^2 + 2A - 1}}, & \text{on } R_2 \cup R_3 \cup R_4, \end{cases}$$

$$W_2 = \begin{cases} 0, & \text{on } R_1 \cup R_2 \cup R_4, \\ \dfrac{A(A-1) - B\sqrt{B^2 + 2A - 1}}{A(B^2 + 2A - 1) + B(A-1)\sqrt{B^2 + 2A - 1}}, & \text{on } R_3. \end{cases}$$

For the case when $B < 0$, we can replace $d\nu(y)$ by $d\nu(-y)$ and apply the above theorem. For the general case when μ is determined by $\alpha, \beta, \gamma, b, a$, we use Equation (8) to derive μ from ν (which is obtained by the above theorem) by translations and dilations.

4. Other results and open questions

The multiplicative renormalization provides a rather simple method for the derivation of orthogonal polynomial generating functions for certain probability measures. But as we have seen from Section 3, it can also be used to derive new probability measures. The computation is usually quite complicated and very tedious. Below we mention some other results and open questions.

1. The case $h(x) = (1-x)^c$ for a general constant c has been studied by S. Namli in his Ph. D. dissertation.[18]
2. Are there other functions $h(x)$ besides e^x and $(1-x)^c$ for which multiplicative renormalization method can be applied?
3. What is the multidimensional generalization of the multiplicative renormalization method?
4. Is an OP-generating function for μ related to some other quantities associated with μ?
5. What is the relationship between MRM-applicability and operations on measures? For example, if μ is MRM applicable, does it follow that its symmetrization $\hat{\mu}$ is also MRM-applicable?

The question 3 is obviously very hard, see the book by Dunkl and Xu.[12] The ideas and results in the papers by Accardi–Kuo–Stan[3,4] and Accardi–Nahni[5] may suggest a suitable formulation for such a method. The answer to question 4, for the case of differential operator, exists in the literature, see, e.g., the paper by Sheffer.[19]

We learned during the conference that N. Demni had used a continued fraction expansion method to obtain some results related to the paper by Kubo–Kuo–Namli.[14,15] See his paper[11] in this volume of proceedings.

Acknowledgments. We are grateful to the referees for their comments and suggestions, which lead to the improvement of this paper.

References

1. Accardi, L.: Meixner classes and the square of white noise, in: *Finite and Infinite Dimensional Analysis in Honor of Leonard Gross, Contemporary Mathematics* **317** (2003) 1–13, Amer. Math. Soc.
2. Accardi, L. and Bożejko, M.: Interacting Fock space and Gaussianization of probability measures, *Infinite Dimensional Analysis, Quantum Probability and Related Topics* **1** (1998) 663–670.
3. Accardi, L., Kuo, H.-H., and Stan, A.: Characterization of probability measures through the canonically associated interacting Fock spaces, *Infinite Dimensional Analysis, Quantum Probability and Related Topics* **7** (2004) 485–505.
4. Accardi, L., Kuo, H.-H., and Stan, A.: Moments and commutators of probability measures, *Infinite Dimensional Analysis, Quantum Probability and Related Topics* **10** (2007) 591–612.

5. Accardi, L. and Nahni, M.: Interacting Fock spaces and orthogonal polynomials in several variables, in: *Non-commutativity, infinite-dimensionality and probability at the crossroads*, (2002) 192–205, World Scientific.
6. Asai, N., Kubo, I., and Kuo, H.-H.: Multiplicative renormalization and generating functions I, *Taiwanese Journal of Mathematics* **7** (2003) 89–101.
7. Asai, N., Kubo, I., and Kuo, H.-H.: Multiplicative renormalization and generating functions II, *Taiwanese Journal of Mathematics* **8** (2004) 593–628.
8. Bożejko, M. and Bryc, W.: On a class of free Lévy laws related to a regression problem, *J. Funct. Anal.* **236** (2006) 59–77.
9. Bożejko, M. and Demni, N.: A short proof of a problem by Asai–Kuo–Kubo and representations of free Meixner laws, *Preprint* (2008).
10. Chihara, T. S.: *An Introduction to Orthogonal Polynomials*. Gordon and Breach, 1978.
11. Demni, N.: Free martingale polynomials for stationary Jacobi processes, (in this volume of proceedings).
12. Dunkl, C. F. and Xu, Y.: *Orthogonal Polynomials of Several Variables*, Cambridge University Press, 2001.
13. Kubo, I.: Generating functions of exponential type for orthogonal polynomials, *Infinite Dimensional Analysis, Quantum Probability and Related Topics* **7** (2004) 155–159.
14. Kubo, I., Kuo, H.-H., and Namli, S.: Interpolation of Chebyshev polynomials and interacting Fock spaces, *Infinite Dimensional Analysis, Quantum Probability and Related Topics* **9** (2006) 361–371.
15. Kubo, I., Kuo, H.-H., and Namli, S.: The characterization of a class of probability measures by multiplicative renormalization, *Communications on Stochastic Analysis* (to appear).
16. Kuo H.-H.: *White Noise Distribution Theory*, CRC Press, Boca Raton, 1996.
17. Meixner, J.: Orthogonale polynomsysteme mit einen besonderen gestalt der erzeugenden funktion, *J. Lond. Math. Soc.* **9** (1934) 6–13.
18. Namli, S.: Multiplicative renormalization method for orthogonal polynomials, Ph.D. dissertation, Louisiana State University, 2007.
19. Sheffer, I. M.: Some properties of polynomials sets of type zero, *Duke Mth. J.* **5** (1939) 590–622.
20. Szegö, M.: *Orthogonal Polynomials*. Coll. Publ. **23**, Amer. Math. Soc., 1975.

QUANTUM L_p AND ORLICZ SPACES

L. E. LABUSCHAGNE

Department of Mathematical Sciences,
Box 392, UNISA, 0003, RSA
E-mail: Labusle@unisa.ac.za

W. A. MAJEWSKI

Institute of Theoretical Physics and Astrophysics, The Gdansk University, Wita
Stwosza 57,
Gdansk, 80-952, Poland
E-mail: fizwam@univ.gda.pl

Let $\mathcal{A}$ ($\mathcal{M}$) be a C^*-algebra (a von Neumann algebra respectively). By a quantumdynamical system we shall understand the pair $(\mathcal{A}, T)$ (($\mathcal{M}, T$)) where $T : \mathcal{A} \to \mathcal{A}$ ($T : \mathcal{M} \to \mathcal{M}$) is a linear, positive (normal respectively), and identity preserving map. In our lecture, we discuss how the techniques of quantum Orlicz spaces may be used to study quantum dynamical systems. To this end, we firstly give a brief exposition of the theory of quantum dynamical systems in quantum L_p spaces. Secondly, we describe the Banach space approach to quantization of classical Orlicz spaces. We will discuss the necessity of the generalization of L_p-space techniques. Some emphasis will be put on the construction of non-commutative Orlicz spaces. The question of lifting dynamical systems defined on von Neumann algebra to a dynamical system defined in terms of quantum Orlicz space will be discussed.

Keywords: (quantum) L_p spaces, (quantum) Orlicz spaces, quantum dynamical systems, C^*-algebras, von Neumann algebras, CP maps.

1. Introduction

To indicate reasons why (quantum) L_p-spaces are emerging in the theory of (quantum) dynamical systems we begin with a particular case of dynamical systems - with stochastic evolution of particle systems. We recall that in the classical theory of particle systems one of the objectives is to produce, describe, and analyze dynamical systems with evolution originating from stochastic processes in such a way that their equilibrium states are given Gibbs states (see Ref. 1). A well known illustration is the so called Glauber dynamics,[2] which may be found in a number of papers. To carry out the analysis of such dynamical systems, it is convenient to use the theory of Markov processes in the context of L_p-spaces. In particular, for the Markov-Feller processes, using the unique correspondence between the process and the corresponding dynamical semigroup, one can give a recipe for the construction of Markov generators for this class of processes (for details see Ref. 1). That correspondence uses the concept of conditional expectation which can be nicely characterized within the L_p-space framework (cf. the Moy paper[3]).

More generally, these Banach spaces, i.e. L_p and their generalizations - Orlicz spaces, are extremely useful in the general description of classical dynamical systems. To support this claim some comments are warranted here. Firstly let $\{\Omega, \Sigma, \mu\}$ be a probability space. We denote by $\mathcal{S}_\mu$ the set of the densities of all the probability measures equivalent to μ, i.e.,

$$\mathcal{S}_\mu = \{ f \in L^1(\mu) : f > 0 \quad \mu - a.s., E(f) = 1 \}$$

$\mathcal{S}_\mu$ can be considered as a set of (classical) states and its natural "geometry" comes from embedding $\mathcal{S}_\mu$ into $L^1(\mu)$. However, it is worth pointing out that the Liouville space technique demands $L^2(\mu)$-space, while employing the interpolation techniques needs other L_p-spaces with $p \geq 1$.

To take one further step, let us consider moment generating functions; so fix $f \in \mathcal{S}_\mu$ and take a real random variable u on $(\Omega, \Sigma, fd\mu)$. Define

$$\hat{u}_f(t) = \int exp(tu) f d\mu, \qquad t \in \mathbb{R}$$

and denote by L_f the set of all random variables such that

(**1**) $\hat{u}_f$ is well defined in a neighborhood of the origin 0,
(**2**) the expectation of u is zero.

One can observe that in this way a nice selection of (classical) observables was made, namely[4] all the moments of every $u \in L_f$ exist and they are the values at 0 of the derivatives of $\hat{u}_f$.

But, it is important to note that L_f is actually the **Orlicz space** based on an exponentially growing function (see Ref. 4). Consequently, one may say that even in classical statistical Physics one could not restrict oneself to merely $L^1(\mu)$, $L^2(\mu)$, $L^\infty(\mu)$ and interpolating $L^p(\mu)$ spaces. In other words, generalizations of L_p-spaces - Orlicz spaces - do appear.

However, contemporary science has been founded on *quantum mechanics*. Therefore, it is quite natural to look for the quantum counterpart of the above approach. Again let us begin with a particle systems with a stochastic evolution. Recently, the quantization of such particle systems was carried out, see Ref. 5–8. The main ingredient of such a quantization, is the concept of a generalized conditional expectation and Dirichlet forms defined in terms of *non-commutative (quantum) L_p-spaces*. The advantage of using quantum L_p-spaces, lies in the fact that when performing the quantization procedure, we can follow the traditional "route" of analysis of dynamical systems, and also in the fact that it is then possible to have one scheme for the quantum counterparts of stochastic dynamics of jump and diffusive-type. In particular, the quantum counterpart of the classical recipe for the construction of quantum Markov generators was obtained. The above scheme is not surprising if we realize that even in the textbook formulation of Quantum Mechanics, states are trace class operators. So, they form a subset of quantum $L_1(\mathcal{B}(\mathcal{H})), Tr)$-space while observables can be identified with self-adjoint elements of $L_\infty(\mathcal{B}(\mathcal{H}), Tr)$-space.

Turning to quantum Orlicz spaces our first remark is that they are a natural generalization of L_p spaces. To provide a simple argument in favor of such a generalization we will follow Streater[9,10]. Let ϱ_0 be a quantum state (a density matrix) and $S(\varrho_0)$ its von Neumann entropy. Assume $S(\varrho_0)$ to be finite. It is an easy observation that in any neighborhood of ϱ_0 (given by the trace norm, so in the sense of quantum L_1-space) there are plenty of states with infinite entropy. This should be considered alongside the thermodynamical rule which tells us that the entropy should be a state function which is increasing in time. Thus we run into serious problems with the explanation of the phenomenon of return to equilibrium. More sophisticated arguments in this direction can be extracted from hypercontractivity of quantum maps and log Sobolev techniques (see Ref. 11 and B. Zegarlinski lecture in Ref. 12.)

The paper is organized as follows: in Section 2 we review some of the standard facts on quantum spin systems. Then quantum L_p-spaces are described (Section 3). In Section 4, we indicate how L_p-space techniques can be used for the construction of quantum stochastic dynamics. Section 5 is devoted to the study of quantum Orlicz spaces.

We want to close this section with a note that the quantum L_p space technique "ideology", presented here, is reproduced from the paper[13] which, to some extent, due to technical problems, is unreadable.

2. Quantum spin systems

In this Section we recall the basic elements of the description of quantum spin systems on a lattice. The best general references are Refs. 14,15. Here, and subsequently, $\mathbb{Z}^d$ stands for the d-dimensional integer lattice. Let $\mathcal{F}$ denote the family of all its finite subsets and let $\mathcal{F}_0$ be an increasing Fisher (or van Hove) sequence of finite volumes invading all of the lattice $\mathbb{Z}^d$. Given a sequence of objects $\{F_\Lambda\}_{\Lambda \in \mathcal{F}_0}$, it will be convenient to denote its limit (in an appropriate topology) as $\Lambda \to \mathbb{Z}^d$ through the sequence $\mathcal{F}_0$ by $\lim_{\mathcal{F}_0} F_\Lambda$.

The basic role in the description of the quantum lattice systems, is played by a $\mathbf{C}^*$- algebra $\mathcal{A}$, with norm $\|\cdot\|$, defined as the inductive limit over finite dimensional complex matrix algebras $\mathbf{M}$. By analogy with the classical commutative spin systems, it is natural to view $\mathcal{A}$ as a noncommutative analogue of the space of bounded continuous functions. For a finite set $X \in \mathcal{F}$, let $\mathcal{A}_X$ denote a subalgebra of operators localised in the set X. We recall that such a subalgebra is isomorphic to $\mathbf{M}^X$. For an arbitrary subset $\Lambda \subset \mathbb{Z}^d$, one defines $\mathcal{A}_\Lambda$ to be the smallest (closed) subalgebra of $\mathcal{A}$ containing $\bigcup \{\mathcal{A}_X : X \in \mathcal{F}, X \subset \Lambda\}$. An operator $f \in \mathcal{A}$ will be called local if there is some $Y \in \mathcal{F}$ such that $f \in \mathcal{A}_Y$. The subset of $\mathcal{A}$ consisting of all local operators will be denoted by $\mathcal{A}_0$. (A detailed account of matricial and operator algebras can be found in Ref. 16.)

Together with the algebra $\mathcal{A}$, we are given a family $\mathbf{Tr}_X$, $X \in \mathcal{F}$, of *normalised partial traces* on $\mathcal{A}$. We mention that the partial traces $\mathbf{Tr}_X$ have all the natural properties of classical *conditional expectations*, i.e. they are (completely) positive, unit preserving projections on the algebra $\mathcal{A}$. There is a unique state $\mathbf{Tr}$ on $\mathcal{A}$, called the normalised trace, such that

$$\mathbf{Tr}\,(\mathbf{Tr}_X f) = \mathbf{Tr}\,(f) \tag{1}$$

for every $X \in \mathcal{F}$, i.e. the normalised trace can be regarded as a (free) Gibbs state in a similar sense as in classical statistical mechanics.

To describe systems with interactions, we need to introduce the notion of an interaction potential. A family $\Phi \equiv \{\Phi_X \in \mathcal{A}_X\}_{X \in \mathcal{F}}$ of selfadjoint operators such that

$$\|\Phi\|_1 \equiv \sup_{i \in \mathbb{Z}^d} \sum_{\substack{X \in \mathcal{F} \\ X \ni i}} \|\Phi_X\| < \infty \tag{2}$$

will be called a (Gibbsian) potential. A potential $\Phi \equiv \{\Phi_X\}_{X \in \mathcal{F}}$ is of *finite range* $R \geq 0$, iff $\Phi_X = 0$ for all $X \in \mathcal{F}$, $diam(X) > R$. The corresponding

Hamiltonian H_Λ is defined by

$$H_\Lambda \equiv H_\Lambda(\Phi) \equiv \sum_{X \subset \Lambda} \Phi_X \qquad (3)$$

In particular, it is an easy observation that *anisotropic and isotropic Heisenberg models (so also Ising model) with nearest-neighbor interactions fall into the considered class of systems!*

Using the Hamiltonian H_Λ, we introduce a density matrix ρ_Λ

$$\rho_\Lambda \equiv \frac{e^{-\beta H_\Lambda}}{\mathbf{Tr} e^{-\beta H_\Lambda}}$$

with $\beta \in (0, \infty)$, and define a finite volume Gibbs state ω_Λ as follows:

$$\omega_\Lambda(f) \equiv \mathbf{Tr}\,(\rho_\Lambda f)$$

It is known, see e.g. Ref. 14, that for $\beta \in (0, \infty)$ the thermodynamic limit state on $\mathcal{A}$

$$\omega \equiv \lim_{\mathcal{F}_0} \omega_\Lambda \qquad (4)$$

exists and is faithful for some exhaustion $\mathcal{F}_0$ of the lattice. In general, a system can possess several such states, so phase transitions are allowed. For a quantum spin system, we can also introduce a natural Hamiltonian dynamics defined in a finite volume as the following automorphism group associated with the potential Φ:

$$\alpha_t^\Lambda(f) \equiv e^{+itH_\Lambda} f e^{-itH_\Lambda} \qquad (5)$$

If the potential $\Phi \equiv \{\Phi_X\}_{X \in \mathcal{F}}$ also satisfies

$$\|\Phi\|_{exp} \equiv \sup_{i \in \mathbb{Z}^d} \sum_{\substack{X \in \mathcal{F} \\ X \ni i}} e^{\lambda|X|} \|\Phi_X\| < \infty \qquad (6)$$

for some $\lambda > 0$, then the limit

$$\alpha_t(f) \equiv \lim_{\mathcal{F}_0} \alpha_t^\Lambda(f), \qquad (7)$$

exists[14] for every $f \in \mathcal{A}_0$. Consequently, the specification of local interactions, leads to a well defined global dynamics, provided that (6) is valid. In other words, the thermodynamic limit

$$(\mathcal{A}_\Lambda, \alpha_t^\Lambda, \omega_\Lambda) \to (\mathcal{A}, \alpha_t, \omega)$$

exists and gives the quantum dynamical system.

3. Non-commutative L_p-spaces

Let $< X, \mu >$ be a measure space, and $p \geq 1$. We denote by $L_p(X, d\mu)$ the set (of equivalence classes) of measurable functions satisfying

$$\|f\|_p \equiv \left(\int_X |f(x)|^p d\mu(x) \right)^{\frac{1}{p}} < \infty.$$

For the pair $(\mathcal{M}, \tau)$ consisting of semifinite von Neumann algebra $\mathcal{M}$ and a trace τ, the analogue of the concept of L_p-spaces ($p \in [1, \infty]$) in the commutative theory, can be introduced as follows : define

$$\mathcal{I}_p = \{ x \in \mathcal{M} \mid \tau(|x|^p) < \infty \}.$$

$\mathcal{I}_p$ is a two sided ideal of $\mathcal{M}$. Further, $\|x\|_p = \tau(|x|^p)^{\frac{1}{p}}$ defines a norm on $\mathcal{I}_p$. The completion of $\mathcal{I}_p$ with respect to the norm $\|\cdot\|_p$ gives Banach $L_p(\mathcal{M}, \tau)$ spaces which can be considered as a generalization of the corresponding spaces defined in the commutative case. It is an easy observation that on setting $\mathcal{M} = \mathcal{B}(\mathcal{H})$ and $\tau = Tr$ (Tr stands for the usual trace on $\mathcal{M}$), one obtains the well known Schatten classes[17]. That is, $L_p(\mathcal{B}(\mathcal{H}), Tr)$ is just the set of compact operators whose singular values are in l_p and the norms of L_p and l_p are equal. Moreover, the family $\{L_p(\mathcal{B}(\mathcal{H}), Tr)\}_{p \geq 1}$ provides a nice example of an abstract interpolation scheme (see Ref. 18).

Using this and the Haagerup theory (Ref. 19; see also[20-25]), we can introduce *quantum L_p spaces for quantum lattice systems*, i.e. for the systems described in the previous Section.

To this end, we firstly note that the quasi-local structure described for quantum lattice systems, can be summarized in the following way:

(1) $\mathcal{A}_0 = \cup_{\Lambda \in \mathcal{F}} \mathcal{A}_\Lambda$ is dense in $\mathcal{A}$.
(2) There exists a family of density operators $\{\varrho_\Lambda \in \mathcal{A}_\Lambda : \varrho_\Lambda > 0, \text{Tr}\varrho_\Lambda = 1\}_{\Lambda \in \mathcal{F}}$ with the compatibility condition $\text{Tr}_{\Lambda_2 \setminus \Lambda_1}\{\varrho_{\Lambda_2}\} = \varrho_{\Lambda_1}$, provided that $\Lambda_1 \subset \Lambda_2$.

We introduce:

- $\|f\|_{L_{p,s}(\omega)} = \lim_\Lambda \|f\|_{L_{p,s}(\omega_\Lambda)}$ for $p \in [1, \infty), s \in [0, 1]$, where $f \in \mathcal{A}$,
- $\|f\|_{L_{p,s}(\omega_\Lambda)} = (\text{Tr}|\varrho_\Lambda^{1-s/p} f \varrho_\Lambda^{s/p}|^p)^{1/p}$,

where $\omega(f) = \lim_{\mathcal{F}_0} \omega_\Lambda(f) \equiv \lim_{\mathcal{F}_0} \text{Tr}\{\varrho_\Lambda f\}$.

One can show that $\|f\|_{L_{p,s}(\omega_\Lambda)}$ is a well defined two-parameter family of norms on $\mathcal{A}$. The same should be done for $\|f\|_{L_{p,s}(\omega)}$ (see Theorem below).

Namely, in[5,7] it was proved:

Theorem 3.1.
For any $p \in [2, \infty)$, $s \in [0,1]$, any local operator $f \in \mathcal{A}_{\Lambda_0}$, $\Lambda_0 \in \mathcal{F}$ and all sets $\Lambda_1, \Lambda_2 \in \mathcal{F}$ such that $\Lambda_0 \subset \Lambda_1 \subset \Lambda_2$, we have
$$||f||_{L_p(\omega_{(\Lambda_2)}, s)} \leq ||f||_{L_p(\omega_{(\Lambda_1)}, s)}.$$
Thus for any $f \in \mathcal{A}_0$ the limit
$$||f||_{L_p(\omega, s)} \equiv \lim_{\mathcal{F}_0} ||f||_{L_p(\omega_{(\Lambda)}, s)}$$
exists and is independent of the countable exhaustion $\mathcal{F}_0$ of the lattice.

For $p \in (1, 2)$ one can use duality to define the correspondings norms[7]:
$$||f||_{L_p(\omega, s)} \equiv sup_{||g||_{L_q(\omega, s)} \leq 1} < g, f >_{\omega, s}$$
where $1/p + 1/q = 1$, $q \in [2, \infty)$ and $< \cdot, \cdot >_{\omega, s}$ is the scalar product associated to the norm $|| \cdot ||_{L_2(\omega, s)}$. Finally, the existence of the norm $|| \cdot ||_{L_1(\omega, s)}$ was established in [5]. Hence quantum L_p-spaces are associated with concrete physical systems:

Corollary 3.2. *To every Gibbs state ω on a C^*-algebra $\mathcal{A}$ defined by a quantum lattice system we can associate an interpolating, two parameter, family of Banach spaces*
$$\{L_p(\omega, s)\}_{p \in [1, \infty), s \in [0,1]}.$$

4. Quantum L_p dynamics

Let $\mathcal{M}$ be a von Neumann algebra generated by $\pi_\omega(\mathcal{A})$, where $\pi_\omega(\cdot)$ is the GNS representation associated with the quantum lattice system $(\mathcal{A}, \omega)$, described in Section 2. By φ_1 we denote the (weak) extension of ω on $\mathcal{M}$. Let $\mathcal{E}_0$ be a conditional expectation, i.e. $\mathcal{E}_0(f^*f) \geq 0$, $\mathcal{E}_0(1) = 1$, $\mathcal{E}_0^2 = \mathcal{E}_0$. We define
$$\varphi_2(\cdot) \equiv \varphi_1 \circ \mathcal{E}_0(\cdot). \tag{1}$$

Suppose that φ_2 is another faithful state on $\mathcal{M}$. Then the Takesaki theorem implies that $\mathcal{E}_0$ commutes with σ_t^2 (the modular automorphism group for $(\mathcal{M}, \varphi_2)$) and hence is symmetric in $(\mathcal{H}_{2, \frac{1}{2}}, < f, g >_{2, \frac{1}{2}} \equiv \varphi_2(\sigma_{\frac{i}{4}}^2(f)^*(\sigma_{\frac{i}{4}}^2(g))))$.

Let $V_t \equiv (D\varphi_1 : D\varphi_2)_t$ be the Radon-Nikodym cocycle. We remind that, in particular, $\sigma_t^1(f) = V_t^* \sigma_t^2(f) V_t$. The main difficulty in carrying out the construction of the Markov generator, is the existence of an analytic extension of $\mathbb{R} \ni t \mapsto V_t \in \mathcal{M}$. The following condition guarantees the desired extension (for details see Ref. 26):

Suppose there exists a positive constant $c \in (0, \infty)$ such that for any $0 \leq f \in \mathcal{M}$ the following inequalities hold:

$$\frac{1}{c}\varphi_1(f) \leq \varphi_2(f) \leq c\varphi_1(f). \tag{2}$$

Then, V_t extends analytically to $-\frac{1}{2} \leq Imz \leq \frac{1}{2}$ and $\xi \equiv V_{t|t=-\frac{i}{2}}$ is a bounded operator in $\mathcal{M}$. Let us note that the above inequalities also guarantee that φ_2 is a faithful state provided that φ_1 has this property.

Now, let us apply the above strategy to a finite system. Fix $X \subset \Lambda \in \mathcal{F}$. Obviously, (2) is satisfied for $\varphi_1(\cdot)(\equiv \varphi_1^\Lambda(\cdot)) = \mathbf{Tr}_\Lambda \varrho_\Lambda(\cdot)$ and $\varphi_2(\cdot)(\equiv \varphi_2^{\Lambda,X}(\cdot)) = \varphi_1 \circ \mathbf{Tr}_X(\cdot)$. Define

$$\mathcal{E}_{X,\Lambda}(a) = \mathbf{Tr}_X(\gamma_{X,\Lambda}^* f \gamma_{X,\Lambda})$$

where $\gamma_{X,\Lambda} = \varrho_\Lambda^{\frac{1}{2}}(\mathbf{Tr}_X \varrho_\Lambda)^{-\frac{1}{2}}$, and $f \in \mathcal{A}_\Lambda$.

One can verify[8] that $\gamma_{X,\Lambda}$ is the analytic extension of the Radon-Nikodym cocycle, and that $\mathcal{E}_{X,\Lambda}$ is a generalized conditional expectation (in the Accardi-Cecchini sense). Moreover[5],

$$P_t^{X,\Lambda} \equiv exp\{t(\mathcal{E}_{X,\Lambda} - id)\}$$

is the well defined Markov semigroup corresponding to the block-spin flip operation. For its construction *only local specifications* $(\varrho_\Lambda, \mathbf{Tr}_X \varrho_\Lambda)$ *are necessary*.

Now we examine (like in the classical case) the question of existence of global dynamics. Denoting $\varphi_2 \equiv \varphi_1 \circ \mathbf{Tr}_X$ and using the same strategy, we have[7]

Theorem 4.1. *Suppose the system is in sufficiently high temperature, $|\beta| < \beta_0$ with interaction Φ fulfilling the condition (6), or that the system is one dimensional at an arbitrary temperature $\beta \in (0, \infty)$ with finite range interactions. Then, for some positive $c \in (0, \infty)$*

$$\frac{1}{c}\varphi_1(f^*f) \leq \varphi_2(f^*f) \leq c\varphi_1(f^*f).$$

Hence, the corresponding Radon-Nikodym cocyles have analytic extension and therefore $\gamma_X \equiv (D\varphi_1 : D\varphi_2)_{|t=-\frac{i\beta}{2}} \in \mathcal{M}$. Hence

$$\mathcal{E}_X(f) \equiv \mathbf{Tr}_X(\gamma_X^* f \gamma_X)$$

defines a generalized conditional expectation which is symmetric in $\mathcal{H}_{\varphi_1}$. (Here $\mathcal{H}_{\varphi_1}$ is just the Hilbert space $L_2(\varphi_1, 1/2)$ constructed on $\mathcal{M}$.)

On the other hand, one has (for details see Ref. 7):

Theorem 4.2. *Let $\mathcal{E}_0$ be a (true) conditional expectation (so not necessary of the form $\mathbf{Tr}_X$). Assume that $\xi \equiv V_{t|t=-\frac{i}{2}}$ is a bounded operator in $\mathcal{M}$ and define*

$$\mathcal{E}(f) \equiv \mathcal{E}_0(\xi^* f \xi).$$

Then, the generalized conditional expectation $\mathcal{E}(\cdot)$ is well defined and it has the following properties:

(1) $\mathcal{E}(1) = 1$,
(2) $\mathcal{E}(f^*f) \geq 0$,
(3) $<\mathcal{E}(f), g>_1 = <f, \mathcal{E}(g)>_1$,

where $<f, g>_1 \equiv \varphi_1\big((\sigma_{\frac{i}{4}}^1(f))^*(\sigma_{\frac{i}{4}}^1(g))\big)$.

Here, again, the generalized conditional expectations are understood in the Accardi-Cecchini sense (cf. Ref. 27–29). Thus we arrive at:

Corollary 4.3. *Theorems 4.1 and 4.2 ensure that the operator given by:*

$$\mathcal{L} \equiv \mathcal{E} - id.$$

is a well defined Markov generator.

Consequently, the (Markov) global quantum stochastic semigroup $P_t \equiv e^{t\mathcal{L}}$ can be constructed (for high temperature region). It is worth pointing out that $P_t|_{\mathcal{M}}$ are completely positive (CP) maps on the von Neumann algebra $\mathcal{M}$ and bounded with respect to $L_2(\varphi_1, \frac{1}{2})$ norm (see Refs. 5,7). So, they give rise to well defined maps on quantum L_2-space. In a similar way, one can perform quantization of other stochastic dynamics[8,12].

However, it is important to note here that we were forced to restrict ourselves to high temperature regions (for lattice systems of dimension larger than 1). As we were not able to overcome this difficulty,[30] one may postulate that besides to the suggestions mentioned in the Introduction, some generalization of quantum L_p spaces could be useful. But to take these hints seriously, one should as a first step study the problem of lifting quantum maps (considered dynamical maps are CP maps on a von Neumann algebra) to well defined maps on quantum Orlicz spaces. This will be done in the next Section.

5. Orlicz spaces

Let us begin with some preliminaries. By the term an *Orlicz function* we understand a convex function $\phi : [0, \infty) \to [0, \infty]$ satisfying $\phi(0) = 0$ and $\lim_{u \to \infty} \phi(u) = \infty$, which is neither identically zero nor infinite valued on all of $(0, \infty)$, and which is left continuous at $b_\phi = \sup\{u > 0 : \phi(u) < \infty\}$. In particular, any Orlicz function must also be increasing.

Let L^0 be the space of measurable functions on some σ-finite measure space (X, Σ, m). The Orlicz space L_ϕ^0 associated with ϕ is defined to be the set

$$L^\phi = \{f \in L^0 : \phi(\lambda|f|) \in L^1 \text{ for some } \lambda = \lambda(f) > 0\}.$$

This space turns out to be a linear subspace of L^0 which becomes a Banach space when equipped with the so-called Luxemburg-Nakano norm

$$\|f\|_\phi = \inf\{\lambda > 0 : \|\phi(|f|/\lambda)\|_1 \leq 1\}.$$

Let ϕ be a given Orlicz function. In the context of semifinite von Neumann algebras $\mathcal{M}$ equipped with an fns trace τ, the space of all τ-measurable operators $\widetilde{\mathcal{M}}$ (equipped with the topology of convergence in measure) plays the role of L^0 (for details see Ref. 23). In the specific case where φ is a so-called Young's function, Kunze[31] used this identification to define the associated noncommutative Orlicz space to be

$$L_\phi^{ncO} = \cup_{n=1}^\infty n\{f \in \widetilde{\mathcal{M}} : \tau(\phi(|f|)) \leq 1\}$$

and showed that this too is a linear space which becomes a Banach space when equipped with the Luxemburg-Nakano norm

$$\|f\|_\phi = \inf\{\lambda > 0 : \tau(\phi(|f|/\lambda)) \leq 1\}.$$

Using the linearity it is not hard to see that

$$L_\phi^{ncO} = \{f \in \widetilde{\mathcal{M}} : \tau(\phi(\lambda|f|)) < \infty \text{ for some } \lambda = \lambda(f) > 0\}.$$

Thus there is a clear analogy with the commutative case.

It is worth pointing out that there is another approach to Quantum Orlicz spaces. Namely, one can replace $(\mathcal{M}, \tau)$ by $(\mathcal{M}, \varphi)$, where φ is a normal faithful state on $\mathcal{M}$ (for details see Ref. 32). However, as we wish to put some emphasis on the universality of quantization, we prefer to follow the Banach space theory approach developed by Dodds, Dodds and de Pagter.[33]

Given an element $f \in \widetilde{\mathcal{M}}$ and $t \in [0, \infty)$, the generalised singular value $\mu_t(f)$ is defined by $\mu_t(f) = \inf\{s \geq 0 : \tau(\mathbb{1} - e_s(|f|)) \leq t\}$ where $e_s(|f|)$ $s \in$

$\mathbb{R}$ is the spectral resolution of $|f|$. The function $t \to \mu_t(f)$ will generally be denoted by $\mu(f)$. For details on the generalised singular values see Ref. 34. (This directly extends classical notions where for any $f \in L^0_\infty$, the function $(0, \infty) \to [0, \infty] : t \to \mu_t(f)$ is known as the decreasing rearrangement of f.) We proceed to briefly review the concept of a Banach Function Space of measurable functions on $(0, \infty)$. (Necessary background is given in Ref. 33.) A function norm ρ on $L^0(0, \infty)$ is defined to be a mapping $\rho : L^0_+ \to [0, \infty]$ satisfying

- $\rho(f) = 0$ iff $f = 0$ a.e.
- $\rho(\lambda f) = \lambda \rho(f)$ for all $f \in L^0_+, \lambda > 0$.
- $\rho(f + g) \leq \rho(f) + \rho(g)$ for all.
- $f \leq g$ implies $\rho(f) \leq \rho(g)$ for all $f, g \in L^0_+$.

Such a ρ may be extended to all of L^0 by setting $\rho(f) = \rho(|f|)$, in which case we may then define $L^\rho(0, \infty) = \{f \in L^0(0, \infty) : \rho(f) < \infty\}$. If now $L^\rho(0, \infty)$ turns out to be a Banach space when equipped with the norm $\rho(\cdot)$, we refer to it as a Banach Function space. If $\rho(f) \leq \liminf_n (f_n)$ whenever $(f_n) \subset L^0$ converges almost everywhere to $f \in L^0$, we say that ρ has the Fatou Property. If less generally this implication only holds for $(f_n) \cup \{f\} \subset L^\rho$, we say that ρ is lower semi-continuous. If further the situation $f \in L^\rho$, $g \in L^0$ and $\mu_t(f) = \mu_t(g)$ for all $t > 0$, forces $g \in L^\rho$ and $\rho(g) = \rho(f)$, we call L^ρ rearrangement invariant (or symmetric). Using the above context Dodds, Dodds and de Pagter[33] formally defined the noncommutative space $L^\rho(\widetilde{\mathcal{M}})$ to be

$$L^\rho(\widetilde{\mathcal{M}}) = \{f \in \widetilde{\mathcal{M}} : \mu(f) \in L^\rho(0, \infty)\}$$

and showed that if ρ is lower semicontinuous and $L^\rho(0, \infty)$ rearrangement-invariant, $L^\rho(\widetilde{\mathcal{M}})$ is a Banach space when equipped with the norm $\|f\|_\rho = \rho(\mu(f))$.

Now for any Orlicz function ϕ, the Orlicz space $L^\phi(0, \infty)$ is known to be a rearrangement invariant Banach Function space with the norm having the Fatou Property, see Theorem 8.9 in Ref. 35. Thus on selecting ρ to be $\|\cdot\|_\phi$, the very general framework of Dodds, Dodds and de Pagter presents us with an alternative approach to realising noncommutative Orlicz spaces.

Note that this approach canonically contains the spaces of Kunze[31] . To see this we recall that any Orlicz function is in fact continuous, non-negative and increasing on $[0, b_\phi)$. The fact that Kunze's approach to noncommutative Orlicz spaces is canonically contained in that of Dodds et al, therefore follows from the observation that if $b_\phi = \infty$, then for any $\lambda > 0$ and any

$f \in \widetilde{\mathcal{M}}$, we have

$$\tau(\phi(\frac{1}{\lambda}|f|)) = \int_0^\infty \phi(\frac{1}{\lambda}\mu_t(|f|))\,dt$$

by [34, 2.8]. More generally we have the following lemma[36]:

Lemma 5.1. *Let ϕ be an Orlicz function and $f \in \widetilde{\mathcal{M}}$ a τ-measurable element. Extend ϕ to a function on $[0,\infty]$ by setting $\phi(\infty) = \infty$. If $\phi(f) \in \widetilde{\mathcal{M}}$, then $\phi(\mu_t(f)) = \mu_t(\phi(|f|))$ for any $t \geq 0$, and $\tau(\phi(|f|)) = \int_0^\infty \phi(\mu_t(|f|))\,dt$.*

It is worth pointing out that the above lemma allows for the possibility that $a_\phi > 0$ and/or $b_\phi < \infty$. It is not difficult to see that if $a_\phi > 0$, then

$$\mathcal{M} \subset \{f \in \widetilde{\mathcal{M}} : \phi(\lambda|f|) \in L_1(\mathcal{M}, \tau) \quad \text{for some} \quad \lambda = \lambda(f) > 0\}.$$

Thus this lemma is not contained in results like Remark 3.3 of Ref. 34, which only hold for those elements of $\widetilde{\mathcal{M}}$ for which $\lim_{t \to \infty} \mu_t(f) = 0$.

Consequently, let us take any Orlicz function ϕ. Then the Orlicz space $L^\phi(0,\infty)$ is a Banach function space with a "good" norm. Thus

$$\|f\|_\phi = \inf\{\lambda > 0 : \int_0^\infty dt \phi(\frac{\mu_t(f)}{\lambda}) \leq 1\} \tag{1}$$

gives the "quantum" Orlicz norm, where $f \in \widetilde{\mathcal{M}}$.

In the next Theorem we collect our results on monotonicity of quantum maps with respect to the Orlicz norm given by the formula (1) (proofs will appear in Ref. 36). However, we need some preliminaries. Firstly, following Arveson[37], we say that a completely positive map $T : \mathcal{M} \to \mathcal{M}$ is pure if, for every completely positive map $T' : \mathcal{M} \to \mathcal{M}$, the property "$T - T'$ is a completely positive map" implies that T' is a scalar multiple of T. Finally, a Jordan $*$-morphism $J : \mathcal{M} \to \mathcal{M}$ is $\epsilon - \delta$ absolutely continuous on the projection lattice of $\mathcal{M}$ with respect to the trace τ,[38] if for any $\epsilon > 0$ there exists a $\delta > 0$ such that for any projection $e \in \mathcal{M}$ we have $\tau(J(e)) < \epsilon$ whenever $\tau(e) < \delta$. We have

Theorem 5.2. *Let $T : \mathcal{M} \to \mathcal{M}$ be a linear positive unital map. Then*

$$\|T(f)\|_\phi \leq C\|f\|_\phi \tag{2}$$

where C is a positive constant, if

(1) T is an inner automorphism, e.g. Hamiltonian type dynamics satisfying Borchers conditions (for exposition on Borchers conditions see e.g. Bratteli, Robinson book[14]).

(2) $T(\cdot) = \sum_1^{N<\infty} W_i^* \cdot W_i$ with $W_i \in \mathcal{M}$.

(3) $T(\cdot)$ *is a pure unital normal CP map.*

(4) T *is a ϵ-δ continuous normal Jordan morphism such that $\tau \circ J \leq \tau$.*

The main idea of the proof is to show that generalized singular values $\mu_t(\cdot)$ are monotonic with respect to the maps T. The rest of the proof follows from the definition of the Orlicz norm (1) and the monotonicity of the Orlicz function.

This theorem ensures the existence of extensions to quantum Orlicz space of a map $T : \mathcal{M} \to \mathcal{M}$ satisfying any of the conditions listed in Theorem 5.2. Consequently, we get the promised possibility of describing quantum dynamical system in terms of Quantum Orlicz spaces; so also in L_p-spaces! This explains why one can expect that dynamical maps defined for quantum L_p spaces may have nice generalizations.

Acknowledgments

The support of Poland-South Africa Cooperation Joint project and (WAM) the support of the grant BW/5400-5-0307-7 is gratefully acknowledged.

References

1. T.M. Ligget, *Interacting particle systems*, Springer Verlag, (1985)
2. R.J. Glauber, *J. Mat. Phys.*, **4**, 294, (1963)
3. S-T. C. Moy, *Pacific J. Math.* **4**, 47-64 (1954)
4. G. Pistone, C. Sempi, *Ann. Stat.* **23**, 1543-1561 (1995)
5. A.W. Majewski, B. Zegarlinski, *Math. Phys. Electronic J.* **1**, Paper 2 (1995)
6. A.W. Majewski, B. Zegarlinski, *Lett. Math. Phys.* **36**, 337 (1996)
7. A.W. Majewski, B. Zegarlinski, *Rev. Math. Phys.* **8**, 689 (1996)
8. A.W. Majewski, B. Zegarlinski, *Markov Proc. and Rel. Fields* **2**, 87 (1996)
9. R. F. Streater, *Open Sys. & Information Dyn.* **11** 359-375, (2004)
10. R. F. Streater, The set of states modeled on an Orlicz space, in the proceedings of the Nottingham conference in honour of R. L. Hudson, July, 2006.
11. R. Olkiewicz, B. Zegarlinski, *J. Funct. Analysis*, **161**, 246-285 (1999)
12. B. Zegarliński, Analysis of classical and quantum interacting particle systems, in *Quantum interacting particle systems (Trento, 2000)*, 241–336, QP–PQ: Quantum Probab. White Noise Anal., **14**,241-336, World Sci. Publ., River Edge, NJ, 2002.
13. W. A. Majewski, On quantum stochastic dynamics. Some recent developments in *Stochastic Analysis and Mathematical Physics*, 255-269 World Scientific (2004)
14. O. Bratteli, D.W. Robinson, *Operator Algebras and Quantum Statistical Mechanics*, Springer Verlag, Vol.I (1979), Vol.II (1981)
15. D. Ruelle, *Statistical Mechanics. Rigorous results.*, Benjamin, 1969

16. R. V. Kadison and J. R. Ringrose, *Fundamentals of the Theory of Operator Algebras I and II*, Pure and Applied Mathematics, vol. 100, Academic Press, New York 1983 and 1986.
17. R. Schatten, *Norms ideals of completely continuous operators*, 2nd Printing, Springer Verlag, 1970
18. M. Reed and B. Simon, *Methods of modern mathematical Physics*, vol. II, Academic Press, 1975.
19. U. Haagerup, L_p-spaces associated with an arbitrary von Neumann algebra in *Algèbres d'opérateurs et leurs applications en physique mathématique*, Colloques internationaux du CNRS, No. 274, Marseille 20-24 juin 1977, 175-184. Éditions du CNRS, Paris 1979
20. H. Araki and T. Masuda, *Publ. R.I.M.S., Kyoto Univ.* **18** (1982), 339-411
21. M. Hilsum, *J. Func. Anal.* **40** (1981), 151-169
22. H. Kosaki, *J. Func. Anal.* **56** (1984), 29-78
23. E. Nelson, *J. Func. Anal.* **15** (1974), 103-116
24. M. Terp, L^p-*spaces associated with von Neumann algebras*. Københavns Universitet, Matematisk Institut, Rapport No. 3 (1981)
25. M. Terp, *J. Op. Theory* **8** (1982), 327-360
26. A. Connes, *Bull. Sc. math.*, 2e série **97** (1973), 253-258
27. L. Accardi, *Phys. Rep.* **77** (1981), 169 - 192.
28. L. Accardi and C. Cecchini, *J. Func. Anal.* **45** 245-273 (1982)
29. M. Takesaki, *J. Func. Anal.* **9** 306-321 (1972)
30. A.W. Majewski, B. Zegarlinski, "On quantum stochastic dynamics for quantum spin systems on a lattice; Low temperature problem" in *Mathematical Results in Statistical Mechanics*, ed. S. Miracle-Solé, J. Ruiz, V. Zagrebnov, World Scientific, 1999
31. W Kunze, *Math Nachr* **147**(1990), 123-138
32. M. H. A. Al-Rashed, B. Zegarlinski, *Studia Mathematica*, **180** 199-209 (2007)
33. PG Dodds, T K.-Y Dodds and B de Pagter, *Math Z* **201**(1989), 583-597.
34. T Fack and H Kosaki, *Pacific J Math* **123**(1986), 269-300.
35. G Bennet and R Sharpley, *Interpolation of Operators*, Academic Press, London, 1988.
36. L.E. Labuschagne, A.W. Majewski, Composition operators on noncommutative Orlicz spaces, in preparation
37. W. B. Arveson, *Acta Math.* **123**, 141-224 (1969)
38. L E Labuschagne, *Expo. Math* **17** (1999), 429–468.

VICIOUS RANDOM WALKERS AND TRUNCATED HAAR UNITARIES

J. I. NOVAK[*]

*Department of Mathematics and Statistics, Queen's University,
Kingston, Ontario, Canada
* E-mail: jnovak@mast.queensu.ca
www.queensu.ca*

We show that certain averages over ensembles of truncated random unitary matrices enumerate configurations of random-turns vicious walkers on $\mathbb{Z}$.

Keywords: random-turns model; truncated random matrices; Colour-Flavour Transformation; symmetric functions.

1. The Random-Turns Model

Vicious random walkers were introduced into statistical mechanics by Fisher[3] in order to model domain walls and wetting in two-dimensional lattice systems, or equivalently the stochastic evolution of interacting particles on a one-dimensional lattice. Fisher introduced two vicious walker models, the "lock-step" model and the "random-turns" model. In this note we study a connection between the random-turns model and random matrices obtained by taking corners of Haar-distributed random unitary matrices.

Consider d walkers (particles) initially positioned at sites $d, d-1, \ldots, 1$ on the integer lattice $\mathbb{Z}$ (note that we label the particles from right to left). At each instant of discrete time, a single randomly chosen walker takes a random positive or negative unit step (a random "turn"), subject only to the condition that no two walkers may occupy the same lattice site simultaneously. Each sequence of configurations of the walkers is a virtual history of particle evolution. Such histories are conveniently pictured as directed systems of non-intersecting lattice paths, as in Figure 1.

Forrester[4] studied a particular asymmetric version of the random-turns model in which the d walkers return to their initial sites $d, d-1, \ldots, 1$ after taking n positive steps followed by n negative steps. He made the

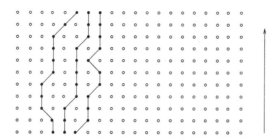

Fig. 1. *A reunion of random-turns vicious walkers.*

following remarkable observation: the number $u_d(n)$ of realizations of the random-turns model of the sort just described is equal to the number of permutations in the symmetric group $S(n)$ which have no increasing subsequence of length greater than d. Recall that a permutation $\sigma \in S(n)$ is said to have an increasing subsequence of length k if their exist indices $1 \leq i_1 < i_2 < \cdots < i_k \leq n$ such that

$$\sigma(i_1) < \sigma(i_2) < \cdots < \sigma(i_k).$$

A well-known earlier result of Rains[9] expresses this latter quantity as an integral $\int_{\mathcal{U}_d} |\operatorname{Trace} U|^{2n} dU$ over the compact group $\mathcal{U}_d$ of $d \times d$ unitary matrices, against Haar measure dU. Thus one has the equality

$$u_d(n) = \int_{\mathcal{U}_d} |\operatorname{Trace} U|^{2n} dU \tag{1}$$

Assuming that the walkers represent d mutually attracting particles, the equilibrium states of the model occur when the walkers are positioned on adjacent lattice site. Thus it is natural to seek a more general integral representation for the number of ways in which d random-turns vicious walkers may reunite at displaced lattice sites $d + q, d - 1 + q, \ldots, 1 + q$ for some $q \geq 0$ by taking $n + dq$ positive steps followed by n negative steps. Such an integral form should reduce to (1) when $q = 0$. Denote this number $u_d(n; q)$.

Our main result is that an integral representation of $u_d(n; q)$ generalizing (1) continues to hold for arbitrary $q \geq 0$ if we enlarge the unitary group $\mathcal{U}_d$ to the semigroup $\mathcal{B}_d$ of contractions of Euclidean space $\mathbb{C}^d$, and deform Haar measure on $\mathcal{U}_d$ to a probability measure $\nu^{(q)}$ on $\mathcal{B}_d$. We should note that the specialization $u_d(n; 0)$ reduces to the original situation investigated by Forrester, while the alternative specialization $u_d(0; q)$ is equivalent to the classical André[1] Ballot Problem.

2. Truncation of random unitary matrices

Let $q \geq 0$ be a non-negative integer, and let $\mathcal{U}_{d+q}$ be the unitary group of dimension $d+q$. A unitary matrix $U \in \mathcal{U}_{d+q}$ may be represented in block form

$$U = \begin{bmatrix} P & Q \\ R & T \end{bmatrix},$$

where P is the $d \times d$ principal submatrix of U. Since $\|P\| \leq \|U\| = 1$ (operator norm), the map

$$U \mapsto P$$

defines a function

$$\mathcal{U}_d \to \mathcal{B}_d$$

called "truncation." Let $\nu^{(q)}$ denote the image of Haar measure on $\mathcal{U}_{d+q}$ under the operation of truncation. Clearly $\nu^{(0)}$ is the Haar measure on $\mathcal{U}_d$, since in the case $q=0$ no truncation is involved.

A random matrix $U_d^{(q)} \in \mathcal{B}_d$ whose distribution is $\nu^{(q)}$ is called a truncated random unitary matrix. The sequence of random matrices $(U_d^{(q)})_{d \geq 1}$ is a parameter-dependent deformation of the classical Circular Unitary Ensemble (CUE) introduced by Dyson. This deformed random matrix ensemble is denoted $\mathrm{CUE}^{(q)}$. We note that for $q \geq d$, the measure $\nu^{(q)}$ is absolutely continuous with respect to Lebesgue measure dZ on the matrix ball $\mathcal{B}_d$ with density proportional to

$$\det(I - Z^*Z)^{q-d} dZ.$$

This is a result of Neretin.[6]

Our main result is the following.

Theorem 2.1. *For any integers $d \geq 1$ and $n, q \geq 0$,*

$$u_d(n;q) = \binom{n+dq}{n} f^{q^d} \int_{\mathcal{U}_{d+q}} |u_{11} + \cdots + u_{dd}|^{2n} dU$$

$$= \binom{n+dq}{n} f^{q^d} \int_{\mathcal{B}_d} |\operatorname{Trace} Z|^{2n} d\nu^{(q)}(Z).$$

The scaling factor f^{q^d} is equal to the number of standard Young tableaux on the $d \times q$-rectangular Young diagram q^d:

$$f^{q^d} = (dq)! \prod_{i=0}^{d-1} \frac{i!}{(q+i)!}.$$

This factor may alternatively be interpreted algebraically as the dimension of a certain irreducible (complex) representation of the symmetric group $S(dq)$, or geometrically as the degree of a certain Grassmannian of hyperplanes in the space $\mathbb{C}^{dq}$.

3. Sketch of proof

The main idea of the proof, which is outlined here and will appear in full detail elsewhere,[7] is to use an integral identity due to Wei and Wettig[10] which connects integration over unitary groups of different dimensions. This identity, which is a version of Zirnbauer's "Colour-Flavour Transform," reads

$$\int_{\mathcal{U}(d+q)} e^{\text{Trace}(X^*UY+Y^*U^*X)} dU$$
$$= H_{q^d} \int_{\mathcal{U}(d)} e^{\text{Trace}(X^*XV^*+VY^*Y)} \det(VY^*X)^{-q} dV,$$

and is valid for any two $(d+q) \times d$ matrices over $\mathbb{C}$ verifying $\det(Y^*X) \neq 0$. Here the scaling factor H_{q^d} is the hook-product of the $d \times q$-rectangular Young diagram, given explicitly as

$$H_{q^d} = \prod_{i=0}^{d-1} \frac{(q+i)!}{i!}.$$

Choosing $X = Y$ to be the rectangular matrix with diagonal entries equal to some real number $t \neq 0$, and all other entries 0, the Colour-Flavour Transform reduces to the identity

$$t^{2dq} \int_{\mathcal{U}_{d+q}} e^{t^2 \text{Trace}(U_d + U_d^*)} dU = H_{q^d} \int_{\mathcal{U}_d} e^{t^2 \text{Trace}(V+V^*)} \det(V^*)^q dV,$$

where we use the notation U_d for the truncation of $U \in \mathcal{U}_{d+q}$. Expanding in power series in t and equating coefficients, we arrive at the identity

$$\binom{n+dq}{n} f^{q^d} \int_{\mathcal{U}_{d+q}} |\text{Trace}\, U_d|^{2n} dU = \int_{\mathcal{U}_d} (\text{Trace}\, V)^{n+dq} (\text{Trace}\, V^*)^n \det(V^*)^q dV.$$

The integral on the right is the scalar product

$$\langle e_1^{n+dq} | e_1^n e_d^q \rangle$$

in the algebra Λ_d of symmetric polynomials in d indeterminates. The scalar product is understood to be the usual Hall scalar product. Expanding the powers of elementary symmetric polynomials as linear combinations of Schur polynomials using the Frobenius identity

$$e_1^N = \sum_{\substack{\lambda \vdash n \\ \ell(\lambda) \leq d}} f^\lambda s_\lambda$$

and using the orthonormality of Schur polynomials with respect to the Hall scalar product (modulo omitted details regarding the adjoint operation of "multiplication by a Schur polynomial") we obtain

$$\langle e_1^{n+dq} | e_1^n e_d^q \rangle = \sum_{\substack{\lambda \vdash n \\ \ell(\lambda) \leq d}} f^{q^d + \lambda} f^\lambda.$$

It is easy to see that this final expression as a sum over Young diagrams has the desired combinatorial interpretation in terms of random-turns walkers (standard Young tableaux are lattice walks in a Weyl chamber, and such a chamber is the configuration space for random-turns vicious walkers[7]).

4. Asymptotics

Probably the most useful feature of the integral representation

$$u_d(n;q) = \binom{n+dq}{n} f^{q^d} \int_{\mathcal{U}_{d+q}} |\operatorname{Trace} U_d|^{2n} dU$$

is the ease with which it yields asymptotic information.

Sommers and Zyckzkowski have proved that the eigenvalues of a random matrix $U_d^{(q)}$ from the ensemble $\mathrm{CUE}^{(q)}$, with $q > 0$, constitute a determinantal point process in the unit disc $\mathbb{D} = \{z \in \mathbb{C} : |z| \leq 1\}$ with kernel

$$K_d^{(q)} = \frac{q}{\pi}(1-|z|^2)^{\frac{q-1}{2}}(1-|w|^2)^{\frac{q-1}{2}} \sum_{j=0}^{d-1} \binom{q+j}{j} z^j \overline{w}^j.$$

One observes from this that the limiting kernel of the spectrum of the scaled random matrix $q^{\frac{1}{2}} U_d^{(q)}$ in the limit $q \to \infty$ is

$$\frac{1}{\pi} e^{-\frac{1}{2}(|z|^2+|w|^2)} \sum_{j=0}^{d-1} \frac{z^j \overline{w}^j}{j!},$$

which is precisely the well-known Ginibre kernel (see e.g. Mehta's book[5]). Thus one may conclude weak convergence

$$q^{\frac{1}{2}} U_d^{(q)} \to G_d$$

in the large q limit, where G_d is the Ginibre matrix whose entries are i.i.d standard complex Gaussians. This is generalizes a classical theorem of E. Borel[2] from the early twentieth century. Borel's theorem states that for a uniformly random point $u = (u_{11}, \ldots, u_{dd}, \ldots, u_{d+q,d+q})$ from the unit sphere S^{d+q-1} in $\mathbb{R}^{d+q}$, the scaled random variables $q^{\frac{1}{2}} u_{11}, \ldots, q^{\frac{1}{2}} u_{dd}$ tend weakly to a family of i.i.d standard Gaussians as $q \to \infty$. The random-matrix version of this result stated above seems to have been first observed by Petz and Reffy,[8] who prove it by a more computational method.

We obtain the asymptotic behaviour of the integral from this weak convergence, namely

$$\int_{\mathcal{U}_{d+q}} |\operatorname{Trace} U_d|^{2n} dU \sim \frac{1}{q^n} \langle \sum_{i=1}^d g_{ii} \rangle = \frac{d^n n!}{q^n}, \quad q \to \infty.$$

Here, the g_{ii}'s are i.i.d. standard complex Gaussians and the angled brackets denote expecation. The rightmost equality follows from the fact that the sum is just a complex Gaussian with mean 0 and variance d. From this integral estimate and Stirling's approximation we immediately obtain the following asymptotic result.

Theorem 4.1. *For fixed integers $d \geq 1$ and $n \geq 0$*

$$u_d(n;q) \sim (2\pi)^{\frac{1-d}{2}} \left(\prod_{i=1}^d i! \right) d^{2n+dq+\frac{1}{2}} q^{\frac{1-d^2}{2}}$$

as $q \to \infty$.

References

1. André, D. "Solution directe du problème résolu par M. Bertrand." *C.R. Acad. Sci. Paris* **105** (1887): 436-437.
2. Borel, E. "Sur le principes de la theorie cinétique des gaz. *Annales de l'École Normale Supérieure* **23** (1906): 9-32.
3. Fisher, M.E. "Walks, walls, wetting, and melting." *Journal of Statistical Physics* **34** (1984): 667-729.
4. Forrester, P.J. "Random walks and random permutations." *Journal of Physics A: Mathematical and General* **34** (2001): L417- L423.
5. Mehta, M.L. *Random Matrices. Third Edition.* Academic Press (2004).
6. Neretin, Y.A. "Hua-type integrals over unitary groups and over projective limits of unitary groups." *Duke Mathematical Journal* **114**(2) (2002): 239-266.
7. Novak, J.I. "Combinatorial aspects of truncated random unitary matrices." *Submitted.*
8. Petz, D., Réffy, J. "On asymptotics of large Haar distributed unitary matrices." *Period. Math. Hungar.* **49** (2004): 103-117.

9. Rains, E. "Increasing subsequences and the classical groups." *Electronic Journal of Combinatorics* **5** (1998): R12.
10. Wei, Y., Wettig, T. "Bosonic color-flavor transformation for the special unitary group." *Journal of Mathematical Physics* **46** (2005).

INCOHERENT QUANTUM CONTROL

A. PECHEN AND H. RABITZ

Department of Chemistry, Princeton University, Princeton, New Jersey 08544, USA
E-mail: apechen@princeton.edu and hrabitz@princeton.edu

Conventional approaches for controlling open quantum systems use coherent control which affects the system's evolution through the Hamiltonian part of the dynamics. Such control, although being extremely efficient for a large variety of problems, has limited capabilities, e.g., if the initial and desired target states have density matrices with different spectra or if a control field needs to be designed to optimally transfer different initial states to the same target state. Recent research works suggest extending coherent control by including active manipulation of the non-unitary (i.e., incoherent) part of the evolution. This paper summarizes recent results specifically for incoherent control by the environment (e.g., incoherent radiation or a gaseous medium) with a kinematic description of controllability and landscape analysis.

Keywords: Incoherent quantum control, control of open quantum systems

1. Introduction

The manipulation of atomic or molecular quantum dynamics commonly uses coherent quantum control, which may be extremely useful for a large variety of problems.[1-12] The dynamical evolution of a closed quantum system under the action of a collection of coherent controls $u = \{u_l(t)\}$ (e.g., Rabi frequencies of the applied laser field) is described by the equation

$$\frac{d\rho_t}{dt} = -i\Big[H_0 + \sum_l Q_l u_l(t), \rho_t\Big], \qquad \rho|_{t=0} = \rho_0 \qquad (1)$$

Here ρ_t is the system density matrix at time t (for an n-level quantum system, the set of all density matrices is $\mathcal{D}_n = \{\rho \in \mathcal{M}_n \,|\, \rho \geq 0, \text{Tr}\,\rho = 1\}$, where $\mathcal{M}_n = \mathbb{C}^{n \times n}$ is the set of $n \times n$ complex matrices), H_0 is the free system Hamiltonian describing evolution of the system in the absence of control fields and each Q_l is an operator describing the coupling of the system to the control field $u_l(t)$.

Coherent control of a closed system induces a unitary transformation of the system density matrix $\rho_t = U_t \rho_0 U_t^\dagger$ and may have some limitations. The first limitation is due to the fact that unitary transformations of an operator preserve its spectrum; thus the spectrum of ρ_t is the same at any t and, for example, a mixed state ρ_0 will always remain mixed.[13] A second limitation is that a control u_{opt} which is optimal for some initial state ρ_0 may be not optimal for another initial state $\tilde\rho_0$ even if ρ_0 and $\tilde\rho_0$ have the same spectrum. This limitation originates from the reversibility of unitary evolution and is due to the fact that $U_t \rho_0 U_t^\dagger \neq U_t \tilde\rho_0 U_t^\dagger$ if $\rho_0 \neq \tilde\rho_0$. To overcame these limitations at least to some degree, control by measurements[14–18] or incoherent control[19–22] may be used. General mathematical notions for the controlled quantum Markov dynamics were formulated.[23]

The necessity to consider incoherent control relies also on the fact that coherent control of quantum systems (e.g., of chemical reactions) in the laboratory is often realized in a medium (solvent) which interacts with the controlled system and plays the role of the environment. Furthermore, then environment may be also affected to some degree by the coherent laser field, thus effectively realizing incoherent control of the system. Moreover, laser sources of coherent radiation at the present time have practical limitations, and some frequencies are very expensive to generate compared to the respective sources of incoherent control (e.g., incoherent radiation as considered in Sec. 2.1 of this work). Thus incoherent control can be used in some cases to reduce the total cost of quantum control.

This paper summarizes recent results specifically for incoherent control by the environment[19] (ICE). A general theoretical formulation for incoherent control is provided in Sec. 2, followed by the examples of control by incoherent radiation (Sec. 2.1) and control through collisions with particles of a medium (e.g., solvent, gas, etc., Sec. 2.2). Relevant known results about controllability and the structure of control landscapes for open quantum systems in the kinematic picture are briefly outlined in Sec. 3.

2. Incoherent control by the environment

The dynamical evolution of an open quantum system under the action of coherent controls in the Markovian regime is described by a master equation

$$\frac{d\rho_t}{dt} = -i\left[H_0 + H_{\text{eff}} + \sum_l Q_l u_l(t), \rho_t\right] + \mathcal{L}\rho_t \qquad (1)$$

The interaction with the environment modifies the Hamiltonian part of the dynamics by adding an effective Hamiltonian term H_{eff} to the free Hamiltonian H_0. Another important effect of the environment is the appearance of the term $\mathcal{L}$ which describes non-unitary aspects of the evolution and is responsible for decoherence. This term in the Markovian regime has the general Gorini-Kossakowski-Sudarshan-Lindblad[24,25] (GKSL) form

$$\mathcal{L}\rho = \sum_i \left(2L_i \rho L_i^\dagger - L_i^\dagger L_i \rho - \rho L_i^\dagger L_i \right)$$

where L_i are some operators acting in the system Hilbert space. The explicit form of the GKSL term depends on the particular type of the environment, on the details of the microscopic interaction between the system and the environment, and on the state of the environment.

The coherent portion of the control in (1) addresses only the Hamiltonian part of the evolution while the GKSL part $\mathcal{L}$ remains fixed (for the analysis of controllability properties for Markovian master equations under coherent controls see for example, Ref. 26). However, the generator $\mathcal{L}$ can also be controlled to some degree. For a fixed system-environmental interaction, the generator $\mathcal{L}$ depends on the state of the environment, which can be either a thermal state at some temperature (including the zero temperature vacuum state) or an arbitrary non-equilibrium state. Such a state is characterized by a (possibly, time dependent) distribution of particles of the environment over their degrees of freedom, which are typically the momentum $\mathbf{k} \in \mathbb{R}^3$ and the internal energy levels parameterized by some discrete index $\alpha \in A$ (e.g., for photons $\alpha = 1, 2$ denotes polarization, for a gas of N-level particles $\alpha = 1, \ldots, N$ denotes the internal energy levels). Denoting the density at time t of the environmental particles with momentum $\mathbf{k}$ and occupying an internal level α by $n_{\mathbf{k},\alpha}(t)$, and the corresponding GKSL generator as $\mathcal{L} = \mathcal{L}[n_{\mathbf{k},\alpha}(t)]$, the equation (1) becomes

$$\frac{d\rho_t}{dt} = -i\left[H_0 + H_{\text{eff}} + \sum_l Q_l u_l(t), \rho_t\right] + \mathcal{L}[n_{\mathbf{k},\alpha}(t)]\rho_t \qquad (2)$$

Here both $u_l(t)$ and $n_{\mathbf{k},\alpha}(t)$ are used as the controls, and for $n_{\mathbf{k},\alpha}(t)$ the optimization is done over $\mathbf{k}, \alpha$ in a time dependent fashion to obtain a desired outcome.

The solution of (2) with the initial condition $\rho|_{t=0} = \rho_0$ for each choice of controls $\{u_l(t)\}$ and $n_{\mathbf{k},\alpha}(t)$ can be represented by a family $P_t\{(u_l), n_{\mathbf{k},\alpha}\}$, $t \geq 0$ of completely positive (CP), trace preserving maps (see Sec. 3 for the explicit definitions) as

$$\rho_t = P_t\{(u_l), n_{\mathbf{k},\alpha}\}\rho_0 \qquad (3)$$

In general, for time dependent controls this family forms not a semigroup but a self-consistent two-parameter family of CP, trace preserving maps

$P_{t,\tau}\{(u_l), n_{\mathbf{k},\alpha}\}$, $t \geq \tau \geq 0$, where each $P_{t,\tau}\{(u_l), n_{\mathbf{k},\alpha}\}$ represents the evolution from τ to t.

The target functional, also called the *performance index*, describes a property of the controlled system which should be minimized during the control and commonly consists of the two terms:

$$J[(u_l), n_{\mathbf{k},\alpha}] = J_1[(u_l), n_{\mathbf{k},\alpha}] + J_2[(u_l), n_{\mathbf{k},\alpha}]$$

The term $J_1[(u_l), n_{\mathbf{k},\alpha}]$, called the *objective functional*, represents the physical system's property which we want to minimize. The term $J_2[(u_l), n_{\mathbf{k},\alpha}]$, called the *cost functional*, represents the penalty for the control fields.

The first general class of objective functionals appears in the problem of minimizing the expectation value of some observable associated to the system at a target time $T > 0$. The system is assumed at the initial time $t = 0$ to be in the state ρ_0. Any observable characterizing the system (e.g., its energy, population of some level, etc.) is represented by some self-adjoint operator O acting in the system Hilbert space, and the corresponding objective functional has the form

$$J_1[(u_l), n_{\mathbf{k},\alpha}] = \mathrm{Tr}\left[\rho_T(u_l, n_{\mathbf{k},\alpha})O\right] \equiv \mathrm{Tr}\left[(P_T\{(u_l), n_{\mathbf{k},\alpha}\}\rho_0)O\right] \quad (4)$$

Here $\rho_T(u_l, n_{\mathbf{k},\alpha}) \equiv P_T\{(u_l), n_{\mathbf{k},\alpha}\}\rho_0$ is the final density matrix of the system evolving from the initial state ρ_0 under the action of u_l and $n_{\mathbf{k},\alpha}$. The physical meaning of this objective functional is that it represents the average measured value of the observable O at the final time T when the system evolves from the initial state ρ_0 under the action of the controls $(u_l), n_{\mathbf{k},\alpha}$.

The second general class of objective functionals appears in the problem of optimal state-to-state transfer. Suppose that initially the system is in a state ρ_0 and the control goal is to steer the system at some target time T into some desired target state ρ_{target}. In this case one seeks controls (u_l) and $n_{\mathbf{k},\alpha}$ which minimize the distance between the states $\rho_T(u_l, n_{\mathbf{k},\alpha})$ and ρ_{target}. The corresponding objective functional has the form

$$J_1[(u_l), n_{\mathbf{k},\alpha}] = \|\rho_T(u_l, n_{\mathbf{k},\alpha}) - \rho_{\text{target}}\| \equiv \|P_T\{(u_l), n_{\mathbf{k},\alpha}\}\rho_0 - \rho_{\text{target}}\| \quad (5)$$

where $\|\cdot\|$ is a suitable matrix norm. Usually the Hilbert-Schmidt norm $\|A\| = \sqrt{\mathrm{Tr}\, A^\dagger A}$ can be used.

The third important class of objective functionals appears in the problem of producing a desired target CP, trace preserving map P_{target}. In this case the objective functional has the form

$$J_1[(u_l), n_{\mathbf{k},\alpha}] = \|P_T\{(u_l), n_{\mathbf{k},\alpha}\} - P_{\text{target}}\| \quad (6)$$

where $\|\cdot\|$ is a suitable norm in the space of all CP, trace preserving maps. In particular, in conventional models of quantum computation the target transformation P_{target} is a unitary gate (e.g., a phase U_ϕ or Hadamard $U_{\mathbb{H}}$ gate, and for these examples $P_{\text{target}} = U_\phi$ or $P_{\text{target}} = U_{\mathbb{H}}$, respectively).[27,28] More general non-unitary target transformations can arise [e.g., in quantum computing with mixed states[29] or for generating controls robust to variations of the initial system's state[30] (see also Sec. 3.1)].

The cost functional J_2 can be chosen to have the form

$$J_2[(u_l), n_{\mathbf{k},i}] = \sum_l \int_0^T dt \alpha_l(t) |u_l(t)|^2 + \max_{0 \leq t \leq T} \sum_i \int d\mathbf{k} \beta_i(\mathbf{k}) n_{\mathbf{k},i}(t)$$

Here each function $\alpha_l(t) \geq 0$ [resp., $\beta_i(\mathbf{k}) \geq 0$] is a weight describing the cost for the control u_l at time t (resp., for the density of particles of the environment with momentum $\mathbf{k}$ and occupying the internal level i). The first term minimizes the energy of the optimal coherent control. The second term minimizes the total density of the environment.

The control functions belong to some sets of admissible controls $(u_l) \in \mathcal{E}$ and $n_{\mathbf{k},\alpha} \in \mathcal{D}$. The following three important problems arise.

Optimal controls. Find, for a given initial state ρ_0 and a target time T, some (or all) controls $(u_l) \in \mathcal{E}$ and $n_{\mathbf{k},\alpha} \in \mathcal{D}$ which minimize the performance index.

Reachable sets. Find, for a given final time $T > 0$ and an initial state ρ_0, the set of all states reachable from ρ_0 up to the time T, i.e., the set

$$\mathcal{R}_T(\rho_0) = \{P_t\{(u_l), n_{\mathbf{k},\alpha}\} \rho_0 \mid t \leq T, (u_l) \in \mathcal{E}, n_{\mathbf{k},\alpha} \in \mathcal{D}\}$$

Landscape analysis. Find, for a given $T > 0$, an initial state ρ_0 and a self-adjoint operator O, all extrema (global and local, and saddles, if any) of the objective functional $J_1[u_l(t), n_{\mathbf{k},\alpha}(t)] = \text{Tr}\left[\{P_T[(u_l), n]\rho_0\}O\right]$ defined by (4) [and similarly for the objective functionals defined by (5) and (6)] or of the corresponding performance index.

2.1. *Incoherent control by radiation*

Non-equilibrium radiation is characterized by its distribution in photon momenta and polarization. For control with distribution of incoherent radiation the magnitude of the photon momentum $|\mathbf{k}|$ can be exploited along with the polarization and the propagation direction in cases where polarization dependence or spatial anisotropy is important (e.g., for controlling a system consisting of oriented molecules bound to a surface).

A thermal equilibrium distribution for photons at temperature T is characterized by Planck's distribution

$$n_\mathbf{k} = \frac{1}{\exp\left(\frac{c\hbar|\mathbf{k}|}{k_\mathrm{B} T}\right) - 1}$$

where c is the speed of light, $\hbar$ and k_B are the Planck and the Boltzmann constants which we set to one below. Non-equilibrium incoherent radiation may have a distribution given as an arbitrary non-negative function $n_{\mathbf{k},\alpha}(t)$. Some practical means to produce non-equilibrium distributions in the laboratory may be based either on filtering thermal radiation or on the use of independent monochromatic sources.

The master equation for an atom or a molecule interacting with a coherent electromagnetic field $E_c(t)$ and with incoherent radiation with a distribution $n_\mathbf{k}(t)$ in the Markovian regime has the form:

$$\frac{d\rho_t}{dt} = -i[H_0 + H_\mathrm{eff} - \mu E_c(t), \rho_t] + \mathcal{L}_\mathrm{Rad}[n_\mathbf{k}(t)]\rho_t \qquad (7)$$

The coherent part of the dynamics is generated by the free system's Hamiltonian $H_0 = \sum_n \varepsilon_n P_n$ with eigenvalues ε_n, forming the spectrum $\mathrm{spec}(H_0)$, and the corresponding projectors P_n, the effective Hamiltonian H_eff resulting from the interaction between the system and the incoherent radiation, dipole moment μ, and electromagnetic field $E_c(t)$.

The GKSL generator $\mathcal{L} = \mathcal{L}_\mathrm{Rad}$ induced by the incoherent radiation with distribution function $n_\mathbf{k}(t)$ has the form (e.g., see Ref. 31)

$$\mathcal{L}_\mathrm{Rad}[n_\mathbf{k}(t)]\rho = \sum_{\omega \in \Omega} [\gamma_\omega^+(t) + \gamma_{-\omega}^-(t)](2\mu_\omega \rho \mu_\omega^\dagger - \mu_\omega^\dagger \mu_\omega \rho - \rho \mu_\omega^\dagger \mu_\omega) \qquad (8)$$

Here the sum is taken over the set of all system transition frequencies $\Omega = \{\varepsilon_n - \varepsilon_m \mid \varepsilon_n, \varepsilon_m \in \mathrm{spec}(H_0)\}$, $\mu_\omega = \sum_{\varepsilon_n - \varepsilon_m = \omega} P_m \mu P_n$, and the coefficients

$$\gamma_\omega^\pm(t) = \pi \int d\mathbf{k}\, \delta(|\mathbf{k}| - \omega)|g_\mathbf{k}|^2 [n_\mathbf{k}(t) + (1 \pm 1)/2]$$

determine the transition rates between energy levels with transition frequency ω. The transition rates depend on the photon density $n_\mathbf{k}(t)$. The form-factor $g_\mathbf{k}$ determines the coupling of the system to the $\mathbf{k}$-th mode of the radiation. Equation (7) together with the explicit structure (8) of the GKSL generator provides the theoretical formulation for analysis of control by incoherent radiation.

The numerical simulations illustrating the capabilities of learning control by incoherent radiation to prepare prespecified mixed states from a pure state is available[19] along with a theoretical analysis of the set of stationary states for the generator $\mathcal{L}_{\text{Rad}}$ for some models.[21] Incoherent control by radiation can extend the capabilities of coherent control by exciting transitions between the system's energy levels for which laser sources are either unavailable at the present time or very expensive compared with the corresponding sources of incoherent radiation. Ref. 22 provides a simple experimental realization of the combined coherent (by a laser) and incoherent (by incoherent radiation emitted by a gas-discharge lamp) control of certain excitations in Kr atoms.

2.2. *Incoherent control by a gaseous medium*

This section considers incoherent control of quantum systems through collisions with particles of a surrounding medium (e.g., a gas or solvent of electrons, atoms or molecules, etc.). This case also includes coherent control of chemical reactions in solvents if the coherent field addresses not only the controlled system but the solvent as well. The particles of the medium in this treatment serve as the control and the explicit characteristic of the medium exploited to minimize the performance index is in general a time dependent distribution of the medium particles over their momenta $\mathbf{k}$ and internal energy levels $\alpha \in A$. This distribution is formally described by a non-negative function $n : \mathbb{R}^3 \times A \times \mathbb{R} \to \mathbb{R}_+$, whose value $n_{\mathbf{k},\alpha}(t)$ (where $\mathbf{k} \in \mathbb{R}^3, \alpha \in A$, and $t \in \mathbb{R}_+$) has the physical meaning of the density at time t of particles of the surrounding medium with momentum $\mathbf{k}$ and in internal energy level α. In this scheme one prepares a suitable, in general non-equilibrium, distribution of the particles in the medium such that the medium drives the system evolution through collisions in a desired way.

It may be difficult to practically create a desired non-equilibrium distribution of medium particles over their momenta. In contrast, a non-equilibrium distribution in the internal energy levels can be relatively easily created, e.g., by lasers capable of exciting the internal levels of the medium particles or through an electric discharge. Then the medium particles can affect the controlled system through collisions and this influence will typically depend on their distribution. A well known example of such control is the preparation of population inversion in a He–Ne gas-discharge laser. In this system an electric discharge passes through the He–Ne gas and brings the He atoms into a non-equilibrium state of their internal degrees of freedom. Then He–Ne collisions transfer the energy of the non-equilibrium state of the He atoms into the high energy levels of the Ne atoms. This process creates a population inversion in the Ne atoms and subsequent lasing. A steady electric discharge can be used to keep the gas of helium atoms in a non-equilibrium state to produce a CW He–Ne laser. This process can serve as an example of incoherent control through collisions by considering the gas of He atoms as the control environment (medium) and the Ne atoms as the system which we want to steer to a desired (excited) state.

Quantum systems controlled through collisions with gas or medium particles in certain regimes can be described by master equations with GKSL generators whose explicit structure is different from the generator $\mathcal{L}_{\rm Rad}$ describing control by incoherent radiation. If the medium is sufficiently dilute, such that the probability of simultaneous interaction of the control system with two or more particles of the medium is negligible, then the reduced dynamics of the system will be Markovian[32,33] and will be determined by two body scattering events between the system and one particle of the medium. Below we provide a formulation for control of quantum systems by a dilute medium, although the assumption of diluteness is not a restriction for ICE, and dense mediums might be used for control as well.

The master equation for a system interacting with coherent fields $u_l(t)$ and with a dilute medium of particles with mass m has the form (2) with the generator $\mathcal{L}[n_{\mathbf{k},\alpha}(t)] = \mathcal{L}_{\rm Medium}[n_{\mathbf{k},\alpha}(t)]$ specified by the distribution function of the medium $n_{\mathbf{k},\alpha}(t)$ and by the T-operator (transition matrix) for the scattering of the system and a medium particle. Below we assume that the particles of the medium are characterized only by their momenta and do not have internal degrees of freedom; otherwise, the state of one particle of the medium should have the form $|\mathbf{k},\alpha\rangle$, where α specifies the internal degrees of freedom. A transition matrix element is $T_{n,n'}(\mathbf{k},\mathbf{k}') = \langle n,\mathbf{k}|T|n',\mathbf{k}'\rangle$, where $|n,\mathbf{k}\rangle \equiv |n\rangle|\mathbf{k}\rangle$ denotes the product state of the system discrete eigenstate $|n\rangle$ (an eigenstate of the system's free Hamiltonian H_0 with eigenvalue ε_n) and a translational state of the system and a medium particle with relative momentum $\mathbf{k}$. If the system is fixed in space (we consider this case below corresponding to the system particle being much more massive than the particles of the surrounding medium) then $|\mathbf{k}\rangle$ is a translation state of a medium particle. The general case of relative system medium particle motion can be considered as well using suitable master equations. We will use the notation $T_\omega(\mathbf{k},\mathbf{k}') := \sum_{m,n:\ \varepsilon_m - \varepsilon_n = \omega} T_{m,n}(\mathbf{k},\mathbf{k}')|m\rangle\langle n|$. The density of particles of the medium at momentum $\mathbf{k}$ is denoted as $n_{\mathbf{k}}(t)$, and the set of all transition frequencies ω of the system among the energy levels of H_0

is denoted as Ω. In this notation the GKSL generator is

$$\mathcal{L}_{\text{Medium}}[n_{\mathbf{k}}(t)]\rho = 2\pi \sum_{\omega \in \Omega} \int d\mathbf{k} n_{\mathbf{k}}(t) \int d\mathbf{k}' \delta\left(\frac{|\mathbf{k}'|^2}{2m} - \frac{|\mathbf{k}|^2}{2m} + \omega\right)$$
$$\times \left[T_\omega(\mathbf{k}',\mathbf{k})\rho T_\omega^\dagger(\mathbf{k}',\mathbf{k}) - \frac{1}{2}\left\{T_\omega^\dagger(\mathbf{k}',\mathbf{k})T_\omega(\mathbf{k}',\mathbf{k}), \rho\right\}\right] \quad (9)$$

where $\{\cdot,\cdot\}$ denotes the anti-commutator. If the medium is at equilibrium with inverse temperature β, then the density has the stationary Boltzmann form $n_{\mathbf{k}}(t) \equiv n_{\mathbf{k}} = C(\beta,n)\exp(-\beta|\mathbf{k}|^2/2m)$. Here the normalization constant $C(\beta,n)$ is determined by the condition $\int d\mathbf{k} n_{\mathbf{k}} = n$, where n is the total density of the medium. The structure of Eq. (9) has been discussed previously for equilibrium media[32,33] and for non-equilibrium stationary media.[34] Non-equilibrium media may be characterized by generally time dependent distributions. Equation (2) with $\mathcal{L}[n_{\mathbf{k}}(t)] \equiv \mathcal{L}_{\text{Medium}}[n_{\mathbf{k}}(t)]$ provides the general formulation for theoretical analysis of control by a coherent field $u_l(t)$ and by a non-equilibrium medium with density $n_{\mathbf{k}}(t)$.

As a simple illustration of such incoherent control, Fig. 1 reproduces the numerical results from Ref. 19 for optimally controlled transfer of a pure initial state of a four-level system into three different mixed target states [i.e., the objective function (5) is chosen]. The control is modelled by collisions with a medium prepared in a static non-equilibrium distribution $n_{|\mathbf{k}|}$ whose form is optimized by learning control using a genetic algorithm (GA)[35] based on the mutation and crossover operations. Since the initial and target states have different spectra, they can not be connected by a unitary evolution induced by coherent control. However, Fig. 1 shows that ICE through collisions can work perfectly for such situations.

3. Kinematic description of incoherent control

Physically admissible evolutions of an n-level quantum system can be represented by CP, trace preserving maps (*Kraus maps*).[36] A map $\Phi : \mathcal{M}_n \to \mathcal{M}_n$ is positive if for any $\rho \in \mathcal{M}_n$ such that $\rho \geq 0$: $\Phi(\rho) \geq 0$. A linear map $\Phi : \mathcal{M}_n \to \mathcal{M}_n$ is CP if for any $l \in \mathbb{N}$ the map $\Phi \otimes \mathbb{I}_l : \mathcal{M}_n \otimes \mathcal{M}_l \to \mathcal{M}_n \otimes \mathcal{M}_l$ is positive (here $\mathbb{I}_l$ denotes the identity map in $\mathcal{M}_l$). A CP map Φ is called trace preserving if for any $\rho \in \mathcal{M}_n$: $\text{Tr}\,\Phi(\rho) = \text{Tr}\,\rho$. The conditions of trace preservation and positivity for physically admissible evolutions are necessary to guarantee that Φ maps states into states. The condition of complete positivity has the following meaning. Consider the elements of $\mathcal{M}_l$ as operators of some l-level ancilla system which does not evolve, i.e., its evolution is represented by the identity mapping $\mathbb{I}_l$. Suppose that the n-level system

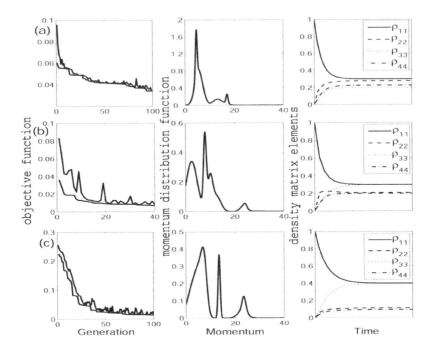

Fig. 1. (From Ref. 19. Copyright (2006) by the American Physical Society.) Results of ICE simulations with a surrounding non-equilibrium medium as the control for target states (a) $\rho_{\text{target}} = \text{diag}(0.3; 0.3; 0.2; 0.2)$, (b) $\rho_{\text{target}} = \text{diag}(0.3; 0.2; 0.3; 0.2)$, and (c) $\rho_{\text{target}} = \text{diag}(0.4; 0.1; 0.4; 0.1)$. Each case shows: the objective function vs GA generation, the optimal distribution vs momentum, and the evolution of the diagonal elements of the density matrix for the optimal distribution. In the plots for the objective function the upper curve is the average value for the objective function and the lower one is the best value in each generation.

does not interact with the ancilla. Then the combined evolution of the total system will be represented by the map $\Phi \otimes \mathbb{I}_l$ and the condition of complete positivity requires that for any l this map should transform all states of the combined system into states, i.e. to be positive.

Any CP, trace preserving map Φ can be expressed using the Kraus operator-sum representation as

$$\Phi(\rho) = \sum_{i=1}^{\lambda} K_i \rho K_i^\dagger$$

where $K_i \in \mathbb{C}^{n \times n}$ are the Kraus operators subject to the constraint $\sum_{i=1}^{\lambda} K_i^\dagger K_i = \mathbb{I}_n$ to guarantee trace preservation. This constraint determines a complex Stiefel manifold $V_n(\mathbb{C}^{\lambda n})$ whose points are $n \times (\lambda n)$ matrices $V = (K_1; K_2; \ldots; K_\lambda)$ (i.e., each V is a column matrix of $K_1, \ldots K_\lambda$) satisfying the orthogonality condition $V^\dagger V = \mathbb{I}_n$.

The explicit evolution $P_t\{(u_l), n_{\mathbf{k},\alpha}\}$ in (3) is unlikely to be known for realistic systems. However, since this evolution is always a CP, trace preserving map, it can be represented in the Kraus form

$$P_t\{(u_l), n_{\mathbf{k},\alpha}\}\rho = \sum_{i=1}^{\lambda} K_i(t, (u_l), n_{\mathbf{k},\alpha})\rho K_i^\dagger(t, (u_l), n_{\mathbf{k},\alpha}),$$

$$\text{where } \sum_{i=1}^{\lambda} K_i^\dagger(t, (u_l), n_{\mathbf{k},\alpha}) K_i(t, (u_l), n_{\mathbf{k},\alpha}\}) = \mathbb{I}_n$$

Assume that any Kraus map can be generated in this way using the available coherent and incoherent controls $\{u_l(t)\}$ and $n_{\mathbf{k},\alpha}(t)$. Then effectively the Kraus operators can be considered as the controls [instead of $\{u_l(t)\}$ and $n_{\mathbf{k},\alpha}(t)$] which can be optimized to drive the evolution of the system in a desired direction. This picture is called *the kinematic picture* in contrast with *the dynamical picture* of Sec. 2. In the next two subsections we briefly outline the controllability and landscape properties in the kinematic picture.

3.1. *Controllability*

Any classical or quantum system at a given time is completely characterized by its state. The related notion of state controllability refers to the ability to steer the system from any initial state to any final state, either at a given time or asymptotically as time goes to infinity, and the important problem in control analysis is to establish the degree of state controllability for a given control system. Assuming for some finite-level system that the set of admissible dynamical controls generates arbitrary Kraus type evolution, the following theorem implies then that the system is completely state controllable.

Theorem 3.1. *For any state $\rho_f \in \mathcal{D}_n$ of an n-level quantum system there exists a Kraus map Φ_{ρ_f} such that $\Phi_{\rho_f}(\rho) = \rho_f$ for all states $\rho \in \mathcal{D}_n$.*

Proof. Consider the spectral decomposition of the final state $\rho_f = \sum_{i=1}^n p_i |\phi_i\rangle\langle\phi_i|$, where p_i is the probability to find the system in the state $|\phi_i\rangle$ ($p_i \geq 0$ and $\sum_{i=1}^n p_i = 1$). Choose an arbitrary orthonormal basis $\{|\chi_j\rangle\}$ in the system Hilbert space and define the operators

$$K_{ij} = \sqrt{p_i}|\phi_i\rangle\langle\chi_j|, \qquad i,j = 1, \ldots, n.$$

The operators K_{ij} satisfy the normalization condition $\sum_{i,j=1}^{n} K_{ij}^{\dagger} K_{ij} = \mathbb{I}_n$ and thus determine the Kraus map $\Phi_{\rho_f}(\rho) = \sum_{i,j=1}^{n} K_{ij} \rho K_{ij}^{\dagger}$. The map Φ_{ρ_f} acts on any state $\rho \in \mathcal{D}_n$ as

$$\Phi_{\rho_f}(\rho) = \sum_{i,j=1}^{n} p_i |\phi_i\rangle\langle\chi_j|\rho|\chi_j\rangle\langle\phi_i| = (\mathrm{Tr}\,\rho) \sum_{i=1}^{n} p_i |\phi_i\rangle\langle\phi_i| = \rho_f$$

and thus satisfies the condition of the Theorem. □

The potential importance of this result is that it shows that there may exist a single incoherent evolution which is capable for transferring all initial states into a given target state, and moreover, the target state can be an arbitrary pure or a mixed state.[30] Thus this theorem shows that non-unitary evolution can break the two general limitations for coherent unitary control described in the second paragraph in the Introduction.

3.2. Control landscape structure

In the kinematic description, under the assumption that any Kraus map can be generated, the objective functional becomes a function on the Stiefel manifold $V_n(\mathbb{C}^{\lambda n})$. In practice, various gradient methods may be used to minimize such an objective function. If the objective function has a local minimum then gradient based optimization methods can be trapped in this minimum and will not provide a true solution to the problem. For such an objective function, if the algorithm stops in some minimum one can not be sure that this minimum is global and therefore this solution may be not satisfactory. This difficulty does not exist if *a priori* information about absence of local minima for the objective function is available as provided by the following theorem for a general class of objective functions of the form $J_1[K_1, \ldots, K_\lambda] = \mathrm{Tr}\,[(\sum_{i=1}^{\lambda} K_i \rho K_i^{\dagger}) O]$ in the kinematic picture.

Theorem 3.2. *For any $n \in \mathbb{N}$, $\rho \in \mathcal{D}_n$, and for any Hermitian $O \in \mathcal{M}_n$ the objective function $J_1[K_1, \ldots, K_\lambda] = \mathrm{Tr}\,[(\sum_{i=1}^{\lambda} K_i \rho K_i^{\dagger}) O]$ on the Stiefel manifold $V_n(\mathbb{C}^{\lambda n})$ does not have local minima or maxima; it has global minimum manifold, global maximum manifold, and possibly saddles whose number and the explicit structure depend on the degeneracies of ρ and O.*

The case $n = 2$ has been considered in detail in Ref. 37, where the global minimum, maximum, and saddle manifolds are explicitly described for each type of initial state ρ. In particular, it is found that the objective function J_1 for a non-degenerate target operator O and for a pure ρ (i.e., such that

$\rho^2 = \rho$) does not have saddle manifolds; for the completely mixed initial state $\rho = \frac{1}{2}\mathbb{I}$, J_1 has one saddle manifold with the value of the objective function $J_{\text{saddle}} = 1/2$; and for any partially mixed initial state J_1 has two saddle manifolds corresponding to the values of the objective function $J_{\text{saddle}}^{\pm} = (1 \pm \|\mathbf{w}\|)/2$, where $\mathbf{w} = \text{Tr}\,[\rho\boldsymbol{\sigma}]$ and $\boldsymbol{\sigma} = (\sigma_x, \sigma_y, \sigma_z)$ is the vector of Pauli matrices (the vector $\mathbf{w}$ is in the unit ball, $\|\mathbf{w}\| \leq 1$ and this vector characterizes the initial state as $\rho = \frac{1}{2}[\mathbb{I}_2 + \langle \mathbf{w}, \boldsymbol{\sigma} \rangle]$). The case of arbitrary n is considered in Ref. 38.

4. Conclusions

This paper outlines recent results for incoherent control of quantum systems through their interaction with an environment. A general formulation for incoherent control through GKSL dynamics is given, followed by examples of incoherent radiation and a gaseous medium serving as the incoherent control environments. The relevant known results on controllability of open quantum systems subject to arbitrary Kraus type dynamics, as well as properties of the corresponding control landscapes, are also discussed.

Acknowledgments

This work was supported by the NSF and ARO. A. Pechen acknowledges also partial support from the RFFI 08-01-00727-a and thanks the organizers of the "28-th Conference on Quantum Probability and Related Topics" (CIMAT-Guanajuato, Mexico, 2007) Prof. R. Quezada Batalla and Prof. L. Accardi for the invitation to present a talk on the subject of this work.

References

1. A. G. Butkovskiy and Y. I. Samoilenko *Control of Quantum-Mechanical Processes and Systems* (Nauka, Moscow, 1984);
   A. G. Butkovskiy and Y. I. Samoilenko *Control of Quantum-Mechanical Processes and Systems* (Kluwer, Dordrecht, 1990) (Engl. Transl.).
2. D. Tannor and S. A. Rice, *J. Chem. Phys.* **83**, 5013 (1985).
3. A. P. Pierce, M. A. Dahleh and H. Rabitz, *Phys. Rev. A* **37**, 4950 (1988).
4. R. S. Judson and H. Rabitz, *Phys. Rev. Lett.* **68**, 1500 (1992).
5. W. S. Warren, H. Rabitz and M. Dahleh, *Science* **259**, 1581 (1993).
6. S. A. Rice and M. Zhao *Optical Control of Molecular Dynamics* (Wiley, New York, 2000).
7. H. Rabitz, R. de Vivie-Riedle, M. Motzkus and K. Kompa, *Science* **288**, 824 (2000).
8. M. Shapiro and P. Brumer *Principles of the Quantum Control of Molecular Processes* (Wiley-Interscience, Hoboken, NJ, 2003).

9. I. A. Walmsley and H. Rabitz, *Physics Today* **56**, 43 (2003).
10. M. Dantus and V. V. Lozovoy, *Chem. Rev.* **104**, 1813 (2004).
11. L. Accardi, S. V. Kozyrev and A. N. Pechen, *QP–PQ: Quantum Probability and White Noise Analysis* vol. **XIX** ed. L. Accardi, M. Ohya and N. Watanabe (World Sci. Pub. Co., Singapore), 1 (2006).
12. D. D'Alessandro *Introduction to Quantum Control and Dynamics* (Chapman and Hall, Boca Raton, 2007).
13. S. G. Schirmer, A. I. Solomon and J. V. Leahy, *J. Phys. A: Math. Gen.* **35**, 4125 (2002).
14. R. Vilela Mendes and V. I. Man'ko, *Phys. Rev. A* **67**, 053404 (2003).
15. A. Mandilara and J. W. Clark, *Phys. Rev. A* **71**, 013406 (2005).
16. A. Pechen, N. Il'in, F. Shuang and H. Rabitz, *Phys. Rev. A* **74**, 052102 (2006).
17. L. Roa, A. Delgado, M. L. Ladron de Guevara and A. B. Klimov, *Phys. Rev. A* **73**, 012322 (2006).
18. F. Shuang, A. Pechen, T.-S. Ho and H. Rabitz, *J. Chem. Phys.* **126**, 134303 (2007).
19. A. Pechen and H. Rabitz, *Phys. Rev. A* **73**, 062102 (2006).
20. R. Romano and D. D'Alessandro, *Phys. Rev. A* **73**, 022323 (2006).
21. L. Accardi and K. Imafuku, *QP–PQ: Quantum Probability and White Noise Analysis* vol. **XIX** ed. L. Accardi, M. Ohya and N. Watanabe (World Sci. Pub. Co., Singapore), 28 (2006).
22. Y. Ding *et al*, *Rev. Sci. Instruments* **78**, 023103 (2007).
23. V. P. Belavkin, *Automatia and Remote Control* **44**, 178 (1983).
24. V. Gorini, A. Kossakowski and E. C. G. Sudarshan, *J. Math. Phys.* **17**, 821 (1976).
25. G. Lindblad, *Comm. Math. Phys.* **48**, 119 (1976).
26. C. Altafini, *J. Math. Phys.* **44**, 2357 (2003).
27. M. Grace, C. Brif, H. Rabitz, I. A. Walmsley, R. L. Kosut and D. A. Lidar, *J. Phys. B: At. Mol. Opt. Phys.* **40**, S103 (2007).
28. M. Grace, C. Brif, H. Rabitz, D. A. Lidar, I. A. Walmsley and R. L. Kosut, *J. Modern Optics* **54**, 2339 (2007).
29. V. E. Tarasov, *J. Phys. A: Math. Gen.* **35**, 5207 (2002).
30. R. Wu, A. Pechen, C. Brif and H. Rabitz, *J. Phys. A: Math. Theor.* **40**, 5681 (2007).
31. L. Accardi, Y. G. Lu and I. V. Volovich *Quantum Theory and Its Stochastic Limit* (Springer, Berlin, 2002).
32. R. Dümcke, *Comm. Math. Phys.* **97**, 331 (1985).
33. L. Accardi, A. N. Pechen and I. V. Volovich, *Infin. Dimens. Anal. Quant. Probab. and Relat. Topics* **6**, 431 (2003).
34. A. N. Pechen, *QP-PQ: Quantum Probability and White Noise Analysis* vol. **XVIII** ed M. Schürmann and U. Franz (World Sci. Pub. Co., Singapore), 428 (2005); *Preprint* http://xxx.lanl.gov/abs/quant-ph/0607134.
35. D. E. Goldberg *Genetic Algorithms in Search, Optimization and Machine Learning* (Addison-Wesley, Reading, MA, 1989).
36. K. Kraus *States, Effects, and Operations* (Springer, Berlin, New York, 1983).
37. A. Pechen, D. Prokhorenko, R. Wu and H. Rabitz *J. Phys. A: Math. Theor.*

41, 045205 (2008).
38. R. Wu, A. Pechen, H. Rabitz, M. Hsieh and B. Tsou, *J. Math. Phys.* **49**, 022108 (2008).

COMPROMISING NON-DEMOLITION AND INFORMATION GAINING FOR QUBIT STATE ESTIMATION*

L. RUPPERT[1,2], A. MAGYAR[1,2] AND K. M. HANGOS[1]

[1] *Process Control Research Group*
Computer and Automation Research Institute,
H-1518 Budapest, POB 63, Hungary
[2] *Department for Mathematical Analysis*
Budapest University of Technology and Economics
H-1521 Budapest, POB 91, Hungary

> An indirect measurement scheme is used in this paper, which allows us to to optimally select the elements of the measurement protocol according to a predefined weighting between the variance of the estimated state parameters and the ratio of the non-demolished copies of the system.

1. Introduction

The state estimation of quantum systems from measurement data is fundamental both in quantum information theory[1,2] and quantum control[3]. Although this problem may be traced back to the seventies [4], the interest in a thorough mathematical analysis of the quantum state estimation procedures has been flourishing recently[5–8].

Determining the state of a quantum mechanical system using projective measurement may be difficult because of two facts: first of all, it is not possible to obtain full information from a single measurement, on the other hand, the second measurement of the same state is not possible because of the destructive nature of the measurement. More precisely, it has been shown that it is impossible to determine the wave function of a single quantum system by using any measuring scheme and protocol including repeated measurements [9].

Therefore, the usual method of quantum tomography is to prepare many

---

*Supported by the Hungarian Research Grant OTKA K67625, and by the Control Engineering Research Group of the Budapest University of Technology and Economics.

systems in the same state, measure them with projective measurements subsequently [10], and does not use the already measured copies for further investigations. In certain physical representations, for example in quantum optics, this is possible to implement, but in many other cases it is not possible to set up several systems in identical states.

A possible way to circumvent the above obstruction is to use an *indirect measurement scheme*, where the 'unknown' quantum system is coupled with a 'measurement' (also called 'probe' or 'ancilla') system and the measurements are only applied on the measurement system [11]. In the literature this method is often termed *weak measurement* [12,13], however, most of these papers use a continuous-time approach to weak measurements.

A large number of the papers apply some kind of feedback either to drive the system into a desired state or to compensate for the 'measurement backaction'. An application of weak measurements in bipartite state purification can be seen in [14], where the authors also use continuous time dynamics. Korotkov and Jordan[15] have shown that "it is possible to fully restore any unknown, pre-measured state, though with probability less than unity."

The aim of this paper is to explore the structure and find a good parametrization of a discrete-time indirect measurement scheme and strategy in the case when both the unknown and the measuring systems are quantum bits. A further aim is to use the measurement scheme as a basis for quantum tomography.

2. Basic notions

Some basic notations regarding the representation of quantum states used in the sequel are presented in this section.

2.1. Bloch representation

Throughout the paper the *Bloch-vector* representation of the states of quantum bits is used, i.e.

$$\rho = \frac{1}{2} \begin{bmatrix} 1 + \theta_3 & \theta_1 - i\theta_2 \\ \theta_1 + i\theta_2 & 1 - \theta_3 \end{bmatrix} = \frac{1}{2}(\sigma_0 + \theta_1\sigma_1 + \theta_2\sigma_2 + \theta_3\sigma_3) \qquad (1)$$

where σ_i stands for the Pauli-matrices, $i = 1, 2, 3$, and θ is the Bloch-vector. This way, the states of a quantum bit can be described with three dimensional real vectors with maximal length 1, i.e. $\theta \in \mathbb{R}^3$, $||\theta|| \leq 1$. Thus, the state space of the system is the unit ball in $\mathbb{R}^3$.

2.2. Dynamics of a single qubit

The Schrödinger picture is used here in discrete time that associates a unitary U to the time-evolution of the system such that

$$\rho(k) = U\rho(k-1)U^* \qquad (2)$$

where $\rho(k)$ is the density matrix of the system at the time instance k, and

$$U = \exp(-ihH(u_x, u_y, u_z)) \qquad (3)$$

with $H(u_x, u_y, u_z) = u_x\sigma_1 + u_y\sigma_2 + u_z\sigma_3$ being the Hamiltonian operator of the system, h is the sampling time, and u_x, u_y, and u_z are the inputs.

Instead of the unitary description of the dynamics (as in Eq. (2)), we will use the so called *T-representation* of the linear mapping $\theta(k-1) \mapsto \theta(k)$ that corresponds to the original state transformation $\rho(k-1) \mapsto \rho(k)$ in (2) with a real 3×3 orthogonal matrix T, such that

$$\theta(k+1) = T(u_x, u_y, u_z)\theta(k), \qquad (4)$$

where the inputs u_x, u_y, and u_z are in the argument of trigonometrical functions, i.e. can effect only the rotational speed of the state vector but not its length. This equation will be generalized to the case of more than one qubit in order to be the basis of the discrete time state equation of the system.

2.3. von Neumann measurement

The most generally used von Neumann measurement is the measurement of the Pauli operators σ_1, σ_2 and σ_3. If one considers the measurement of the observable σ_1, the possible outcomes are the different eigenvalues of the observable, i.e. ± 1. The probabilities of the different outcomes are

$$p_{+1} = \text{Prob}(+1) = \text{Tr}\rho E_{+1} = \tfrac{1}{2}(1+\theta_1)$$
$$p_{-1} = \text{Prob}(-1) = \text{Tr}\rho E_{-1} = \tfrac{1}{2}(1-\theta_1)$$

respectively, where the spectral decomposition of σ_1 is $\sigma_1 = E_{+1} - E_{-1}$.

To represent the above measurement as a *stochastic disturbance* it is important to know what are the eigenstates of the measurement. Measuring σ_1, the state after measurement is

$$\bar{\rho} = \frac{E_{\pm 1}\rho E_{\pm 1}}{\text{Tr}E_{\pm 1}\rho E_{\pm 1}},$$

depending on the actual outcome. In the Bloch vector representation, these states are

$$\theta_{+1} = \begin{bmatrix} +1 & 0 & 0 \end{bmatrix}^T, \theta_{-1} = \begin{bmatrix} -1 & 0 & 0 \end{bmatrix}^T.$$

3. Indirect measurement applied to qubits

In order to compute the effect of an indirect measurement for the coupled unknown-measurement qubit system, it is necessary to write up the dynamics of coupled qubit pairs in their Bloch-vector representation.

In what follows, the 'unknown' qubit is denoted by the subscript S and the 'measurement' qubit or the *probe system* is denoted by M.

3.1. *Dynamics of coupled qubit pairs*

Let us denote the Bloch representation of the unknown system and the probe (measurement device) as

$$\rho_S(k) = \frac{1}{2}(I + \theta_S(k)\sigma^S) \ , \ \rho_M(k) = \frac{1}{2}(I + \theta_M(k)\sigma^M) \qquad (1)$$

where θ_S and θ_M are 3 dimensional real vectors, σ^S and σ^M are symbolic vectors constructed from the Pauli operators acting on the Hilbert spaces $\mathcal{H}^S$ and $\mathcal{H}^M$, with $\mathcal{H}^S = \mathcal{H}^M = \mathbb{C}^2$.

The state of the composite system is represented as a 4×4 density matrix $\rho_{S+M}(k)$. The state of the composite system after the interaction is given by

$$\rho_{S+M}(k+1) = U_{S+M}\rho_{S+M}(k)U^*_{S+M} \qquad (2)$$

where U_{S+M} is the overall system evolution unitary. Note, that we shall use (2) with $\rho_{S+M} = \rho_S \otimes \rho_M$ in the following because of our special measurement strategy. Since we are interested in the dynamical change of the system S, the first reduced density matrix should only be considered:

$$\rho_S(k+1) = \text{Tr}_M \rho_{S+M}(k+1). \qquad (3)$$

In order to have a simple parametrization of the interaction (coupling) between the unknown and measurement qubits, the Cartan decomposition[16,17] of the discrete time evolution unitary U_{S+M} is used in the form

$$U_{S+M} = L_1 e^{ah} L_2 \qquad (4)$$

where L_1 and L_2 are in $SU(2) \otimes SU(2)$ and $a \in \mathbf{a}$ with

$$\mathbf{a} = \mathrm{i}\ span\{\sigma_1^S \otimes \sigma_1^M, \sigma_2^S \otimes \sigma_2^M, \sigma_3^S \otimes \sigma_3^M\} \qquad (5)$$

Because both L_1 and L_2 are in a product form, they describe the product of the local dynamical effects L_i^S and L_i^M ($i = 1, 2$), and the interaction is parameterized by three real parameters a_1, a_2 and a_3.

Therefore, the dynamical equation of qubit S in (3) becomes

$$\rho_S(k+1) = L_1^S \text{Tr}_M e^{ah} \left(\tilde{\rho}_S(k) \otimes \tilde{\rho}_M(k) \right) e^{a^*h} L_1^{S*} \tag{6}$$

where $L_1 = L_1^S \otimes L_1^M$, $L_2 = L_2^S \otimes L_2^M$ both time dependent, and $\tilde{\rho}_S = L_2^S \rho_S L_2^{S*}$, $\tilde{\rho}_M = L_2^M \rho_M L_2^{M*}$. In order to simplify the forthcoming computations, we consider the case *with no local dynamics*, when $L_i^S = L_i^M = I$ ($i = 1, 2$).

A simple example is a case when

$$U_{S+M} = e^{-ih(a_1 \sigma_1^S \otimes \sigma_1^M)} \tag{7}$$

i.e. the qubits are interacting only in the x direction. Computing the dynamics of the system in Bloch representation we obtain

$$\theta_S(k+1) = \begin{bmatrix} 1 & 0 & 0 \\ 0 & \cos(2a_1 h) & -\sin(2a_1 h)\theta_{M1} \\ 0 & \sin(2a_1 h)\theta_{M1} & \cos(2a_1 h) \end{bmatrix} \theta_S(k) = T(a_1, h, \theta_M)\theta_S(k) \tag{8}$$

if there were no measurements performed.

3.2. *Measurement protocol*

Indirect measurement means, that the projective measurements are performed on the probe system in state θ_M attached to the one we are interested in (θ_S). In the composite system (in state ρ_{S+M}) an indirect measurement corresponds to the observables of the form $I \otimes A_M$, where A_M is a self-adjoint operator on the Hilbert space of system M. For the sake of simplicity, it is assumed, that A_M is a Pauli spin operator. The *measurement*

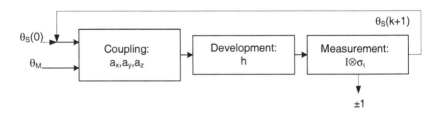

Fig. 1. Signal flow diagram of indirect measurement

protocol is shown in Fig. 1. At each time instant of the discrete time set, the measurement qubit is coupled to the unknown system S. They evolve according to the bipartite dynamics (6) for the sampling time h, and at the end of the sampling interval, a von Neumann measurement is performed on the measurement qubit. At the next time instant, the previous steps are repeated.

3.3. Parameters of the protocol

The above setting of the indirect measurement allows us to adjust various parameters of the measurement protocol. These can be used to make an optimal compromise between the information gained from the measurement and the demolition caused by the measurement back-action.

The coupling parameters a_1, a_2, a_3 of the Cartan decomposition (4-5) determine how (in terms of strength and direction) the measurement system is coupled to the unknown one. The sampling time h amplifies this effect and appears as a multiplicative factor to the coupling parameters.

The state of the measurement qubit (θ_M) can be different at each time instant, which allows us in the future to introduce a feedback to the measurement protocol.

It is important to note that one can make a 'no information - no demolition' situation by setting the coupling parameters to zero, and a 'maximal information - complete demolition' situation, too. Examples of such extreme cases will be given in the next section, see Remark 4.1.

4. A simple example for indirect measurement

A simple special case of an indirect measurement is investigated analytically here to find the effect of the protocol parameters. As we shall se later, this case can be used to selectively estimate one of the components of the unknown qubit's Bloch vector, similarly to the so-called standard measurement scheme[18] for single qubits. A straightforward modification of the measurement setup leads to the estimators of the other two Bloch vector components.

4.1. Measurement setup

Consider the case when the qubits are interacting only in the y direction for time h (sampling time). Afterwards, an indirect measurement is performed, i.e. a von Neumann measurement of the observable $I \otimes \sigma_1$ on the composite system

$$e^{-ih(a_2\sigma_2^S\otimes\sigma_2^M)}\cdot(\rho_S\otimes\rho_M)\cdot e^{-ih(a_2\sigma_2^S\otimes\sigma_2^M)^*}.$$

For the sake of simplicity, choose h and a_i in such a way, that $2a_ih = \frac{\pi}{2}$. For example, the above setting corresponds to the parameters

$$h = \frac{1}{10}, \quad a_2 = \frac{5}{2}\pi, \quad a_1 = a_3 = 0.$$

The probabilities of the different outcomes of $I\otimes\sigma_1$'s measurement are

$$\text{Prob}(+1) = \tfrac{1}{2}(1+\theta_{S2}\theta_{M3})$$
$$\text{Prob}(-1) = \tfrac{1}{2}(1-\theta_{S2}\theta_{M3}). \tag{1}$$

Now the probabilities depend on the both the state θ_S and the measurement qubit θ_M. The states after the measurement are

$$\theta_S^{+1} = \begin{bmatrix} \frac{\theta_{S3}\theta_{M2}+\theta_{S1}\theta_{M1}}{1+\theta_{S2}\theta_{M3}} \\ \frac{\theta_{S2}+\theta_{M3}}{1+\theta_{S2}\theta_{M3}} \\ \frac{\theta_{S3}\theta_{M1}-\theta_{S1}\theta_{M2}}{1+\theta_{S2}\theta_{M3}} \end{bmatrix}, \quad \theta_S^{-1} = \begin{bmatrix} \frac{\theta_{S3}\theta_{M2}-\theta_{S1}\theta_{M1}}{1-\theta_{S2}\theta_{M3}} \\ \frac{\theta_{S2}-\theta_{M3}}{1-\theta_{S2}\theta_{M3}} \\ \frac{-\theta_{S3}\theta_{M1}-\theta_{S1}\theta_{M2}}{1-\theta_{S2}\theta_{M3}} \end{bmatrix} \tag{2}$$

if $+1$ or -1 was the result, respectively. This case is useful for state parameter estimation since the probabilities and the new states depend on both θ_S and θ_M. This means that we both gain information from the measurements and retrieve information in the new states after the measurement.

4.2. *Properties*

Let us concentrate on the estimate of the second unknown state co-ordinate, i.e. we want to describe the change of θ_{S2} (notation: $x = \theta_{S2}$) during the measurements. Let us further assume that θ_{M3} is constant (denoted by c) and $\theta_{M1} = \theta_{M2} = 0$.

Remark 4.1. If $c=1$ then we get the standard measurement scheme:

$$\text{Prob}(\pm 1) = \frac{1}{2}(1+x^{\pm}) \quad , \quad x^{\pm 1} = \pm 1$$

It is easy to see from Eq. (2) that this would be a totally invasive measurement, i.e. the information about the true state would be lost, thus we assume $|c| < 1$.

Proposition 4.1. *If we measure first +1, and thereafter -1 (or vice versa), then the state of θ_{S2} ($=x$) will not change.*

Proof: First from x it will be $x' = \frac{x+c}{1+cx}$ then from x' it will be $x'' = \frac{x'-c}{1-cx'} = \frac{\frac{x+c}{1+cx}-c}{1-c\frac{x+c}{1+cx}} = \cdots = x$. The reverse goes similarly.

Corollary 4.1. *All of the possible cases of states can be ordered in a line such a way, that after each measurement we jump in the neighboring state on the left or right side.*

Proposition 4.2. *If we measure first +1, and thereafter -1 (or vice versa), then the probability of these outcomes doesn't depend on x.*

Proof: First from x it will be $x' = \frac{x+c}{1+cx}$ with probability: $P = \frac{1}{2}(1+cx)$ then from x' it will be x with probability: $Q = \frac{1}{2}(1-cx') = \frac{1}{2}(1-c\frac{x+c}{1+cx}) = \frac{1}{2}\frac{1-c^2}{1+cx}$. So the probability of this outcome is $P \cdot Q = \frac{1-c^2}{4}$. The reverse goes similarly.

Corollary 4.2. *If two outcome sequences contain the same number of $+1$ and -1 measurement outcomes, then their probabilities are the same.*

Let us introduce the following notations:

p_n the probability that from n measurements all outcomes are $+1$s
x_n the resulting state from n measurement when all outcomes are $+1$s

By definition $p_0 := 1$.

Corollary 4.3. *With this notation, the probability that there are k times $+1$ and l times -1 outcomes ($k > l$) in the sequence can be computed as:*

$$\left(\frac{1-c^2}{4}\right)^l \cdot p_{k-l}$$

The state after this sequence of outcomes will be x_{k-l}, and we can represent the sequence of the measurement outcomes as a Markov-process.

Proposition 4.3. *p_k is a linear function of x.*

Proof: The proof goes by induction. Let $q_k := 2^k p_k$ and $x_k := \frac{y_k}{q_k}$, where (we will proof) q_k and y_k are simple polynomials.
If $k = 0$ then $q_0 = p_0 = 1$ and $y_0 = x$.
Next let us suppose that both q_k and y_k are simple polynomials, and p_k is linear in x. Then

$$p_{k+1} = p_k \cdot \frac{1}{2}(1+cx_k) = \frac{q_k}{2^k} \cdot \frac{1}{2}\left(1+c\frac{y_k}{q_k}\right) = \frac{1}{2^{k+1}} \cdot (q_k + cy_k)$$

On the other hand $p_{k+1} = \frac{q_{k+1}}{2^{k+1}}$, so

$$q_{k+1} = q_k + cy_k \tag{3}$$

Furthermore

$$x_{k+1} = \frac{x_k + c}{1 + cx_k} = \frac{\frac{y_k}{q_k} + c}{1 + c\frac{y_k}{q_k}} = \frac{y_k + cq_k}{q_k + cy_k} = \frac{y_k + cq_k}{q_{k+1}}$$

On the other hand $x_{k+1} = \frac{y_{k+1}}{q_{k+1}}$, therefore

$$y_{k+1} = y_k + cq_k \tag{4}$$

Finally we conclude that q_k and y_k are really simple polynomials, and from the recursion we can see that both q_k and y_k are linear in x, so p_k is linear in x, too.

Remark 4.2. The proof gives us a recursive calculation for x_k and p_k, so we can build up easily a stochastic model based on the above 3 propositions, and develop a state estimation strategy.

5. Towards optimal quantum state estimation by indirect measurements

Let us suppose that we have N identical copies of the composite quantum system (the two coupled qubits, S and M). We shall use the following *measurement strategy*:

(1) Perform 2 subsequent measurements (a *measurement pair*) with a pre-specified $c = \theta_{M3}$ and compute the maximum-likelihood (ML) estimate of x.
(2) Retain the systems on which the measured outcomes were $+1$ and -1 (in any order) for further studies, because the are not affected by the measurements, i.e. their $\theta_{S2} = x$ is left unchanged (see Proposition 4.1).

Note that the above implies $n = 2$ for the results in Subsection 4.2. Now we investigate how the selection of c (the initial state of the measurement system) affects the variance of the estimate (we want it to be small), and the ratio of the unaffected system copies (we want this to be large).

Denote the number of the $(+1, +1)$ outcome by N_+, and the probability that a measurement pair results in this outcome by $p_+ = p_2 = \frac{1+c^2+2cx}{4}$. Similarly, the number of the $(-1,-1)$ outcome is denoted by N_-, and its probability is by $p_- = \frac{1+c^2-2cx}{4}$. Then the number of the non-effective

$((+1, -1)$ or $(-1, +1))$ outcomes is $N_0 = N - N_+ - N_-$, and its probability is $p_0 = \frac{1-c^2}{2}$. Then the likelihood function of a measurement pair is the following polynomial distribution:

$$P = \frac{N!}{N_+! \, N_-! \, N_0!} p_+^{N_+} p_-^{N_-} p_0^{N_0}.$$

The maximum likelihood estimate of x is obtained from this by taking the logarithm of P above, and maximizing it with respect to x:

$$\hat{x}_{ML}(N_+, N_-, c) = \frac{1+c^2}{2c} \frac{N_+ - N_i}{N_+ + N_-}. \quad (1)$$

This estimate is well-defined if at least one of N_+ or N_- is positive, that holds with probability one when number of measurements goes to infinity. On the other hand, this estimation is asymptotically unbiased.

5.1. *The variance and the non-demolition probability*

In the case of the investigated measurement setup (see section 4.1), the variance **V** of the Maximum Likelihood estimator (1) is as follows:

$$\mathbf{V}(c, x) = \sum_{i=1}^{N} Var\left(\frac{1+c^2}{2c} \frac{N_+ - N_-}{N_+ + N_-} \mid N_+ + N_- = i\right) \cdot P(N_+ + N_- = i) =$$

$$= \sum_{i=1}^{N} \left(\frac{1+c^2}{2c}\right)^2 \frac{1}{i^2} Var\left(N_+ - N_- \mid N_+ + N_- = i\right) \cdot P(N_+ + N_- = i)$$

where $Var(.)$ denotes the variance of a random variable.

Let X_j be a random variable that takes the value $+1$ if the outcome of the measurement pair is $(+1, +1)$, and -1 when the outcome is $(+1, +1)$. Then $X_j = 1$ with probability $\frac{p_+}{p_+ + p_-}$, and $X_j = -1$ with probability $\frac{p_-}{p_+ + p_-}$. These are the conditional properties of being $+1, +1$ and $-1, -1$, if we know that the two outcome is the same. Then

$$Var\left(N_+ - N_- \mid N_+ + N_- = i\right) = Var\left(\sum_{j=1}^{i} X_j\right) = i \cdot Var(X_1)$$

From simple calculation we obtain:

$$Var(X_1) = 1 - \left(\frac{p_+ - p_-}{p_+ + p_-}\right)^2 = 1 - \left(\frac{2cx}{1+c^2}\right)^2$$

Therefore, the variance of the Maximum Likelihood estimator is:

$$\mathbf{V}(c,x) = \left(\frac{1+c^2}{2c}\right)^2 \left[1 - \left(\frac{2cx}{1+c^2}\right)^2\right] \sum_{i=1}^{N} \frac{1}{i} \cdot P(N_+ + N_- = i)$$

$$\sum_{i=1}^{N} \frac{1}{i} \cdot P(N_+ + N_- = i) = \mathbb{E}\left(\frac{1}{N_+ + N_-}\right) \approx \frac{1}{N(p_+ + p_-)}$$

where $\mathbb{E}$ denotes the mean value, and $\approx$ stands for asymptotic equality. Thus we obtain

$$\mathbf{V}(c,x) \approx \frac{1}{N} \frac{(c+1/c)^2 - 4x^2}{2(1+c^2)}$$

The other important aim would be to minimize the disturbed system instances, i.e. the cases when the outcomes were $(+1,+1)$, or $(-1,-1)$. The probability of having such outcomes is

$$p(c,x) = \left(1 + c^2\right)/2,$$

5.2. Optimal measurement strategy

If one wants to have a compromising strategy, then a possible way is to minimize the expression

$$\Psi(c,x) = A \cdot \mathbf{V}(c,x) + (1 - A) \cdot p(c,x),$$

where $A \in \mathbb{R}^+$ is a normalized parameter ($1 \geq A \geq 0$) which determines our trade-off strategy. If $A \approx 1$, then the aim is accuracy, while in the case of $A \approx 0$ we aim at minimal demolition. Figure 2 shows the substantial part of the loss function $\Psi(c,x)$ over the domain $(1- \leq x \leq 1), 0.2 \leq c \leq 1)$. Note that the function is symmetric to the $c = 0$ line, but it is indefinite at $c - 0$. It is seen that there is a definitely optimal value $c \approx 0.6$ for the initial state of the measurement qubit in case $A = 0.1$ that is the same for every x. In the case of $A = 0.9$, however, the minimum is taken at $c = 1$, i.e. at the complete demolition situation.

It is worth noting that the theoretical (analytical) analysis of $\Psi(c,x)$ is also possible but it requires the determination of the roots of a 4th order polynomial. This is a computationally hard problem in a parametric case, this motivated us to use the simple numerical analysis above.

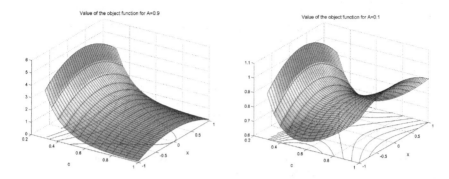

Fig. 2. The optimal measurement qubit state for different A values: more information ($A = 0.9$, left) versus more non-demolished system ($A = 0.1$, right)

6. Conclusions and Future work

An indirect measurement scheme is used in the paper which allows us to optimally select the elements of the measurement protocol according to a pre-defined weighting between the variance of the estimated state parameters and the ratio of the non-demolished copies of the system.

References

1. M. Nielsen and I. Chuang, *Quantum Computation and Quantum Information* (Cambridge University Press, Cambridge, 2000).
2. D. Petz, *Quantum Information Theory and Quantum Statistics* (Springer, Berlin and Heidelberg, 2008).
3. V. Belavkin, *Automatica and Remote Control* **44**, 178 (1983).
4. C. W. Helstrøm, *Quantum decision and estimation theory* (Academic Press, New York, 1976).
5. M. Paris and J. Reháček, *Quantum state estimation, Lect. Notes Phys.* **649** (Springer, Berlin, 2004).
6. E. Bagan, M. Ballester, R. Gill, A. Monras and R. Munoz-Tapia, *Physical Review A* **73**, p. 032301 (2006).
7. J. A. Bergou, U. Herzog and M. Hillery, Discrimination of quantum states, in *Quantum State Estimation, Lect. Notes Phys.* **649**, eds. M. Paris and J. Reháček (Springer, London, 2004) pp. 417–465.
8. M. Hayashi and K. Matsumoto, Asymptotic performance of optimal state estimation in quantum two level system, arXiv:quant-ph/0411073v2, (2006).
9. G. D'Ariano and H. Yuen, *Physical Review Letters* **76**, 2832 (1996).
10. J. Reháček, B. Englert and D. Kaszlikowski, *Physical Review A* **70**, p. 052321 (2004).
11. A. Jordan, B. Trauzettel and G. Burkard, Weak measurement of quantum dot spin qubits, arXiv:0706.0180v1[cond-mat.mes-hall], (2007).

12. G. Smith, A. Silberfarb, I. Deutsch and P. Jessen, Efficient quantum state estimation by continuous weak measurement and dynamical control, arXiv:quant-ph/0606115v1, (2006).
13. L. Diósi, Weak measurement in bipartite quantum mechanics, arXiv:quant-ph/0505075v1, (2007).
14. C. Hill and J. Ralph, Weak measurement and rapid state reduction in bipartite quantum systems, arXiv:quant-ph/0610156v2, (2006).
15. A. Korotkov and A. Jordan, *Physical Review Letters (arXiv:quant-ph/0610156v2)* **97**, p. 166805 (2006).
16. N. Khaneja and S. J. Glaser, Cartan decomposition of $su(2^n)$, constructive controllability of spin systems and universal quantum computing, arXiv:quant-ph/0010100v1, (2000).
17. R. Romano and D. D'Alessandro, Incoherent control and entanglement for two-dimensional coupled systems, arXiv:quant-ph/0510020v1, (2005).
18. D. Petz, K. Hangos and A. Magyar, *Jounal of Physics A: Mathematical and General* **40**, 7955 (2006).

POSITIVITY OF FREE CONVOLUTION SEMIGROUPS

M. SCHÜRMANN & S. VOß

Department of Mathematics and Computer Science, University of Greifswald
Friedrich-Ludwig-Jahn-Straße 15 a, 17487 Greifswald, Germany
e-mail: schurman@uni-greifswald.de

We show that the continuous convolution semigroups of states on dual groups (in the sense of D. Voiculescu) are precisely the convolution exponentials of conditionally positive linear functionals.

Keywords: free probability, freeness, Lévy processes, quantum stochastic processes

1. Introduction

A Lévy process on a (topological) group G is a stochastic process $X_t : E \to G$ taking values in G defined on some probability space E with independent, stationary increments $X_s^{-1} X_t$, $s \leq t$, and such that the process starts at the unit element e of G and X_t converges to e weakly for $t \searrow 0$ ⟩. The distributions of X_t form a convolution semigroup of probability measures on G which determines the process up to stochastic equivalence. When G is a Lie group the formula of Hunt for the generator of this convolution semigroup gives a description of Lévy processes on G. The 'dual' picture of this is obtained by replacing the group G and the probability space E by appropriate commutative *-algebras of functions on G and E respectively. The probability measure on E is replaced by a state on the *-algebra of functions on E. In a non-commutative or quantum version of Lévy processes groups are replaced by Hopf algebras and the notion of independence is the 'tensor' independence; cf.[7] It is also possible to consider free Lévy processes; cf.[2,3] Then tensor independence is replaced by free independence (also called freeness) in the sense of Voiculescu.[9] In[2] Hopf algebras are replaced by Voiculecu's dual groups (which can be formally thought of as Hopf *-algebras on which the usual tensor product is replaced by the free product). Again Lévy processes on dual groups are determined by their 1-dimensional distributions which form a free convolution semigroup of states on the underlying dual group $\mathcal{B}$. This convolution semigroup is a convolution exponential of its generator which is a conditionally positive, hermitian linear functional on $\mathcal{B}$ vanishing at the unit element $\mathbf{1}$ of $\mathcal{B}$. Conversely, in this paper we prove that the convolution exponential of such a linear functional always is a convolution semigroup of states. This establishes, in the dual group case, *Schoenberg's correspondence* between conditionally positive, hermitian linear functionals on a dual group and convolution semigroups of states on it; cf.[7] for the bialgebra case. By an inductive limit construction which can be done in complete analogy to the construction in[7] Corollary 1.9.7 in the tensor case, each convolution semigroup of states on $\mathcal{B}$ gives rise to a free Lévy process on $\mathcal{B}$. This means that our result gives a complete description of the free Lévy processes on a dual group up to (quantum) stochastic equivalence. In a forthcoming paper we will show how to construct a free Lévy process on a full Fock space from its generator; cf.[8] for the tensor case.

The idea of the proof of Schoenberg's correspondence for dual groups given in this paper is to construct first the Gelfand-Naimark-Segal representations of the convolution semigroup which then automatically yields its positivity. The GNS-representation is given on a full Fock space by approximating it by a sequence of convolutions of full Fock space creation, anihilation and preservation operators where this approximation is only to be understood in terms of distributions. What is actually approximated are the states of the convolution semigroup by taking the vacuum expectations of the approximating sequence of operators on full Fock space. In more detail, first a basic lemma (Prop. 3.2) on convolution semigroups on a coalgebra is stated[8] which says that the convolution exponential of a linear functional γ on the coalgebra is the limit of the nth convolution powers of $\delta + \frac{\gamma}{n} + R_n$ where δ denotes the counit and where R_n is dominated by a constant times $\frac{1}{n^2}$. Using the reduction to bialgebras arguments of,[2] this lemma is lifted to the case of convolution semigroups on dual groups (Prop. 3.3). Starting from a generator ψ, the GNS construction for ψ, analoguous to the bialgebra case, gives an inner product space and an 'additive Lévy process' on the full Fock space over this inner product space. For an interval $[0, t]$ and the sequence of equidistant partitions of $[0, t]$, a sequence of convolutions of increments of the additive Lévy process associated with the partitions, by the above mentioned lemma, converges to the convolution semigroup in the vacuum state (see Prop. 3.4). This is an approximation of the convolution semigroup by states, and thus proves the positivity of the convolution semigroup.

2. Preliminaries and statement of the results

All vector spaces are over the complex field $\mathbb{C}$. An algebra is a (complex) associative algebra. An algebra is called unital if it has a unit element $\mathbf{1}$. For two algebras $\mathcal{A}_{1/2}$ we define their *free* product $\mathcal{A}_1 \sqcup \mathcal{A}_2$ by the following universal property. There are algebra embeddings $\iota_{1/2} : \mathcal{A}_{1/2} \to \mathcal{A}_1 \sqcup \mathcal{A}_2$ such that the following holds. For each algebra $\mathcal{C}$ and each pair of algebra homomorphisms $j_{1/2} : \mathcal{A}_{1/2} \to \mathcal{C}$ there is a unique algebra homomorphism $j_1 \sqcup j_2 : \mathcal{A}_1 \sqcup \mathcal{A}_2 \to \mathcal{C}$ with $j_{1/2} = (j_1 \sqcup j_2) \circ \iota_{1/2}$. By this universal property the triplet $(\mathcal{A}_1 \sqcup \mathcal{A}_2, \iota_1, \iota_2)$ is uniquely determined up to isomorphisms. A realization of the free product of algbebras is obtained as follows. Denote by $\mathbb{A}$ the set of all alternating finite sequences of 1 and 2, i.e.

$$\mathbb{A} = \{(\epsilon_1, \ldots, \epsilon_n) \mid n \in \mathbb{N}, \epsilon_i = 1, 2, \epsilon_i \neq \epsilon_{i+1}\}.$$

As a vector space $\mathcal{A}_1 \sqcup \mathcal{A}_2$ is equal to the vector space direct sum $\bigoplus_{\epsilon \in \mathbb{A}} \mathcal{A}_\epsilon$ with $\mathcal{A}_\epsilon = \mathcal{A}_{\epsilon_1} \otimes \cdots \otimes \mathcal{A}_{\epsilon_n}$ for $\epsilon = (\epsilon_1 \ldots, \epsilon_n) \in \mathbb{A}$. The embeddings $\iota_{1/2}$ are given by identification of $\mathcal{A}_{1/2}$ and $\mathcal{A}_{(1/2)}$. The multiplication in $\mathcal{A}_1 \sqcup \mathcal{A}_2$ is given by

$$(a_1 \otimes \cdots \otimes a_n)(b_1 \otimes \cdots \otimes b_m) = \begin{cases} a_1 \otimes \cdots \otimes a_n \otimes b_1 \otimes \cdots \otimes b_m & if \ \epsilon_n \neq \delta_1 \\ a_1 \otimes \cdots \otimes a_n b_1 \otimes \cdots \otimes b_m & if \ \epsilon_n = \delta_1 \end{cases}$$

for $a_1 \otimes \cdots \otimes a_n \in \mathcal{A}_\epsilon$ and $b_1 \otimes \cdots \otimes b_m \in \mathcal{A}_\delta$, and we have

$$(j_1 \sqcup j_2)(a_1 \otimes \cdots \otimes a_n) = j_{\epsilon_1}(a_1) \cdots j_{\epsilon_n}(a_n).$$

The free product $\sqcup$ is the coproduct in the category of algebras; cf.[4] Similarly, the free product $\mathcal{A}_1 \sqcup_1 \mathcal{A}_2$ of unital algebras $\mathcal{A}_{1/2}$ can be introduced via the analogous universal property, and a realization of $\mathcal{A}_1 \sqcup_1 \mathcal{A}_2$ is the unital algebra $\mathbb{C}\mathbf{1} \oplus \mathcal{A}_1 \sqcup \mathcal{A}_2$ divided by the relations $\mathbf{1} = \mathbf{1}_1 = \mathbf{1}_2$ where $\mathbf{1}_{1/2}$ denotes the units in $\mathcal{A}_{1/2}$. For an algebra $\mathcal{A}$ denote by $\widetilde{\mathcal{A}}$ the unital algebra $\mathbb{C}\mathbf{1} \oplus \mathcal{A}$. Then in a natural way

$$\widetilde{\mathcal{A}_1} \sqcup_1 \widetilde{\mathcal{A}_2} \cong \widetilde{\mathcal{A}_1 \sqcup \mathcal{A}_2}. \tag{1}$$

Finally, notice that the tensor product of algebras is the coproduct in the category of unital commutative algebras.

A *dual semigroup* (cf.[10]) is a unital *-algebra equipped with mappings

$$\Delta : \mathcal{B} \to \mathcal{B} \sqcup_1 \mathcal{B}$$
$$\delta : \mathcal{B} \to \mathbb{C}$$

such that Δ, δ are unital *-algebra homomorphisms and

$$(\Delta \sqcup_1 \text{id}) \circ \Delta = (\text{id} \sqcup_1 \Delta) \circ \Delta$$
$$(\delta \sqcup_1 \text{id}) \circ \Delta = \text{id} = (\text{id} \sqcup_1 \delta) \circ \Delta.$$

We call $\mathcal{B}$ a *dual group* if there exists a *-algebra homomorphism $A : \mathcal{B} \to \mathcal{B}$ satisfying $M \circ (A \sqcup_1 \text{id}) \circ \Delta = \delta \mathbf{1} = M \circ (\text{id} \sqcup_1 A) \circ \Delta$. The map Δ is called the comultiplication, δ the counit and A the antipode of $\mathcal{B}$. Notice that for $\mathcal{B}_0 := \text{kern}\,\delta$ we have $\Delta \mathcal{B}_0 \subset \mathcal{B}_0 \sqcup \mathcal{B}_0$ where we use $\mathcal{B} = \mathbb{C}\mathbf{1} \oplus \mathcal{B}_0$ and $\mathcal{B} \sqcup_1 \mathcal{B} = \mathbb{C}\mathbf{1} \oplus \mathcal{B}_0 \sqcup \mathcal{B}_0$ which is (1). Denote by Δ_0 the restriction of Δ to $\mathcal{B}_0$, i.e. $\Delta_0 : \mathcal{B}_0 \to \mathcal{B}_0 \sqcup \mathcal{B}_0$. Then we can define a dual semigroup (cf.[2]) to be a *-algebra $\mathcal{B}_0$ equipped with a (not necessarily unital) *-algebra hommorphism $\Delta_0 : \mathcal{B}_0 \to \mathcal{B}_0 \sqcup \mathcal{B}_0$ such that

$$(\Delta_0 \sqcup \text{id}) \circ \Delta_0 = (\text{id} \sqcup \Delta_0) \circ \Delta_0$$
$$p_1 \circ \Delta_0 = \text{id} = p_2 \circ \Delta_0$$

where $p_{1/2}$ denote the projections from $\mathcal{B}_0 \sqcup \mathcal{B}_0 = \mathcal{B}_0 \oplus \mathcal{B}_0 \oplus (\mathcal{B}_0 \otimes \mathcal{B}_0) \oplus \cdots$ to the first and the second copy of $\mathcal{B}_0$ respectively. A linear functional φ on a unital algebra $\mathcal{B}$ is called *normalized* if $\varphi(\mathbf{1}) = 1$. A linear functional φ on a unital *-algebra $\mathcal{B}$ is called *positive* if $\varphi(b^*b) \geq 0$ for all $b \in \mathcal{B}$. A *state* is a normalized, positive linear functional. A linear functional $\psi : \mathcal{B} \to \mathbb{C}$ on a dual semigroup $\mathcal{B}$ is called *conditionally positive* if $\psi(b^*b) \geq 0$ for all $b \in \mathcal{B}_0$ and *hermitian* if $\psi(b^*) = \overline{\psi(b)}$ for all $b \in \mathcal{B}$. A *generator* on $\mathcal{B}$ is a hermitian, conditionally positive linear functional on $\mathcal{B}$ with $\psi(\mathbf{1}) = 0$. For two normalized linear functionals φ_1, φ_2 on unital algebras $\mathcal{B}_1, \mathcal{B}_2$ the *free product* $\varphi_1 \cdot \varphi_2$ of φ_1 and φ_2 is the normalized linear funtional on $\mathcal{B}_1 \sqcup_1 \mathcal{B}_2$ with

$$\varphi_1 \cdot \varphi_2(b_1 \cdots b_m) = 0 \tag{2}$$

if $m \geq 1$, $(\epsilon_1, \ldots, \epsilon_m) \in \mathbb{A}$, $b_i \in \mathcal{B}_{\epsilon_i}$, $\varphi_{\epsilon_i}(b_i) = 0$, $i = 1, \ldots, n$. The following axioms of a 'natural' product are satisfied (cf.[1,2,5,6])

$$\varphi_1 \cdot \varphi_2 \circ \iota_{1/2} = \varphi_{1/2} \tag{3}$$
$$\varphi_1 \cdot (\varphi_2 \cdot \varphi_3) = (\varphi_1 \cdot (\varphi_2 \cdot \varphi_3)) \tag{4}$$
$$(\varphi_1 \circ j_1) \cdot (\varphi_2 \circ j_2) = (\varphi_1 \cdot \varphi_2) \circ (j_1 \sqcup j_2) \tag{5}$$

where $j_{1/2} : \mathcal{C}_{1/2} \to \mathcal{B}_{1/2}$ are unital algebra homomorphisms, $\mathcal{C}_{1/2}$ unital algebras.

For two normalized linear functionals φ_1, φ_2 on a dual semigroup $\mathcal{B}$ we define their convolution product by

$$\varphi_1 \star \varphi_2 := (\varphi_1 \cdot \varphi_2) \circ \Delta. \tag{6}$$

A *continuous convolution semigroup* on a dual semigroup is a family $(\varphi_t)_{t\geq 0}$ of linear functionals φ_t on $\mathcal{B}$ such that

$$\varphi_s \star \varphi_t = \varphi_{s+t} \ \forall s,t \in \mathbb{R}_+$$
$$\lim_{t\to 0+} \varphi_t(b) = \delta(b) \ \forall b \in \mathcal{B}.$$

The following is the main result of the present paper. It it a corollary to Theorem 2.2 below which will be proved in Section 3.

Theorem 2.1. *(a) Let $(\varphi_t)_{t\geq 0}$ be a continuous convolution semigroup of states on the dual semigroup $\mathcal{B}$. Then, for each $b \in \mathcal{B}$,*

$$\frac{1}{t}(\varphi_t(b) - \delta(b)) \tag{7}$$

converges to a limit $\psi(b)$ as $t \to 0+$ and ψ is a generator on $\mathcal{B}$.
(b) Let ψ be a generator on $\mathcal{B}$. Then there exists a unique continuous convolution semigroup of states $(\varphi_t)_{t\geq 0}$ on $\mathcal{B}$ such that

$$\psi(b) = \lim_{t\to 0+} \frac{1}{t}(\varphi_t(b) - \delta(b)). \tag{8}$$

For a vector space $\mathcal{V}$ we denote by $\mathrm{T}(\mathcal{V})$ the *tensor algebra* over $\mathcal{V}$ which is characterized by the following universal property. For a unital algebra $\mathcal{A}$ and a linear map $R : \mathcal{V} \to \mathcal{A}$ there is a unique unital algebra homomorphism $\mathrm{T}(R) : \mathrm{T}(\mathcal{V}) \to \mathcal{A}$ such that $\mathrm{T}(R) \circ \iota = R$ where ι is a fixed vector space embedding of $\mathcal{V}$ into $\mathrm{T}(\mathcal{V})$. A realization of $\mathrm{T}(\mathcal{V})$ is given by

$$\mathbb{C}\mathbf{1} \oplus \mathcal{V} \oplus (\mathcal{V} \otimes \mathcal{V}) \oplus (\mathcal{V} \otimes \mathcal{V} \otimes \mathcal{V}) \oplus \cdots \tag{9}$$

with ι the obvious embedding of $\mathcal{V}$. Multiplication is given by the tensor product. We denote by $\mathrm{T}_0(\mathcal{V})$ the non-unital version, i.e. $\mathrm{T}_0(\mathcal{V}) = \mathcal{V} \oplus (\mathcal{V} \otimes \mathcal{V}) \oplus (\mathcal{V} \otimes \mathcal{V} \otimes \mathcal{V}) \oplus \cdots$. Similarly, the symmetric tensor algebra $\mathrm{S}(\mathcal{V})$ over $\mathcal{V}$ is characterized by the analogous universal property but with unital algebras replaced by unital commutative algebras. A realization is again given by (9) but with the tensor product replaced by the symmetric tensor product. We have $\mathrm{T}(\mathcal{V}_1 \oplus \mathcal{V}_2) \cong \mathrm{T}(\mathcal{V}_1) \sqcup_1 \mathrm{T}(\mathcal{V}_2)$, $\mathrm{T}_0(\mathcal{V}_1 \oplus \mathcal{V}_2) \cong \mathrm{T}_0(\mathcal{V}_1) \sqcup \mathrm{T}_0(\mathcal{V}_2)$ and $\mathrm{S}(\mathcal{V}_1 \oplus \mathcal{V}_2) \cong \mathrm{S}(\mathcal{V}_1) \otimes \mathrm{S}(\mathcal{V}_2)$ in a natural way. If $\mathcal{V}$ carries an involution (i.e. antilinear, selfinverse mapping) $v \mapsto v^*$ then $\mathrm{T}(\mathcal{V})$ and $\mathrm{S}(\mathcal{V})$ become *-algebras in the obvious ways.

It was shown in[2] that for each pair $\mathcal{B}_1, \mathcal{B}_2$ of *-algebras there exists a linear mapping

$$\sigma_{\mathcal{B}_1,\mathcal{B}_2} : \mathcal{B}_1 \sqcup \mathcal{B}_2 \to \mathrm{S}(\mathcal{B}_1) \otimes \mathrm{S}(\mathcal{B}_2) \tag{10}$$

such that

$$\widetilde{\varphi_1} \cdot \widetilde{\varphi_2} \lceil \mathcal{B}_1 \sqcup \mathcal{B}_2 = (S(\varphi_1) \otimes S(\varphi_2)) \circ \sigma_{\mathcal{B}_1, \mathcal{B}_2}. \quad (11)$$

Here $\widetilde{\varphi_{1/2}}$ denotes the normalized linear functionals on $\widetilde{\mathcal{B}_{1/2}} = \mathbb{C}1 \oplus \mathcal{B}_{1/2}$ with $\widetilde{\varphi_{1/2}} \lceil \mathcal{B}_{1/2} = \varphi_{1/2}$, where we again use $\widetilde{\mathcal{B}_1 \sqcup_1 \mathcal{B}_2} = \mathbb{C}1 \oplus \mathcal{B}_1 \sqcup \mathcal{B}_2$. Moreover, for a dual semigroup $\mathcal{B}$, the space $S(\mathcal{B}_0)$ becomes a commutative *-bialgebra with comultiplication and counit given by $S(\sigma \circ \Delta_0)$ and $S(0)$, and

$$\varphi_1 \star \varphi_2 = S(\varphi_1 \lceil \mathcal{B}_0) \star S(\varphi_2 \lceil \mathcal{B}_0) \lceil \mathcal{B} \quad (12)$$

where the second $\star$ is with respect to the comultiplication $S(\sigma \circ \Delta_0)$. Here $\mathcal{B}$ is identified with $\mathbb{C}1 \oplus \mathcal{B}_0 \subset S(\mathcal{B}_0)$. For a continuous convolution semigroup $(\varphi_t)_{t \geq 0}$ of normalized linear functionals on $\mathcal{B}$ we have that $(S(\varphi_t \lceil \mathcal{B}_0))_{t \geq 0}$ is a convolution semigroup of normalized linear functionals on $S(\mathcal{B}_0)$. Moreover, $\lim_{t \to 0+} S(\varphi_t \lceil \mathcal{B}_0)(x) = S(0)(x)$ for all $x \in S(\mathcal{B}_0)$ and this convolution semigroup is continuous, too. It is well-known (see[2]) that the fundamental theorem on coalgebras guarantees the existence of

$$\Psi(x) := \lim_{t \to 0+} \frac{1}{t}(S(\varphi_t \lceil \mathcal{B}_0)(x) - S(0)(x)) \quad (13)$$

for all $x \in S(\mathcal{B}_0)$ and that

$$S(\varphi_t \lceil \mathcal{B}_0)(x) = \sum_{k=0}^{\infty} \frac{t^k \Psi^{\star k}(x)}{k!}. \quad (14)$$

Moreover, $\Psi = D(\psi)$ with $\psi = \Psi \lceil \mathcal{B}$ and $D(\psi)(1) = 0$,

$$D(\psi)(b_1 \otimes \cdots \otimes b_m) = \begin{cases} 0 & \text{if } m \neq 1 \\ \psi(b_1) & \text{if } m = 1 \end{cases} \quad (15)$$

$m \in \mathbb{N}, b_1, \ldots, b_m \in \mathcal{B}_0$. We will write

$$S(\varphi_t \lceil \mathcal{B}_0) = \exp_\star(tD(\psi))$$

and have

$$\varphi_t = \exp_\star(tD(\psi)) \lceil \mathcal{B}. \quad (16)$$

Conversely, given a linear funtional ψ on $\mathcal{B}$ with $\psi(1) = 0$ we have that (16) is a continuous convolution semigroup of normalized linear functionals on $\mathcal{B}$. It follows that Theorem 2.1 is proved if we can show

Theorem 2.2. *For a linear functional ψ on $\mathcal{B}$ the following are equivalent*
(i) ψ is a generator.
(ii) $\exp_\star(tD(\psi)) \lceil \mathcal{B}$ is a continuous convolution semigroup of states.

3. Proof of Theorem 2.2

In the sequel, we write $\exp_\star \psi$ for $\exp_\star D(\psi) \lceil \mathcal{B}$. For linear functionals $\psi_1, \ldots, \psi_k$ on $\mathcal{B}$, $\psi_i(1) = 0$, we put

$$(\psi_1 \uplus \cdots \uplus \psi_k)(b) = \frac{\partial^k}{\partial t_1 \cdots \partial t_n}(\delta + t_1 \psi_1) \star \cdots \star (\delta + t_k \psi_k)(b)|_{t_1 = \cdots = t_k = 0} \quad (1)$$

and $\psi^{\uplus 0} = \delta$, $\psi^{\uplus n} = \psi \uplus \cdots \uplus \psi$ (n times). It is not difficult to see that

Proposition 3.1.

$$(\exp_\star \psi)(b) = \sum_{k=0}^{\infty} \frac{\psi^{\uplus k}}{k!}(b), \quad b \in \mathcal{B}. \quad (2)$$

Let $(\mathcal{C}, \Delta, \delta)$ be a coalgebra. Using the fundamental theorem on coalgebras, one proves (see[8])

Proposition 3.2. *Let γ be a linear functional on $\mathcal{C}$, and let $R_n : \mathcal{C} \to \mathbb{C}$, $n \in \mathbb{N}$, be linear and such that for each $b \in \mathcal{C}$ there exists a constant $C_b > 0$ with $|R_n(b)| \leq \frac{1}{n^2} C_b$. Then*

$$\lim_{n \to \infty} (\delta + \frac{\gamma}{n} + R_n)^{\star n}(b) = (\exp_\star \gamma)(b) \; \forall b \in \mathcal{C}.$$

Proposition 3.3. *Let $\mathcal{B}$ be a dual semigroup. Let γ be a linear functional on $\mathcal{B}$, $\gamma(1) = 0$, and let $R_n : \mathcal{B} \to \mathbb{C}$, $n \in \mathbb{N}$, be linear, $\mathbb{R}_n(1) = 0$, such that for each $b \in \mathcal{B}$ there exists a constant $C_b > 0$ with $|R_n(b)| \leq \frac{1}{n^2} C_b$. Then*

$$\lim_{n \to \infty} (\delta + \frac{\gamma}{n} + R_n)^{\star n}(b) = (\exp_\star \gamma)(b) \; \forall b \in \mathcal{B}$$

Proof. It follows from the assumptions that

$$S(\delta + \frac{\gamma}{n} + R_n) = S(0) + \frac{D(\gamma)}{n} + T_n \quad (3)$$

with T_n linear, $T_n(1) = 0$, and such that for each $x \in S(\mathcal{B}_0)$ there is a constant $\tilde{C}_x > 0$ with $|T_n(x)| \leq \frac{1}{n^2} \tilde{C}_x$ for all $x \in S(\mathcal{B}_0)$. An application of Proposition (3.2) to the right hand side of (3) yields the result. □

For an inner product space K define the full Fock space over K to be the inner product space

$$\Gamma(K) := \mathbb{C} \oplus K \oplus (K \otimes K) \oplus (K \otimes K \otimes K) \oplus \cdots = \bigoplus_{n=0}^{\infty} K^{\otimes n} \quad (4)$$

where the direct sum is in the algebraic sense. The vector $\Omega = (1, 0, \ldots)$ in $\Gamma(K)$ is called the *vacuum vector*. For an inner product space K we denote by $\mathcal{L}^a(K)$ the *-algebra of adjointable linear mappings from K to K. For $u \in K$, $r \in \mathcal{L}^a(K)$, define $A^*(u), A(u), \Lambda(r) \in \mathcal{L}^a(\Gamma(K))$ by

$$A^*(u)u_1 \otimes \cdots \otimes u_n = u \otimes u_1 \otimes \cdots \otimes u_n, n \geq 1; \; A(u)\Omega = u$$

$$A(u)u_1 \otimes \cdots \otimes u_n = \langle u, u_1 \rangle u_2 \otimes \cdots \otimes u_n, n \geq 2; A(u)u_1 = \langle u, u_1 \rangle \Omega; \; A(u)\Omega = 0$$

$$\Lambda(r)u_1 \otimes \ldots \otimes u_n = (ru_1) \otimes u_2 \otimes \cdots u_n, n \geq 1; \; \Lambda(r)\Omega = 0$$

For a linear subspace M in K with $K = M \oplus M^\perp$ denote by $\mathcal{A}(M)$ the *-subalgebra of $\mathcal{L}^a(\Gamma(K))$ generated by the operators $A^*(u)$, $A(u)$ and $\Lambda(r \oplus 0)$, $u \in M$, $r \in \mathcal{L}^a(M)$. Then it is well-known (see e.g.[11]) that for $K = K_1 \oplus \cdots \oplus K_l$, in the vacuum state, the algebras $\mathcal{A}(K_1), \ldots, \mathcal{A}(K_l)$ are freely independent unital *-subalgebras of $\mathcal{L}^a(\Gamma(K))$, i.e.

$$\langle \Omega, A_1 \cdots A_m \Omega \rangle = 0$$

for each choice of $m \in \mathbb{N}$, $(\epsilon_1, \ldots, \epsilon_m) \in \mathbb{A}$, and $A_i \in \mathcal{A}(K_{\epsilon_i})$ with $\langle \Omega, A_i \Omega \rangle = 0$, $i = 1, \ldots, m$.

Consider the embedding

$$\mathrm{E} : \mathcal{B}_0 \sqcup \mathcal{B}_0 \to \mathrm{T}(\mathcal{B}_0) \sqcup_1 \mathrm{T}(\mathcal{B}_0) = \mathbb{C}\mathbf{1} \oplus \mathrm{T}_0(\mathcal{B}_0) \sqcup \mathrm{T}_0(\mathcal{B}_0)$$

of vector spaces given by the inclusion $\mathcal{B}_0 \subset \mathrm{T}_0(\mathcal{B}_0)$. Then $(\mathrm{T}(\mathcal{B}_0), \mathrm{T}(\mathrm{E} \circ \Delta_0), \mathrm{T}(0))$ forms a dual semigroup. Let ψ be a generator on $\mathcal{B}$. By putting

$$\mathcal{N}_\psi := \{b \in \mathcal{B} \,|\, \psi((b - \delta(b)\mathbf{1})^*(b - \delta(b)\mathbf{1})) = 0\}$$

$$\eta : \mathcal{B} \to \mathcal{B}/\mathcal{N}_\psi \text{ the canonical mapping}$$

$$\rho(b)\eta(c) = \eta(bc) - \eta(b)\delta(c), \; b, c \in \mathcal{B}$$

we obtain an inner product space $D = \mathcal{B}/\mathcal{N}\psi$, a surjective linear mappping $\eta : \mathcal{B} \to D$, and a *-representation ρ of $\mathcal{B}$ on D such that

$$\rho(b)\eta(c) - \eta(bc) + \eta(b)\delta(c) = 0 \tag{5}$$

$$\delta(b)\psi(c) - \psi(bc) + \psi(b)\delta(c) = -\langle \eta(b^*), \eta(c) \rangle. \tag{6}$$

This means that η is a 1-cocycle with respect to the $\mathcal{B}$-bimodule structure $b.\xi.c = \rho(b)(\xi)\delta(c)$ of D and $(b, c) \mapsto -\langle \eta(b^*), \eta(c) \rangle$ is the coboundary of ψ with respect to the $\mathcal{B}$-bimodule structure $b.\xi.c = \delta(b)\delta(c)\xi$ of D; cf.[7] Next put

$$I_{st}(b) = A_{st}(\eta(b^*)) + \Lambda_{st}(\rho(b)) + A^*_{st}(\eta(b)) + (t-s)\psi(b), \tag{7}$$

$b \in \mathcal{B}_0$, where

$$A_{st}(\xi) = A(\chi_{[s,t]} \otimes \xi)$$
$$\Lambda_{st}(\alpha) = \Lambda(\chi_{[s,t]} \otimes \alpha)$$
$$A^*_{st}(\xi) = A^*(\chi_{[s,t]} \otimes \xi)$$

$\xi \in D$, $\alpha \in \mathcal{L}^a(D)$. Then the quantum random variables

$$T(I_{st}) : T(\mathcal{B}_0) \to \mathcal{L}^a(\Gamma(L^2(\mathbb{R}_+, D)))$$

(where $L^2(\mathbb{R}_+, D) = \{f : \mathbb{R}_+ \to D \mid f \text{ measurable}, \int_0^\infty \|f(t)\|^2 dt < \infty\}$) are freely independent for disjoint intervals, i.e. for $0 \le t_1 \le \cdots \le t_{n+1}$ the *-algebras $T(I_{t_1,t_2})(T(\mathcal{B}_0)), \ldots, T(I_{t_n,t_{n+1}})(T(\mathcal{B}_0))$ are freely independent in the vacuum state of $\mathcal{L}^a(\Gamma(L^2(\mathbb{R}_+, D)))$. This means that

$$\langle \Omega, T(I_{0,\frac{t}{n}}(x_1) \sqcup_1 \cdots \sqcup_1 T(I_{\frac{n-1}{n}t,t}(x_n)\Omega \rangle = \left(\Phi_{\frac{t}{n}} \star \ldots \star \Phi_{\frac{t}{n}}\right)(x_1 \otimes \cdots \otimes x_n)$$

for all $x_1, \ldots, x_n \in T(\mathcal{B}_0)$ where

$$\Phi_{\frac{t}{n}}(x) = \langle \Omega, T(I_{0,\frac{t}{n}})(x)\Omega \rangle$$

is the distribution of $T(I_{0,\frac{t}{n}})$ and where $\Phi_{\frac{t}{n}} \star \ldots \star \Phi_{\frac{t}{n}}$ denotes the n-fold free product of $\Phi_{\frac{t}{n}}$ on $T(\mathcal{B}_0) \sqcup_1 \cdots \sqcup_1 T(\mathcal{B}_0) = \mathbb{C}\mathbf{1} \oplus \tilde{T}_0(\mathcal{B}_0) \sqcup \cdots \sqcup T_0(\mathcal{B}_0)$. It follows that

$$\Phi_{\frac{t}{n}}^{\star n}(x) = \langle \Omega, T(I_{0,\frac{t}{n}}) \star \cdots \star T(I_{\frac{n-1}{n}t,t})(x)\Omega \rangle, x \in T(\mathcal{B}_0).$$

Denote by $M : T(\mathcal{B}_0) \to \mathcal{B}$ the *-algebra homomorphism $T(\kappa)$ where κ is the embedding of the vector space $\mathcal{B}_0$ into the unital algebra $\mathcal{B} = \mathbb{C}\mathbf{1} \oplus \mathcal{B}_0$.

Proposition 3.4. *Let* $\Phi_t(x) = \langle \Omega, T(I_{0t})(x)\Omega \rangle$ *be the distribution of* $T(I_{0t})$, $t > 0$. *Then there are constants* $C_x > 0$, $x \in T(\mathcal{B}_0)$, *such that*

$$\left|(\Phi_t - (T(0) + t\,\psi \circ M))(x)\right| \le t^2 C_x \qquad (8)$$

for all t in some bounded interval of $\mathbb{R}_+$.

Proof. It suffices to prove the existence of C_x for x of the form $x = b_1 \otimes \cdots \otimes b_k$, $b_i \in \mathcal{B}_0$, $k \in \mathbb{N}$. We have for $k = 1$

$$\Phi_t(b) = \langle \Omega, I_{0t}(b)\Omega \rangle = t\,\psi(b), \qquad (9)$$

and (8) holds. For $k \ge 2$

$$\Phi_t(b_1 \otimes \cdots \otimes b_k) = \langle \Omega, T(I_{0t})(b_1 \otimes \cdots \otimes b_k)\Omega \rangle$$
$$= \langle \Omega, I_{0t}(b_1) \cdots I_{0t}(b_n)\Omega \rangle$$

is a polynomial in t of degree $\le n$. The constant term of this polynomial is 0, the linear term equals

$$\langle \Omega, A_{0t}(\eta(b_1^*))\Lambda_{0t}(\rho(b_2)) \cdots \Lambda_{0t}(\rho(b_{n-1}))A_{0t}^*(\eta(b_n))\Omega \rangle, \qquad (10)$$

and all remaining terms are of order ≥ 2. This holds because each term with more than 2 creation/annihilation operators is 0 or of order ≥ 2. Since (10) is equal to $t\psi(b_1\cdots b_n)$ it follows that

$$\Phi_t(b_1\otimes\cdots\otimes b_k) = t\psi(b_1\cdots b_n) + t^2\, L_t(x) \tag{11}$$

with $L_t(x)$ bounded for fixed x and for t in some bounded interval which proves the proposition. □

Now by Proposition 3.3

$$\lim_{n\to\infty}\Phi_{\frac{t}{n}}^{\star n}(x) = \exp_\star(t\,\psi\circ M)(x),\ x\in \mathrm{T}(\mathcal{B}_0)$$

where convolution is with respect to $\mathrm{T}(\mathrm{E}\circ\Delta_0)$.

Proposition 3.5. *For two normalized linear functionals φ_1 and φ_2 on $\mathcal{B}$ we have*

$$(\varphi_1\circ M)\star_{\mathrm{T}(\mathrm{E}\circ\Delta_0)}(\varphi_2\circ M) = (\varphi_1\star_\Delta\varphi_2)\circ M. \tag{12}$$

Proof. We have

$$(M\sqcup_1 M)\circ\mathrm{T}(\mathrm{E}\circ M) = \Delta\circ M$$

because both sides are algebra homomorphisms and for $b\in\mathcal{B}_0$

$$(M\sqcup_1 M)\circ\mathrm{T}(\mathrm{E}\circ\Delta_0)(b) = (M\sqcup_1 M)\Delta_0 b$$
$$= \Delta_0 b = \Delta b = \Delta(Mb).$$

Using (5), we obtain

$$(\varphi_1\circ M)\star(\varphi_2\circ M) = \bigl((\varphi_1\circ M)\cdot(\varphi_2\circ M)\bigr)\circ\mathrm{T}(\mathrm{E}\circ\Delta_0)$$
$$= (\varphi_1\cdot\varphi_2)\circ(M\sqcup_1 M)\circ\mathrm{T}(\mathrm{E}\circ\Delta_0)$$
$$= (\varphi_1\cdot\varphi_2)\circ\Delta\circ M$$
$$= (\varphi_1\star\varphi_2)\circ M.\qquad\square$$

It follows from this proposition that

$$(\psi\circ M)\uplus_{\mathrm{T}(\mathrm{E}\circ\Delta_0)}\cdots\uplus_{\mathrm{T}(\mathrm{E}\circ\Delta_0)}(\psi_k\circ M) = (\psi_1\uplus_\Delta\cdots\uplus_\Delta\psi_k)\circ M$$

and thus

$$\exp_\star(t\,\psi\circ M)(x) = \sum_{k=0}^{\infty}\frac{(t\,\psi\circ M)^{\uplus k}}{k!}$$
$$= \exp_\star(t\,\psi)(Mx).$$

Now we have, using that $T(I_{0,\frac{t}{n}}) \star \cdots \star T(I_{\frac{n-1}{n}t,t})$ is a *-algebra homomorphism,

$$\begin{aligned}
\exp_\star(t\,\psi)(b^*b) &= \exp_\star(t\,\psi \circ M)(b^* \otimes b) \\
&= \lim_{n\to\infty} \Phi^{\star n}_{\frac{t}{n}}(b^* \otimes b) \\
&= \lim_{n\to\infty} \langle \Omega, T(I_{0,\frac{t}{n}}) \star \cdots \star T(I_{\frac{n-1}{n}t,t})(b^* \otimes b)\Omega \rangle \\
&= \lim_{n\to\infty} \|T(I_{0,\frac{t}{n}}) \star \cdots \star T(I_{\frac{n-1}{n}t,t})(b)\Omega\|^2 \geq 0
\end{aligned}$$

which proves that (i) implies (ii) in Theorem 2.2. Since (ii) implies (i) by differentiating, this completes the proof of Theorem 2.2 and thus of Theorem 2.1.

References

1. A. Ben Ghorbal, M. Schürmann: Non-commutative notions of stochastic independence. Math. Proc. Cambridge Philos. Soc. **133**, 531-561 (2002)
2. A. Ben Ghorbal, M. Schürmann: Quantum Lévy processes on dual groups. Math. Z. **251**, 147-165 (2005)
3. P. Biane: Processes with free increments. Math. Z. **227**, 143-174 (1998)
4. U. Franz: Lévy processes on quantum groups and dual groups. In: Schürmann, M., Franz, U. (eds.) Quantum independent increment processes II. (Lect. Notes Math., vol. 1866). Berlin Heidelberg New York, Springer 2006
5. N. Muraki: The five independences as quasi-universal products. Infin. Dimens. Anal. Quantum Probab. Rel. Top. **4**, 39-58 (2001)
6. N. Muraki: The five independences as natural products. Greifswald Preprint Series Math. no. 3/2002, 2002
7. M. Schürmann: White Noise on Bialgebras. Springer Lect. Notes Math., vol. 1544. New York Berlin Heidelberg, Springer 1993
8. M. Schürmann, M. Skeide, S. Volkwardt: Transformations of Lévy processes. Greifswald Preprint Series Math. no. 13/2007, 2007
9. D. Voiculescu: Symmetries of some reduced free product C*-algebras. In: Araki, H., Moore, C.C., Stratila, S., Voiculescu, D. (eds.) Operator algebras and their connection with toplogy and ergodic theory. Proceedings, Busteni 1983. (Lect. Notes Math., vol. 1132). Berlin Heidelberg New York, Springer 1985
10. D. Voiculescu: Dual algebaric structures on operator algebras related to free products. J. Operator Theory **17**, 85-98 (1987)
11. D. Voiculescu, K. Dykema, A. Nica: Free random variables. CRM Monograph Series no. 1, Amer. Math. Soc., Providence 1992

QUASI-FREE STOCHASTIC INTEGRALS AND MARTINGALE REPRESENTATION

W J. SPRING

Quantum Information and Probability Group,
School of Computer Science, University of Hertfordshire,
Hatfield, Herts, AL10 9AB, UK
*E-mail: j.spring@Herts.ac.uk

We discuss the construction of quantum stochastic integrals over the positive plane. Non-commutative representations for the quasi-free CAR and CCR settings are presented extending results obtained previously for martingales over the two parameter plane.

Keywords: Martingale; Representation; Isometry; Quantum Stochastic Integral.

1. Introduction

Following the work commenced by the London based BSW group (Barnett, Streater and Wilde) it was natural to conjecture the possibility of a corresponding two parameter theory and martingale representation for both the Clifford and quasi-free settings. Quantum analogues[1–4] of the classical Ito integral[5] and the Wong and Zakai integral[6] have been explored over the positive plane. In this paper we consider martingale representations for the CAR and CCR quasi-free setting. A 'quantum stochastic base' $(\mathcal{H}, \mathcal{A}, g, (\mathcal{A}_z), Z)$ is introduced in which $\mathcal{H}$ is an underlying Hilbert space, $\mathcal{A}$ a von Neumann algebra, g a guage, $(\mathcal{A}_z)$ a filtration of associated von Neumann algebras and Z an associated parameter set. We note that in place of the von Neumann algebra $\mathcal{A}$ and its corresponding filtration $(\mathcal{A}_z)$ one may work with C*-algebras $\mathcal{U}$ and their corresponding filtration $(\mathcal{U}_z)$ or the Hilbert space $\mathcal{H}$ and its underlying filtration $(\mathcal{H}_z)$.

2. Quasi-Free Construction

2.1. *The Quasi-Free CAR Construction*

In terms of a stochastic base for the CAR construction we work with $(\mathcal{F}(\mathcal{H}_R), \mathcal{A}, \omega, (\mathcal{A}_z), \mathbb{R}_+^2)$ in which $\mathcal{F}(\mathcal{H}_R)$ represents $\mathcal{F}_0(\mathcal{H}_R) \otimes \mathcal{F}_0(\mathcal{H}_R)$, the tensor product of the anti-symmetric Fermi-Fock space $\mathcal{F}_0(\mathcal{H}_R)$ with itself,[3] in which $\mathcal{H}_R = L^2(R)$ and $R \subseteq \mathbb{R}_+^2$, a closed square with lower left corner fixed at the origin and sides parallel to the axes. Following[9] we define creation and annihilation operators b^* and b satisfying the CAR (Canonical Anticommutation Relations) properties, on $\mathcal{F}(\mathcal{H}_R)$ via

$$b^*(f) = b_0^*((1-\rho)^{1/2} f) \otimes \mathbb{I} + \Gamma(-1) \otimes b_0(\rho^{1/2}\overline{f})$$

and

$$b(f) = b_0((1-\rho)^{1/2} f) \otimes \mathbb{I} + \Gamma(-1) \otimes b_0^*(\rho^{1/2}\overline{f})$$

with $\Gamma(-1)\Omega_0 = \Omega_0$ on $\mathcal{H}_0 = \mathbb{C}$, $\otimes^n(-1)$ on $\otimes^n \mathcal{H}$ and ρ a measurable function on R with $0 < \rho < 1$. The von Neumann algebra $\mathcal{A} = \mathcal{U}''$ may be defined via $\mathcal{U}$ the C^*-algebra generated by the fermion creation and annihilation operators as f varies in $L^2(R)$. For our gauge we work with $\omega : \mathcal{U} \longrightarrow \mathbb{C}$ by $\omega(u) = (u\Omega, \Omega)$, defining a guage-invariant quasi-free state on $\mathcal{U}$ in which

$$w(b^*(f)) = w(b(g)) = 0$$

and

$$w(b^*(f)b(g)) = (\rho f, g)_{L^2(R)}$$

Various filtrations $\mathcal{U}_z, \mathcal{A}_z$ and $\mathcal{H}_z$ may now be developed (with $z \in R$) in $\mathcal{U}, \mathcal{A}$ and $\mathcal{H}$ respectively together with ω-invariant conditional expectations on $\mathcal{U}$ and $\mathcal{A}$ and orthogonal projections on $\mathcal{H}$. We will work with $\mathcal{H}$. As in the Clifford case,[1] with $\psi(z)$, the families $\{b^\#(\chi_{R_z} u) : z \in R\}$ form centred martingales in which $b^\#$ denotes b or b^*, and R_z, denotes a rectangle with lower left corner fixed at the origin, sides parallel to the axes and upper right corner at $z \in \mathbb{R}_+^2$. For further details we defer to.[2,3]

2.2. *The Quasi-Free CCR Construction*

For the CCR construction we take $\mathcal{F}(\mathcal{H}_R)$ to represent the tensor product of the symmetric Boson-Fock space over $\mathcal{H}_R = L^2(R)$ with itself. Creation and annihilation operators $a^*(f)$ and a(f) satisfying the CCR properties are defined on $\mathcal{F}(\mathcal{H}_R)$ by

$$a^*(f) = a_0^*((1+\tau)^{1/2}f) \otimes \mathbb{I} + \mathbb{I} \otimes a_0(\tau^{1/2}\overline{f})$$

and

$$a(f) = a_0((1+\tau)^{1/2}f) \otimes \mathbb{I} + \mathbb{I} \otimes a_0^*(\tau^{1/2}\overline{f})$$

with τ a measurable function on $\mathbb{R}_+^2$ such that $\tau \in L^\infty_{loc}(\mathbb{R}_+^2)$, $\tau(z) > 0$ and a_0^*, a_0 denote the Fock space creation and annihilation operators. The guage invariant quasi-free state ω is defined as for the CAR case above but with Ω denoting the vector $\Omega_0 \otimes \Omega_0$ in which Ω_0 denotes the boson-Fock no-particle vector rather than the fermi-Fock no-particle vector, and $f, g \in \mathcal{D}(\tau^{1/2}) = \{f : \tau^{1/2}f \in L^2(\mathbb{R}_+^2)\}$. Filtrations for this discussion will be denoted by $(\mathcal{F}(\mathcal{H}_R)_z)$ the closure of the unital polynomial *-algebra generated the boson creation and annihilation operators $a^*(f)$ and $a(f)$ on $\mathcal{F}(\mathcal{H}_R)$ as f varies in $L^2(\mathbb{R}_+^2)$ with support in R_z. As above, the families $\{a^\#(\chi_{R_z}u)\Omega : z \in R\}$ in which $a^\#$ denotes a or a^* form centred martingales. For further details we defer to.[2]

Remark For properties of the CAR and CCR quasi-free state ω we refer the interested reader to Powers and Størmer, Brattelli and Robinson and earlier papers.[2,3,8,10]

3. Quantum Stochastic Integrals

We consider Quasi-free stochastic integrals for $\mathcal{H}$. Such integrals may also be realised for $\mathcal{U}$ and $\mathcal{A}$. Points in the parameter space $\mathbb{R}_2^+$ in which $A \prec B$ or $A \wedge B$ form[a] posets in $\mathbb{R}_+^2$,[6] and lead to two different types of stochastic integral, referred to as type I and type II integrals. Type I quantum stochastic integrals are quantum analogues of the Itô integral.[5] We define these in terms of elementary adapted processes of the form $h(z) = \chi_\Delta a$. Here Δ is a rectangle in R, $a \in \mathcal{H}_{z'}$ and $z' = inf\Delta$. The type I integral over R_z is defined to be

$$\iint_{R_z} db_{z'}^\# h(z') = b^\#(\chi_\Delta \chi_{R_z})a$$

We extend by linearity to $\mathcal{H}$-valued simple adapted processes.

---

[a] $A \prec B$ denotes points $z = (x,y) \in A$, $z' = (x',y') \in B$ such that $x \leq x'$ and $y \leq y'$ whilst $A \wedge B$ denotes points $z = (x,y) \in A$, $z' = (x',y') \in B$ such that $x \leq x'$ and $y \geq y'$

Type II integrals are quantum analogues of the Wong-Zakai integral.[6] These are defined on elementary adapted processes of the form $h(z, z') = \chi_{\Delta_i}\chi_{\Delta_j}a$ with $\Delta_i \wedge \Delta_j$ as

$$\iint\limits_{R_z}\iint\limits_{R_z} db^\#_{z'} db^\#_{z''} h(z', z'') = b^\#(\chi_{\Delta_i}\chi_{\Delta_{R_z}}) b^\#(\chi_{\Delta_j}\chi_{\Delta_{R_z}}) a$$

and again we extend these by linearity to $\mathcal{H}$-valued simple adapted processes.

The quasi-free CCR stochastic integrals are similarly defined in terms of $a^\#$.[2]

The type I and type II integrals for both the CAR and CCR cases each result in six different stochastic integrals, two type I and four type two integrals. Each of the six integrals satisfy isometry conditions and extend via isometry to completions of the $\mathcal{H}$-valued simple adapted processes. Each of the six integrals are also pairwise orthogonal, orthogonal to Ω and generate families of martingales. For further details we defer to.[2,3]

Remark In contrast to the quasi-free type I and type II stochastic integrals[1] the Clifford stochastic integrals result in one (rather than two) type I integral and one (rather than four) type II integral which as in the quasi-free case extend via isometry to a completion of the simple adapted processes. Martingale and orthogonality properties for the integrals follow and a representation theorem for $L^2(\mathcal{A})$ valued martingales in terms of these integrals is established.[4]

4. Representation

Following the results outlined above it is natural to consider the possibility of extending the representation theorem for martingales as quantum stochastic integrals to the quasi-free setting.

Theorem (Quasi-Free CAR and CCR Representations) Let $\{X_z | z \in R\}$ denote a $\mathcal{H}$-valued martingale. Then there exist unique α, $f_1, \ldots f_6$ (one from each of the orthogonal completions described in the last section) such that

$$X = \alpha\Omega + \sum_{i=1}^{2} \iint\limits_{R_z} db^\#_{z'} f_i(z') + \sum_{j=3}^{6} \iint\limits_{R_z}\iint\limits_{R_z} db^\#_{z'} db^\#_{z''} f_j(z', z'')$$

Proof

(CAR case) Let $b^\#(C)$ denote $b^\#(\chi_C)$ for $C \subseteq \mathbb{R}_+^2$ and χ_C the characteristic function on C. We consider products of the form

$$X = b^\#(\chi_{C_1}) b^\#(\chi_{C_2}) \ldots b^\#(\chi_{C_n}) \Omega$$

these being dense in $\mathcal{H}$. Using CAR's and linearity of the $b^\#$ such products may be rearranged into a product of the form

$$X = b^\#(A_1) b^\#(A_2) \ldots b^\#(A_p) b^\#(B_1) \ldots b^\#(B_q) Y$$

with $Y \in \mathcal{H}_z$ and the $\{A_i\}_{i=1}^p$, $\{B_j\}_{j=1}^q$ being rectangular subdomains of $R \subseteq \mathbb{R}_+^2$, each 'proud' of R_z such that $\forall 1 \leq i \leq p$, and $1 \leq j \leq q$ $A_i \wedge B_j$, each of the $A_i's$ have the same maximum y value, each the $B_j's$ have the same maximum x value and each subdomain is mutually exclusive to each of the other subdomains.

Using horizontal cuts for the $A_i's$ and vertical cuts for the $B_j's$ we divide each domain in half. At the nth cut we obtain $X = 4^n$ products of the same type as X together with a sum of products with less than pq proud domains which we will refer to as Z. It follows that

$$\|X - Z\|^2 = \sum_{i=1}^{2^n} \sum_{j=1}^{2^n} \prod_{k=1}^{p} \omega(b^{\#*}(A_k^i) b^\#(A_k^i)) \prod_{l=1}^{q} \omega(b^{\#*}(B_l^j) b^\#(B_l^j)) \omega(Y^* Y)$$

$$= \sum_{i=1}^{2^n} \sum_{j=1}^{2^n} \prod_{k=1}^{p} (\rho' \chi_{A_k^i}, \chi_{A_k^i}) \prod_{l=1}^{q} (\rho' \chi_{B_l^j}, \chi_{B_l^j}) \omega(Y^* Y)$$

$$< \sum_{i=1}^{2^n} \sum_{j=1}^{2^n} \prod_{k=1}^{p} |A_k^i| \prod_{l=1}^{q} |B_l^j| \omega(Y^* Y)$$

$$= 2^{-(p+q-2)n} \prod_{k=1}^{p} |A_k| \prod_{l=1}^{q} |B_l| \omega(Y^* Y) \longrightarrow 0 \text{ as } n \longrightarrow \infty$$

Here $\rho' = \rho$ or $1 - \rho$ depending upon $b^\#$

Induction on p and q, establishes X as a limit of type II integrals, and so type II by completion. Isometry and orthogonality on the completion ensure existance and uniqueness for the four $f_i's$ corresponding to each of the type II integrals.

Alternative configurations of the 'proud domains' each lead to sums of type I and/or type II integrals with the CAR condition generating $\alpha'\Omega$. Isometry and orthogonality ensure existance and uniqueness for the α and $f_1, \ldots, f_6$ in the respective completion.

(CCR Case) Proceeding as outlined above we obtain

$$\begin{aligned}
||X - Z||^2 &= \sum_{i=1}^{2^n}\sum_{j=1}^{2^n}\prod_{k=1}^{p} \omega(a^{\#^*}(A_k^i)a^{\#}(A_k^i))\prod_{l=1}^{q} \omega(a^{\#^*}(B_l^j)a^{\#}(B_l^j))\omega(Y^*Y) \\
&= \sum_{i=1}^{2^n}\sum_{j=1}^{2^n}\prod_{k=1}^{p}(\tau'\chi_{A_k^i},\chi_{A_k^i})\prod_{l=1}^{q}(\tau'\chi_{B_l^j},\chi_{B_l^j})\omega(Y^*Y) \\
&< \sum_{i=1}^{2^n}\sum_{j=1}^{2^n}\prod_{k=1}^{p}\|\tau'\|_\infty |A_k^i|\prod_{l=1}^{q}|B_l^j|\omega(Y^*Y) \\
&= 2^{-(p+q-2)n}\prod_{k=1}^{p}\|\tau'\|_\infty |A_k|\prod_{l=1}^{q}|B_l|\omega(Y^*Y) \longrightarrow 0 \text{ as } n \longrightarrow \infty
\end{aligned}$$

Here $\tau' = \tau$ or $1 + \tau$ depending upon $a^{\#}$.

The result now follows for $\{X_z\}$ a $\mathcal{H}$ valued martingale in both the CAR and CCR settings by considering expressions for X_z and $X_{z'}$ with $z \prec z'$ and projecting onto $\mathcal{H}_z$. The resulting expressions establish unique $\alpha, f_1, \ldots, f_6$ from their respective completions via isometry and orthogonality.

$\square$

5. Concluding Remarks

Further research has been carried out in both the Clifford and Quasi-Free settings the results of which will appear elsewhere. The author would like to thank in particular Ivan F.Wilde for helpful discussions and collaboration on earlier papers.

References

1. W. J. Spring and I. F. Wilde, Rep. Math. Phys. **42**, 389 (1998).
2. W. J. Spring and I. F. Wilde, Rep. Math. Phys. **49**, 63 (2002).
3. W. J. Spring and I. F. Wilde, Quant. Prob. and White Noise Analysis **15**, (2003).
4. W. J. Spring, Foundations of Quantum Probability and Physics **4**, (2006).
5. K. Itô, Memoirs of the American Mathematical Society **4**, 1 (1951).
6. E. Wong and M. Zakai, Z. Wahrscheinlichkeitstheorie und Verw. Gebiete **29**, 109 (1974).
7. J. M. Cook, Trans. Amer. Math. Soc. **74**, 222 (1953).
8. O. Bratteli and D. W. Robinson, *Operator Algebras and Quantum Statistical Mechanics II* (Springer, New York, 1981).
9. D. E. Evans, Commun. Math. Phys. **70**, 53 (1979).
10. R. Powers and E. Størmer, Commun. Math. Phys. **16**, 1 (1970).

A METHOD OF RECOVERING THE MOMENTS OF A PROBABILITY MEASURE FROM COMMUTATORS

A. I. STAN

Department of Mathematics
The Ohio State University at Marion
1465 Mount Vernon Avenue
Marion, OH 43302, U.S.A.
E-mail: stan.7@osu.edu

J. J. WHITAKER

Department of Mathematical Sciences
Shawnee State University
Portsmouth, OH 45501, U.S.A
E-mail: jwhitaker@shawnee.edu

A simple method of recovering the moments of a probability measure, on $\mathbb{R}$, from the commutator between its annihilation and creation operators, the commutator between the annihilation and preservation operator, and the first moment is presented.

Keywords: Szegö–Jacobi parameters, annihilation, preservation, and creation operators

1. Introduction

It was shown in Ref. 1 that the moments of a probability measure, on $\mathbb{R}^d$, where d is a fixed natural number, can be recovered from its preservation operators, and the commutators between the annihilation and creation operators. Moreover, some properties of a probability measure μ, on $\mathbb{R}^d$, having finite moments of all orders, like polynomial symmetry and factorizability, were shown in Ref. 2 to be equivalent to some properties of the preservation, annihilation, and creation operators. More precisely, the condition that μ is polynomially symmetric about a vector $c = (c_1, c_2, \ldots, c_d) \in \mathbb{R}^d$, i.e.,

$$E[(x_1 - c_1)^{i_1}(x_2 - c_2)^{i_2} \cdots (x_d - c_d)^{i_d}] = 0, \tag{1}$$

for all i_1, i_2, ..., i_d non–negative integers, such that $i_1+i_2+\cdots+i_d$ is odd, where E denotes the expectation with respect to μ, was shown in Ref. 2 to be equivalent to the fact that for all $i \in \{1, 2, ..., d\}$, $a^0(i) = c_i I$, where $a^0(i)$ denotes the preservation operator generated by the operator X_i of multiplication by the coordinate random variable x_i, and I is the identity operator of the space F of all polynomial functions. Here $x = (x_1, x_2, ..., x_d)$ denotes a generic vector (in fact outcome) of the sample space $\mathbb{R}^d$. It was shown in Ref. 1 that if μ is polynomially symmetric about c, then there exists a symmetric probability measure ν about c, on $\mathbb{R}^d$, that has the same moments as μ. The fact that ν is symmetric about c means that for any Borel subset $B \subset \mathbb{R}^d$, $\nu(c-B) = \nu(c+B)$, where $B+v := \{x+v \mid x \in B\}$, for all $v \in \mathbb{R}^d$. It was also shown in Ref. 2 that μ is polynomially factorizable, i.e.,

$$E[x_1^{i_1} x_2^{i_2} \cdots x_d^{i_d}] = E[x_1^{i_1}] E[x_2^{i_2}] \cdots E[x_d^{i_d}],$$

for all i_1, i_2, ...i_d non–negative integers, if and only if for any $i \neq j$, any operator from the set $\{a^-(i), a^0(i), a^+(i)\}$ commutes with any operator from the set $\{a^-(j), a^0(j), a^+(j)\}$, where $a^-(k)$ and $a^+(k)$ denote the annihilation and creation operators generated by the operator X_k of multiplication by x_k, for all $1 \le k \le d$. We know (see Ref. 1) that if μ is polynomially factorizable, then there exists a product probability measure $\nu = \nu_1 \otimes \nu_2 \otimes \cdots \otimes \nu_d$, where $\nu_1, \nu_2, ..., \nu_d$ are probability measures on $\mathbb{R}$, having finite moments of all orders, such that for all non–negative integers $i_1, i_2, ... i_d$, the monomial $x_1^{i_1} x_2^{i_2} ... x_d^{i_d}$ has the same expectation with respect to both μ and ν. It is not hard to see that due to the commutation relationship $[X_i, X_j] := X_i X_j - X_j X_i = 0$, and the fact that all preservation operators are symmetric operators, while each creation operator $a^+(i)$ is the adjoint of the annihilation operator $a^-(i)$, the condition that for any $i \neq j$, any operator from the set $\{a^-(i), a^0(i), a^+(i)\}$ commutes with any operator from the set $\{a^-(j), a^0(j), a^+(j)\}$, is equivalent to the simpler fact that for any $i \neq j$, $[a^-(i), a^+(j)] = 0$ and $[a^-(i), a^0(j)] = 0$. The condition $[a^-(i), a^0(j)] = 0$ does not fit into the theme of the paper Ref. 1, since there all probability measures are studied in terms of the commutators $[a^-(i), a^+(j)] = 0$ and the action of the preservation operators $a^0(k)$ alone, where $(i, j, k) \in \{1, 2, ..., d\}^3$. The fact that the polynomially factorizable probability measures are characterized in terms of two families of commutators, and not in terms of one family of commutators and the preservation operators alone, has made us rethink about the results from Ref. 2 and reformulate all of them in terms of two families of operators, namely the commutators between the annihilation and creation operators, and the commutators between the annihilation and preservation operators. In doing so, we came out with a simple method of recovering the moments of a probability measure from these two families of commutators, assuming that the first order moments $E[x_1]$, $E[x_2]$, ..., $E[x_d]$ are known. This method works very nicely for some classic probability distributions on $\mathbb{R}$, but unfortunately for probability measures on $\mathbb{R}^d$, that are not product measures, the computations involved seem to be quite complicated. For this reason, we will focus in this paper only on the one dimensional case $d = 1$, and show how this method of commutators can be successfully employed in a classic example. It is also worth to understand very well the one–dimensional case first before moving on to higher dimensions. In section 2 we give the definition of the annihilation, preservation and creation operators, of any probability measure on $\mathbb{R}$, having finite moments of all orders. In section 3 we present a method of recovering the moments from commutators and the first order moment. Finally in section 4 we show an application of this method to a concrete example.

2. Background

Let μ be a probability measure on $\mathbb{R}$, having finite moments of all orders. Applying the Gram–Schmidt orthogonalization procedure to the sequence of monomial random variables $1, x, x^2, \ldots$, one can construct a sequence of orthogonal polynomials: $f_0(x)$, $f_1(x)$, $f_2(x)$, ... If the support of the probability measure μ is a finite set, then this sequence of orthogonal polynomials has only finitely many non–zero terms. Otherwise, all these orthogonal polynomials will be different from zero and we can choose them in such a way that their leading coefficient is equal to 1. It is well known (see for example Ref. 3,4) that there exist two sequences $\{\alpha_n\}_{n\geq 0}$ and $\{\omega_n\}_{n\geq 1}$ of real numbers, such that, for all $k \geq 0$, the following relation holds:

$$xf_n(x) = f_{n+1}(x) + \alpha_n f_n(x) + \omega_n f_{n-1}(x).$$

When $n = 0$, we define $f_{-1}(x) := 0$ (the null polynomial) and $\omega_0 := 0$. The terms of the sequences $\{\alpha_n\}_{n\geq 0}$ and $\{\omega_n\}_{n\geq 1}$ are called the *Szegö–Jacobi parameters of μ*. We define the space $\mathcal{H} := \oplus_{n\geq 0} \mathbb{C}f_n$ and call $\mathcal{H}$ the *chaos space of μ*. We can now define three densely defined linear operators on $\mathcal{H}$, a^-, a^0, and a^+, called the *annihilation, preservation*, and *creation* operators, respectively, by presenting their actions on each orthogonal polynomial. For each $n \geq 0$, we define $a^- f_n(x) := \omega_n f_{n-1}(x)$, $a^0 f_n(x) := \alpha_n f_n(x)$, and $a^+ f_n(x) := f_{n+1}(x)$. It is easy to see now, from (1), that the operator X

of multiplication by x (that means the operator that maps any polynomial random variable $g(x)$ into $xg(x)$) is the sum of these three operators. Hence:

$$X = a^- + a^0 + a^+. \tag{1}$$

We also define the *number operator* $\mathcal{N}$ as the linear operator for which $\mathcal{N} f_n(x) := n f_n(x)$, for all $n \geq 0$. The domain of a^-, a^0, a^+, and $\mathcal{N}$ is understood to be the space F of all polynomial functions $g(x)$ of one real variable x with complex coefficients.

If A and B are two operators defined on the same space, then we define their commutator $[A, B]$, as:

$$[A, B] := AB - BA. \tag{2}$$

It is easy to see, by induction on m, that:

$$[A, B^m] = \sum_{i=0}^{m-1} B^{m-1-i} [A, B] B^i, \tag{3}$$

for all natural numbers m. One can think of (3) as being an analogue of the product rule of differentiation from Calculus.

Since the operators $\mathcal{N}$ and a^0 are diagonalized in the same basis $\{f_n(x)\}_{n \geq 1}$, it follows that they commute. Thus:

$$[\mathcal{N}, a^0] = 0. \tag{4}$$

On the other hand, for all $n \geq 0$,

$$\begin{aligned}[a^-, \mathcal{N}] f_n(x) &= a^- \mathcal{N} f_n(x) - \mathcal{N} a^- f_n(x) \\ &= a^- [n f_n(x)] - \mathcal{N} [\omega_n f_{n-1}(x)] \\ &= n \omega_n f_{n-1}(x) - (n-1) \omega_n f_{n-1}(x) \\ &= \omega_n f_{n-1}(x) \\ &= a^- f_n(x).\end{aligned}$$

Thus we obtain that:

$$[a^-, \mathcal{N}] = a^-. \tag{5}$$

Taking the adjoint in both sides of (5) we conclude that:

$$[\mathcal{N}, a^+] = a^+. \tag{6}$$

3. A commutator method

We formulate now the following method of computing the moments of a probability measure, on $\mathbb{R}$, from its commutators $[a^-, a^+]$ and $[a^-, a^0]$, and first moment $E[x]$.

Method. Let us assume that μ is a probability measure on $\mathbb{R}$, having finite moments of all orders, whose commutators $[a^-, a^+]$, $[a^-, a^0]$, and first moment $E[x]$ are known. Let $\langle \cdot\,,\,\cdot \rangle$ denote the inner product generated by μ. It is clear that x^n is obtained by applying the operators X, of multiplication by x, n times to the constant polynomial (random variable) 1. Thus we have:

$$E[x^n] = \langle X^n 1, 1 \rangle$$
$$= \langle X \circ X^{n-1} 1, 1 \rangle.$$

Step 1 Replace the first X by $a^- + a^0 + a^+$ and get:

$$E[x^n] = \langle (a^- + a^0 + a^+) X^{n-1} 1, 1 \rangle$$
$$= \langle a^- X^{n-1} 1, 1 \rangle + \langle a^0 X^{n-1} 1, 1 \rangle + \langle a^+ X^{n-1} 1, 1 \rangle.$$

Observe that:

$$\langle a^+ X^{n-1} 1, 1 \rangle = \langle X^{n-1} 1, a^- 1 \rangle$$
$$= 0,$$

since $a^- 1 = 0$. Remark also that:

$$\langle a^0 X^{n-1} 1, 1 \rangle = \langle X^{n-1} 1, a^0 1 \rangle$$
$$= \langle X^{n-1} 1, E[x] 1 \rangle$$
$$= E[x] \langle X^{n-1} 1, 1 \rangle$$
$$= E[x] E[x^{n-1}],$$

since $a^0 1 = E[x] 1$. Thus we obtain:

$$E[x^n] = \langle a^- X^{n-1} 1, 1 \rangle + E[x] E[x^{n-1}].$$

Step 2 Swap (commute) a^- with X^{n-1} using the simple formula $a^- X^{n-1} = [a^-, X^{n-1}] + X^{n-1} a^-$ and rule (3). Since $a^- 1 = 0$, we get:

$$\begin{aligned}
E[x^n] &= \langle a^- X^{n-1} 1, 1 \rangle + E[x] E[x^{n-1}] \\
&= E[x] E[x^{n-1}] + \langle [a^-, X^{n-1}] 1, 1 \rangle + \langle X^{n-1} a^- 1, 1 \rangle \\
&= E[x] E[x^{n-1}] + \langle \sum_{i=0}^{n-2} X^{n-2-i} [a^-, X] X^i 1, 1 \rangle + 0 \\
&= E[x] E[x^{n-1}] + \sum_{i=0}^{n-2} \langle X^{n-2-i} [a^-, X] X^i 1, 1 \rangle.
\end{aligned}$$

Therefore, we obtain:

$$E[x^n] = E[x] E[x^{n-1}] + \sum_{i=0}^{n-2} \langle X^{n-2-i} [a^-, X] X^i 1, 1 \rangle. \tag{1}$$

Step 3 Replace $[a^-, X]$ by $[a^-, a^+] + [a^-, a^0]$, since:

$$\begin{aligned}
[a^-, X] &= [a^-, a^- + a^0 + a^+] \\
&= [a^-, a^-] + [a^-, a^0] + [a^-, a^+] \\
&= [a^-, a^0] + [a^-, a^+].
\end{aligned}$$

Go back to Step 2 if necessary and continue this algorithm until all annihilation operators disappear, obtaining in the end a recursive relation expressing $E[x^n]$ in terms of the lower order moments $E[x^{n-1}]$, $E[x^{n-2}]$, ..., $E[x]$, 1.

4. Gaussian and Poisson probability distributions

Let us find all probability measures on $\mathbb{R}$, having finite moments of all orders, whose Szegö–Jacobi parameters are $\alpha_n = bn + c$ and $\omega_n = dn$, for all $n \geq 0$, where b, c, and d are real numbers, such that $d > 0$. The condition $d > 0$ follows from the fact the ω_n must be positive for all $n \geq 1$ (Favard's theorem). We avoid the trivial case $d = 0$ which forces b to be also zero and implies that $\mu = \delta_{\{c\}}$, i.e., the Dirac delta measure at c. Observe that $\alpha_n = bn + c$, for all $n \geq 0$, is equivalent to the fact that $a^0 = b\mathcal{N} + cI$, where I denotes the identity operator. Thus we obtain:

$$\begin{aligned}
[a^-, a^0] &= b[a^-, \mathcal{N}] + c[a^0, I] \\
&= ba^-.
\end{aligned}$$

For all $n \geq 0$, we have:

$$\begin{aligned}
{[a^-, a^+]f_n(x)} &= a^-a^+f_n(x) - a^+a^-f_n(x) \\
&= a^-f_{n+1}(x) - a^+[dnf_{n-1}(x)] \\
&= d(n+1)f_n(x) - dnf_n(x) \\
&= df_n(x).
\end{aligned}$$

Therefore,

$$[a^-, a^+] = dI.$$

Finally,

$$\begin{aligned}
E[x] &= \alpha_0 \\
&= c.
\end{aligned}$$

Therefore, we can reformulate our problem as:

Problem *Find the probability measure on $\mathbb{R}$, having finite moments of all orders, for which:*

$$[a^-, a^+] = dI, \tag{1}$$
$$[a^-, a^0] = ba^-, \tag{2}$$
$$E[x] = c. \tag{3}$$

Solution. Let $n \geq 1$, be a fixed natural number. Applying formula (1) of the commutator method we get:

$$\begin{aligned}
E[x^n] &= E[x]E[x^{n-1}] + \sum_{i=0}^{n-2} \langle X^{n-2-i}[a^-, X]X^i 1, 1\rangle \\
&= cE[x^{n-1}] + \sum_{i=0}^{n-2} \langle X^{n-2-i}\left([a^-, a^+] + [a^-, a^0]\right) X^i 1, 1\rangle \\
&= cE[x^{n-1}] + \sum_{i=0}^{n-2} \langle X^{n-2-i}\left(dI + ba^-\right) X^i 1, 1\rangle \\
&= cE[x^{n-1}] + d\sum_{i=0}^{n-2} \langle X^{n-2} 1, 1\rangle + b\sum_{i=0}^{n-2} \langle X^{n-2-i} a^- X^i 1, 1\rangle \\
&= cE[x^{n-1}] + d(n-1)E[x^{n-2}] + b\sum_{i=0}^{n-2} \langle X^{n-2-i} a^- X^i 1, 1\rangle.
\end{aligned}$$

Let us observe that in the last sum we can take $i \geq 1$, since for $i = 0$, $X^{n-2-i}a^- X^i 1 = X^{n-2}a^- 1 = 0$. Swap now a^- and X^i, and use rule (3) again, to obtain:

$$E[x^n] = cE[x^{n-1}] + d(n-1)E[x^{n-2}] + b\sum_{i=1}^{n-2}\langle X^{n-2-i}a^- X^i 1, 1\rangle$$

$$= cE[x^{n-1}] + d(n-1)E[x^{n-2}] + b\sum_{i=1}^{n-2}\langle X^{n-2-i}[a^-, X^i]1, 1\rangle$$

$$+ b\sum_{i=1}^{n-2}\langle X^{n-2-i} X^i a^- 1, 1\rangle$$

$$= cE[x^{n-1}] + d(n-1)E[x^{n-2}]$$
$$+ b\sum_{i=1}^{n-2}\sum_{j=0}^{i-1}\langle X^{n-2-i} X^{i-1-j}[a^-, X] X^j 1, 1\rangle$$

$$= cE[x^{n-1}] + d(n-1)E[x^{n-2}]$$
$$+ b\sum_{0 \leq j < i \leq n-2}\langle X^{n-3-j}\left(dI + ba^-\right) X^j 1, 1\rangle$$

$$= cE[x^{n-1}] + d(n-1)E[x^{n-2}] + bd\sum_{0 \leq j < i \leq n-2} E[x^{n-3}]$$
$$+ b^2 \sum_{0 \leq j < i \leq n-2}\langle X^{n-3-j}a^- X^j 1, 1\rangle.$$

Since the cardinality of the set $\{(i,j) \mid 0 \leq j < i \leq n-2\}$ is $\binom{n-1}{2}$, we obtain:

$$E[x^n] = cE[x^{n-1}] + d(n-1)E[x^{n-2}] + bd\binom{n-1}{2}E[x^{n-3}]$$
$$+ b^2 \sum_{0 \leq j < i \leq n-2}\langle X^{n-3-j}a^- X^j 1, 1\rangle.$$

Repeating this procedure (that means observing that in the last sum we can take $j \geq 1$, then swapping a^- and X^j, and so on), we can push a^- more and more to the right, until there is no power of X left in between a^- and the vacuum polynomial 1. It is easy to see that in the end we get:

$$E[x^n] = cE[x^{n-1}] + d\sum_{k=1}^{n-1}\binom{n-1}{k}b^{k-1}E[x^{n-1-k}], \qquad (4)$$

for all $n \geq 2$. We analyze now two cases:

Case 1. If $b = 0$, then the recursive relation (4) becomes:
$$E[x^n] = cE[x^{n-1}] + d(n-1)E[x^{n-2}], \tag{5}$$
for all $n \geq 2$. Let us try to see whether we can find a probability measure μ that is absolutely continuous with respect to the Lebesgue measure dx on $\mathbb{R}$, for which the recursive relation (5) holds. Assuming that μ is a continuous probability distribution, then if g denotes the density function of μ, it follows from (5), that for all $n \geq 2$, we have:
$$\int_\mathbb{R} x^{n-1}(x-c)g(x)dx = d\int_\mathbb{R} (n-1)x^{n-1}g(x)dx$$
$$= d\int_\mathbb{R} \left(x^{n-1}\right)' g(x)dx,$$
for all $n \geq 2$ (this formula works even for $n = 1$ since $E[x] = c$). Let us try to see whether we can take g to be a Schwartz function, that means a smooth function that decreases faster than any polynomial is growing at $\pm\infty$. If this happens, then integrating by parts, in the last relation, we would get:
$$\int_\mathbb{R} x^{n-1}(x-c)g(x)dx = -\int_\mathbb{R} x^{n-1}dg'(x)dx,$$
for all $n \geq 1$. It follows now that the function $dg'(x) + (x-c)g(x)$ is orthogonal to all monomials x, x^2, ..., and thus by a leap of faith, we can try to find g, such that:
$$g'(x) = -\frac{1}{d}(x-c)g(x), \tag{6}$$
for all $x \in \mathbb{R}$. Solving this differential equation, we conclude that there exists a real constant k, such that:
$$g(x) = ke^{-\frac{1}{2d}(x-c)^2}, \tag{7}$$
for all $x \in \mathbb{R}$. Since $\int_\mathbb{R} g(x)dx = 1$, we can see that $k = \frac{1}{\sqrt{2\pi d}}$, and thus μ is the probability distribution of a Gaussian random variable with mean c and variance d. Since the problem of moments has a unique solution when the moments are equal to the moments of a Gaussian random variable, we conclude that μ must be a normal probability distribution.

Case 2. If $b \neq 0$, then it follows from (4) that:
$$E[x^n] = cE[x^{n-1}] + \frac{d}{b}\sum_{k=1}^{n-1}\binom{n-1}{k}b^k E[x^{n-1-k}].$$

This is equivalent to:

$$E[x^n] = \left(c - \frac{d}{b}\right) E[x^{n-1}] + \frac{d}{b} E\left[(x+b)^{n-1}\right], \qquad (8)$$

for all $n \geq 1$. For all $k \geq 0$, let $M_k := \max\{|E[x^i]|/i! \mid 0 \leq i \leq k\}$. We can also conclude from (4) that:

$$|E[x^n]| \leq |c||E[x^{n-1}]| + \frac{|d|}{|b|} \sum_{k=1}^{n-1} \binom{n-1}{k} |b|^k |E[x^{n-1-k}]|$$

$$\leq |c|M_{n-1}(n-1)! + \frac{|d|}{|b|} \sum_{k=1}^{n-1} \binom{n-1}{k} |b|^k M_{n-1}(n-1-k)!$$

$$= |c|M_{n-1}(n-1)! + \frac{|d|}{|b|} M_{n-1}(n-1)! \sum_{k=1}^{n-1} \frac{|b|^k}{k!}$$

$$\leq |c|M_{n-1}(n-1)! + \frac{|d|}{|b|} M_{n-1}(n-1)! \sum_{k=1}^{\infty} \frac{|b|^k}{k!}$$

$$= \left[|c| + \frac{|d|}{|b|}\left(e^{|b|} - 1\right)\right] M_{n-1}(n-1)!,$$

for all $n \geq 1$. Let $A := |c| + \frac{|d|}{|b|}(e^{|b|} - 1)$. We have:

$$\frac{|E[x^n]|}{n!} \leq \frac{A}{n} M_{n-1},$$

for all $n \geq 1$. Thus there exists a natural number N_0, such that for all $n \geq N_0$, we have $|E[x^n]|/n! \leq M_{n-1}$. Hence:

$$M_n = \max\{|E[x^n]|/n!, M_{n-1}\}$$
$$= M_{n-1},$$

for all $n \geq N_0$. Thus we conclude that $M_{N_0-1} = M_{N_0} = M_{N_0+1} = \cdots$. Therefore, there exists a constant $M \geq 1$, such that for all $n \geq 0$,

$$|E[x^n]| \leq M n!. \qquad (9)$$

(We can actually take $M := M_{N_0}$.) In particular for $n = 2m$, where m is a non-negative integer, we have:

$$E[x^{2m}] \leq M(2m)!. \qquad (10)$$

Using now Schwarz' or Jensen's inequality we conclude that, for all $m \geq 0$,

$$E[|x|^m] \leq \sqrt{E[x^{2m}]}$$
$$\leq \sqrt{M(2m)!}$$
$$\leq 2^m \sqrt{M} m!,$$

since $\binom{2m}{m} \leq \sum_{j=0}^{2m} \binom{2m}{m} = 2^{2m}$. Hence, for all $m \geq 0$, we have:

$$E[|x|^m] \leq 2^m \sqrt{M m!}. \tag{11}$$

Inequality (11) implies, via the convexity of the function $f(x) = x^m$ on $[0, \infty)$, that:

$$E[|x+b|^m] \leq E\left[(|x|+|b|)^m\right]$$
$$= 2^m E\left[\left(\frac{1}{2}|x| + \frac{1}{2}|b|\right)^m\right]$$
$$\leq 2^m E\left[\frac{1}{2}|x|^m + \frac{1}{2}|b|^m\right]$$
$$\leq 2^{m-1}[2^m \sqrt{M m!} + |b|^m].$$

Thus we obtain:

$$E[|x+b|^m] \leq \frac{1}{2} 4^m \sqrt{M m!} + \frac{1}{2}|2b|^m, \tag{12}$$

for all $m \geq 0$. The bounds (11) and (12) allow us, after multiplying both sides of the recursive relation (8) by $t^{n-1}/(n-1)!$, to sum up from $n=1$ to ∞, and interchange the series with the expectation, thanks to dominated convergence theorem, where t belongs to a small interval $(-\epsilon, \epsilon)$ about zero, and obtain:

$$E\left[\sum_{n=1}^{\infty} x \frac{(tx)^{n-1}}{(n-1)!}\right] = \left(c - \frac{d}{b}\right) E\left[\sum_{n=1}^{\infty} \frac{(tx)^{n-1}}{(n-1)!}\right] + \frac{d}{b} E\left[\sum_{n=1}^{\infty} \frac{[t(x+b)]^{n-1}}{(n-1)!}\right].$$

The last relation is equivalent to:

$$E\left[xe^{tx}\right] = \left(c - \frac{d}{b}\right) E\left[e^{tx}\right] + \frac{d}{b} E\left[e^{t(x+b)}\right].$$

for all t in a neighborhood of zero. Let $\varphi(t) := E[e^{tx}]$ (i.e., φ is the Laplace transform of the polynomial random variable x). Then, using again the dominated convergence theorem, we can easily see that $E\left[xe^{tx}\right]$ is the derivative of φ with respect to t, for all t in a neighborhood of zero. Thus, we conclude that the Laplace transform φ of x, satisfies the differential equation:

$$\varphi'(t) = \left[\frac{d}{b}(e^{bt} - 1) + c\right] \varphi(t), \tag{13}$$

for all t near zero. Since $\varphi(0) = 1$, we conclude that:

$$\varphi(t) = e^{(d/b^2)(e^{bt}-1)+(c-d/b)t}. \tag{14}$$

This means that:

$$E\left[e^{t(x-c+d/b)}\right] = e^{(d/b^2)(e^{bt}-1)}, \quad (15)$$

for all t sufficiently close to zero. We recognize that the right–hand side of (15) represents the Laplace transform of the constant b times a Poisson random variable Y with parameter $\lambda := d/b^2$. Thus μ is the probability distribution of the shifted re–scaled Poisson random variable $bY + c - d/b$.

Final Comments The problem of finding the Szegö–Jacobi parameters of a probability distribution on $\mathbb{R}$ from its moments, and its converse are classic ones. They have been studied by many mathematicians. One way to study these problems is to use the technique of re–normalization (see Ref. 5–8). Our commutator method brings a new insight into the converse problem (i.e., the problem of finding the moments from the Szegö–Jacobi parameters) by disregarding the actual form of the orthogonal polynomials $\{f_n(x)\}_{n\geq 0}$, and exploiting the structure of the Lie algebra generated by the annihilation, preservation, and creation operators.

References

1. L. Accardi, H.-H. Kuo, and A. I. Stan, Moments and commutators of probability measures, *Infin. Dimens. Anal. Quantum Probab. Relat. Top.*, **10**, No. 4, (2007) 591–612.
2. L. Accardi, H.-H. Kuo, and A. I. Stan, Characterization of probability measures through the canonically associated interacting Fock spaces, *Infin. Dimens. Anal. Quantum Probab. Relat. Top.*, **7**, No. 4, (2004) 485–505.
3. T. S. Chihara, *An Introduction to Orthogonal Polynomials* (Gordon & Breach, New York, 1978).
4. M. Szegö, *Orthogonal Polynomials*, Coll. Publ. **23** (Amer. Math. Soc., 1975).
5. N. Asai, I. Kubo, and H.-H. Kuo, Multiplicative renormalization and generating functions I, *Taiwanese J. Math.* **7**, (2003) 89–101.
6. N. Asai, I. Kubo, and H.-H. Kuo, Multiplicative renormalization and generating functions II, *Taiwanese J. Math.* **8**, (2004) 593–628.
7. N. Asai, I. Kubo, and H.-H. Kuo, Generating functions of orthogonal polynomials and Szegö–Jacobi parameters, *Prob. Math. Stat.* **23**, (2003) 273–291.
8. I. Kubo, H.-H. Kuo, and S. Namli, Interpolation of Chebyshev polynomials and interacting Fock spaces, *Infin. Dimens. Anal. Quantum Probab. Relat. Top.*, **9**, No. 3, (2006) 361–371.

DESCRIPTION OF DECOHERENCE BY MEANS OF TRANSLATION-COVARIANT MASTER EQUATIONS AND LÉVY PROCESSES

B. VACCHINI

Dipartimento di Fisica dell'Università di Milano and INFN, Sezione di Milano
Via Celoria 16, 20133, Milan, Italy

Translation-covariant Markovian master equations used in the description of decoherence and dissipation are considered in the general framework of Holevo's results on the characterization of generators of covariant quantum dynamical semigroups. A general connection between the characteristic function of classical Lévy processes and loss of coherence of the statistical operator describing the center of mass degrees of freedom of a quantum system interacting through momentum transfer events with an environment is established. The relationship with both microphysical models and experimental realizations is considered, focusing in particular on recent interferometric experiments exploring the boundaries between classical and quantum world.

Keywords: Lévy processes; decoherence; quantum dynamical semigroups

1. Introduction

A natural standpoint about quantum mechanics, which is however not the one usually considered in textbooks written for physics students, is to look at it as a new probability theory, different and reacher than the classical one[1]. This point of view becomes mandatory or at least very fruitful if one is faced with more advanced research topics, such as the description of open quantum systems or quantum information and communication theory (for a general reference see[2,3]). In these fields tools and concepts obtained relying on a probabilistic approach, also working in direct analogy with classical probability theory, have become of paramount importance. An example in this direction is given by quantum dynamical semigroups, which provide the quantum generalization of classical Markov semigroups. The subject has been the object of active research in the mathematics, physics and chemistry community over decades by now, but it is still of great interest. In particular covariance properties of such mappings under translations have been considered in detail only recently. Besides a mathematical characterization[4-7] also the actual physical relevance[8-10] of such covariant quantum dynamical semigroups has been considered.

In the present contribution we will focus mainly on the application of such translation-covariant quantum dynamical semigroups to the study and the description of the phenomenon called decoherence in the physics literature[11,12]. By such a term a whole variety of situations is meant, all having in common a loss of typical quantum interference capability, arising as a dynamical consequence of interaction of the system of interest with some other, typically much bigger, system. The phenomenology of decoherence is ubiquitous when considering open quantum systems, but its actual quantitative study requires very special experimental conditions, which can be realized e.g. in interferometric setups for massive particles, observing loss of interference fringes as a consequence of external disturbance, arising because the approximation of isolation of the system is no more realistic. For a quantitative study of the phenomenon it is in fact crucial that such decoherence effects can actually be engineered, so that their strength is under the control of the experimenter.

The paper is organized as follows. In Sect. 2 we briefly sketch the formal expression of the generator of a translation-covariant quantum dynamical semigroup. In Sect. 3 we show how such a general structure in a suitable limit can account for decoherence behaviors quantitatively described by means of the characteristic function of a classical Lévy process. In Sect. 4 we further explore how a particular physical example of realization of such generators applies to the description of decoherence in both position and momentum space, finally mentioning possible extension of the formalism in Sect. 5.

2. Translation-covariant master equations

Provided memory effects can be neglected, quantum dynamical semigroups[13,14] give a general setting for the description of the dynamics of an open quantum system[15]. In the physical literature major efforts have been devoted to the derivation or phenomenological assessment of possible generators of such quantum dynamical semigroups, so called master equations. The typical benchmark is the Lindblad structure of such generators, which goes back to the work of Gorini, Kossakowski and Sudarshan[16] and of Lindblad[17], holding true for a generator given by a bounded mapping. Attention was later devoted to possible constraints on the structure of such generators arising as a consequence of symmetry (see e.g.[18] for references).

In this respect the results of Holevo for symmetry under translations are of particular importance because of the many possible physical applications, especially in connection with typical quantum phenomena such as decoherence.

We first consider the general expression of formal generators of translation-covariant quantum dynamical semigroups as obtained by Holevo[4-6,19]. The covariance of the mapping corresponds to the requirement that its action has to commute with the unitary representation of translations on the Hilbert space of interest. The physical system we are going to consider is the centre of mass of a particle in free space, so that $\mathcal{H} = L^2(\mathbb{R}^3)$. Let $\mathcal{L}'$ be the mapping describing the dynamics in Heisenberg picture, thus acting on an observable A. In order to be covariant $\mathcal{L}'$ has to satisfy the requirement

$$\mathcal{L}'\left[e^{i\mathbf{A}\cdot\mathbf{P}/\hbar}\mathsf{A}e^{-i\mathbf{A}\cdot\mathbf{P}/\hbar}\right] = e^{i\mathbf{A}\cdot\mathbf{P}/\hbar}\mathcal{L}'[\mathsf{A}]e^{-i\mathbf{A}\cdot\mathbf{P}/\hbar} \quad \forall \mathbf{A} \in \mathbb{R}^3, \qquad (1)$$

where P denotes the momentum operator of the massive particle. The general structure of generator complying with this requirement is given by the formal operator expression

$$\mathcal{L}'[\mathsf{A}] = \frac{i}{\hbar}[\mathsf{H}(\mathsf{P}),\mathsf{A}] + \mathcal{L}_G[\mathsf{A}] + \mathcal{L}_P[\mathsf{A}], \qquad (2)$$

where the symbols G and P denote a Gaussian and a Poisson component, the names arising from the connection with the classical Lévy-Khintchine formula. One has in particular for the Gaussian component

$$\mathcal{L}_G[\mathsf{A}] = \frac{i}{\hbar}\left[\mathsf{Y}_0 + \frac{1}{2i}\sum_{k=1}^{3}\left(\mathsf{Y}_k L_k(\mathsf{P}) - L_k^\dagger(\mathsf{P})\mathsf{Y}_k\right),\mathsf{A}\right]$$
$$+ \frac{1}{\hbar}\sum_{k=1}^{3}\left[(\mathsf{Y}_k + L_k(\mathsf{P}))^\dagger \mathsf{A}(\mathsf{Y}_k + L_k(\mathsf{P}))\right.$$
$$\left. - \frac{1}{2}\left\{(\mathsf{Y}_k + L_k(\mathsf{P}))^\dagger(\mathsf{Y}_k + L_k(\mathsf{P})),\mathsf{A}\right\}\right]$$

where $\mathsf{Y}_j = \sum_{i=1}^{3} a_{ji}\mathsf{X}_i$ with $a_{ji} \in \mathbb{R}$ for $j = 0,1,2,3$, that is to say it is a linear combination of the three position operators of the test particle, appearing at most quadratically, while for the Poisson component

$$\mathcal{L}_P[\mathsf{A}] = \int d\mu\,(\boldsymbol{Q}) \sum_j \left[\mathsf{L}_j^\dagger\,(\mathsf{P};\boldsymbol{Q})\,\mathrm{e}^{-i\boldsymbol{Q}\cdot\mathsf{X}/\hbar}\mathsf{A}\mathrm{e}^{i\boldsymbol{Q}\cdot\mathsf{X}/\hbar}\mathsf{L}_j\,(\mathsf{P};\boldsymbol{Q}) \right.$$

$$\left. -\frac{1}{2}\left\{ \mathsf{L}_j^\dagger\,(\mathsf{P};\boldsymbol{Q})\,\mathsf{L}_j\,(\mathsf{P};\boldsymbol{Q}),\mathsf{A}\right\}\right]$$

$$+ \int d\mu\,(\boldsymbol{Q}) \sum_j \left[\omega_j\,(\boldsymbol{Q})\,\mathsf{L}_j^\dagger\,(\mathsf{P};\boldsymbol{Q})\left(\mathrm{e}^{-i\boldsymbol{Q}\cdot\mathsf{X}/\hbar}\mathsf{A}\mathrm{e}^{i\boldsymbol{Q}\cdot\mathsf{X}/\hbar} - \mathsf{A}\right)\right.$$

$$\left. + \left(\mathrm{e}^{-i\boldsymbol{Q}\cdot\mathsf{X}/\hbar}\mathsf{A}\mathrm{e}^{i\boldsymbol{Q}\cdot\mathsf{X}/\hbar} - \mathsf{A}\right)\mathsf{L}_j\,(\mathsf{P};\boldsymbol{Q})\,\omega_j^*\,(\boldsymbol{Q})\right]$$

$$+ \int d\mu\,(\boldsymbol{Q})\,|\omega_j\,(\boldsymbol{Q})|^2 \sum_j \left[\mathrm{e}^{-i\boldsymbol{Q}\cdot\mathsf{X}/\hbar}\mathsf{A}\mathrm{e}^{i\boldsymbol{Q}\cdot\mathsf{X}/\hbar} - \mathsf{A} - \frac{i}{\hbar}\frac{[\mathsf{A},\boldsymbol{Q}\cdot\mathsf{X}]}{1+Q^2/Q_0^2}\right].$$

Such expressions can cover a huge variety of physical situations, accounting for both dissipative and decoherence effects. Some rough insight can be gained considering the dummy integration label $\boldsymbol{Q}$ as a momentum. The dynamics of the open system, in our case the centre of mass of a tracer particle, is thus described by an interaction only characterized by the momentum transfers between system and environment, taking place e.g. as a consequence of collisions, thus complying with translational invariance. The unitary operators $\exp(i\boldsymbol{Q}\cdot\mathsf{X}/\hbar)$ appearing in the Poisson part describe in fact a momentum kick, with rates which are not only given by functions of the momentum transfer $\boldsymbol{Q}$ itself, but also depend on the momentum operator P, thus becoming dynamic quantities. This is in particular necessary in order to correctly describe phenomena like energy transfer and approach to equilibrium. The Gaussian part corresponds to a dynamics arising as a consequence of a big number of small momentum transfers, leading to a diffusive behavior.

An interesting limiting situation appears if we neglect dissipative effects and therefore the dynamics of the momentum operator, apart from its appearance in the free kinetic term, so that, also switching to the preadjoint mapping in Schrödinger picture, and assuming $\mu(\boldsymbol{Q})$ to be absolutely continuous with respect to the Lebesgue measure, the two contributions can be written

$$\mathcal{L}_G[\rho] = -\frac{i}{\hbar}\sum_{i=1}^{3} \mathsf{b}_i[\mathsf{X}_i,\rho] - \sum_{i,j=1}^{3} \frac{1}{2}\mathsf{D}_{ij}[\mathsf{X}_i,[\mathsf{X}_j,\rho]] \tag{3}$$

$$\mathcal{L}_P[\rho] = \int d\boldsymbol{Q}|\lambda(\boldsymbol{Q})|^2\left[\mathrm{e}^{i\boldsymbol{Q}\cdot\mathsf{X}/\hbar}\rho\mathrm{e}^{-i\boldsymbol{Q}\cdot\mathsf{X}/\hbar} - \rho - \frac{i}{\hbar}\frac{[\boldsymbol{Q}\cdot\mathsf{X},\rho]}{1+Q^2/Q_0^2}\right] \tag{4}$$

where $\mathbf{b} \in \mathbb{R}, \boldsymbol{D} \geq 0$, and the integration measure satisfies the Lévy condition

$$\int d\boldsymbol{Q} |\lambda(\boldsymbol{Q})|^2 \frac{Q^2}{1+Q^2} < \infty. \tag{5}$$

It is very convenient to write the contributions given by Eq. (3) and Eq. (4) in the position representation, leading to the simple expression

$$\langle \boldsymbol{X} | \mathcal{L}_G[\rho] + \mathcal{L}_P[\rho] | \boldsymbol{Y} \rangle = -\Psi(\boldsymbol{X}-\boldsymbol{Y}) \langle \boldsymbol{X} | \rho | \boldsymbol{Y} \rangle, \tag{6}$$

where according to Eq. (3) and Eq. (4) we have introduced the function

$$\Psi(\boldsymbol{X}-\boldsymbol{Y}) = \frac{i}{\hbar} \mathbf{b} \cdot (\boldsymbol{X}-\boldsymbol{Y}) + \frac{1}{2}(\boldsymbol{X}-\boldsymbol{Y})^T \cdot \boldsymbol{D} \cdot (\boldsymbol{X}-\boldsymbol{Y}) \tag{7}$$
$$- \int d\boldsymbol{Q} |\lambda(\boldsymbol{Q})|^2 \left[e^{i\boldsymbol{Q}\cdot(\boldsymbol{X}-\boldsymbol{Y})/\hbar} - 1 - \frac{i}{\hbar} \frac{\boldsymbol{Q}\cdot(\boldsymbol{X}-\boldsymbol{Y})}{1+Q^2/Q_0^2} \right],$$

only depending on the difference $\boldsymbol{X}-\boldsymbol{Y}$ due to translational invariance. The action of the contributions given by Eq. (3) and Eq. (4) in the position representation is therefore very simple, it only amounts to multiplying the matrix elements of the statistical operator by a function of the particular form (7), whose general properties as we shall see naturally account for a description of decoherence.

3. Decoherence and Lévy processes

The master equation corresponding to Eq. (3) and Eq. (4) can be easily solved in the position representation, giving a dynamics which only changes the initial statistical operator by a multiplicative time dependent factor

$$\langle \boldsymbol{X} | \rho_t | \boldsymbol{Y} \rangle = e^{-t\Psi(\boldsymbol{X}-\boldsymbol{Y})} \langle \boldsymbol{X} | \rho_0 | \boldsymbol{Y} \rangle. \tag{1}$$

A key point is now the observation that Eq. (7) actually gives the general expression of the characteristic exponent appearing in the characteristic function of a Lévy process, corresponding to the celebrated Lévy-Khintchine formula[20]. As a consequence the function

$$\Phi(t, \boldsymbol{X}-\boldsymbol{Y}) = e^{-t\Psi(\boldsymbol{X}-\boldsymbol{Y})} \tag{2}$$

gives the general possible expressions for the characteristic function of a classical Lévy process, different processes, e.g. Gaussian, Poisson, compound Poisson or Lévy stable processes arising corresponding to the different possible values of $\mathbf{b}, \boldsymbol{D}$ and of the positive weight $|\lambda(\boldsymbol{Q})|^2$ in the measure. These different Lévy processes intuitively correspond to the different ways according to which momentum is transferred to the test particle as a consequence of interaction with the environment. Thus for example a Poisson process corresponds to a situation in which the different possible interaction events are characterized by a fixed momentum transfer, given by the height of the jumps in the Poisson process. More generally a physically realistic situation involves a compound Poisson process, characterized by the fact that the momentum transfer in the single interaction events is not a deterministic quantity, but it is itself described by a probability density, depending on the detail of the microscopic interaction mechanism, according to which the Poisson process is composed.

The function $\Phi(t, \boldsymbol{X} - \boldsymbol{Y})$ is a characteristic function, so that it has the following interesting properties, explaining why Eq. (1) generally gives a well defined master equation describing loss of coherence in the position representation:

- $\Phi(t, 0) = 1$
- $|\Phi(t, \boldsymbol{X} - \boldsymbol{Y})| \leqslant 1$
- $\Phi(t, \boldsymbol{X} - \boldsymbol{Y})$ is positive definite
- $\Phi(t, \boldsymbol{X} - \boldsymbol{Y}) \longrightarrow 0$ for $t \to \infty$
- $\Phi(t, \boldsymbol{X} - \boldsymbol{Y}) \longrightarrow 0$ for $(\boldsymbol{X} - \boldsymbol{Y}) \to \infty$, provided there exists a probability density.

These properties typical of characteristic functions[21] automatically entail that the diagonal matrix elements in the position representation are not affected with elapsing time, thus preserving normalization of the statistical operator, while the off-diagonal matrix elements are generally suppressed as expected due to decoherence. Furthermore for a fixed spatial distance $\boldsymbol{X} - \boldsymbol{Y}$ the off-diagonal matrix elements in the position representation are fully suppressed for long enough interaction times, while for a fixed interaction time these off-diagonal matrix elements only go to zero if the associated process admits a proper probability density, which is not the case e.g. for a compound Poisson process. Depending on the particular process describing the random momentum transfers in each scattering event different characteristic functions appear, corresponding to different behaviors in the suppression of the off-diagonal matrix elements for large spatial separations. The function $|\Phi(t, \boldsymbol{X} - \boldsymbol{Y})|$, which is responsible for the loss of visibility in interferometric experiments testing decoherence, for a fixed interaction time t might monotonically decrease to zero for growing values of $\boldsymbol{X} - \boldsymbol{Y}$, or also oscillate and reach asymptotically a finite value corresponding to a residual coherence. These quite different behaviors, corresponding to a more or less effective decoherence effect, are all encoded in the possible expressions of the characteristic function Φ. Application of this formalism to actually realized experiments has been considered in[10]. Typical experiments testing decoherence in a quantitative way involve an interferometer for massive particles (such as fullerenes[22,23] or atoms[24,25]), in which the interfering particle is exposed to some environment during the time of flight, such as a background gas, a laser field or even the internal degrees of freedom of the interfering particle itself.

4. Decoherence in momentum and position for a massive tracer particle

The general structure of translation-covariant quantum dynamical semigroups allows for the description of decoherence effects provided one considers the behavior in time of the so called coherences, that is to say the off-diagonal matrix elements of the statistical operator in a given basis, selected by the dynamics itself or by the observation which can be performed on the open system. For the considered massive particle interacting with some environment the natural basis are given by momentum or position. In order to describe both phenomena we obviously cannot neglect the momentum dynamics as implicitly done going over from Eq. (2) to Eq. (3) and Eq. (4). We therefore need a physical example of realization of the general structure Eq. (2), as given by the quantum version of the classical linear Boltzmann equation[8,26–29]. Such a master equation describes the dynamics of a quantum test particle interacting through collisions with a homogeneous gas, thus providing a quantum counterpart of the classical linear Boltzmann equation. For the case of a scattering cross section $\sigma(\boldsymbol{Q})$ only depending on the momentum transfer the equation can be written

$$\mathcal{L}[\rho] = \frac{n_{\text{gas}}}{m_*^2} \int d\boldsymbol{Q} \sigma(\boldsymbol{Q}) \left[e^{i\boldsymbol{Q}\cdot\mathsf{X}/\hbar} \sqrt{S(\boldsymbol{Q},\mathsf{P})} \rho \sqrt{S(\boldsymbol{Q},\mathsf{P})} e^{-i\boldsymbol{Q}\cdot\mathsf{X}/\hbar} \right. \quad (1)$$
$$\left. - \frac{1}{2} \{S(\boldsymbol{Q},\mathsf{P}), \rho\} \right],$$

with n_{gas} the density of gas particles with mass m, M the mass of the test particle, $m_* = mM/(m+M)$ the reduced mass, $S(\boldsymbol{Q},\mathsf{P})$ a two-point correlation function of the gas known as dynamic structure factor and explicitly given by

$$S(\boldsymbol{Q},\boldsymbol{P}) = \sqrt{\frac{\beta m}{2\pi}} \frac{1}{Q} \exp\left(-\frac{\beta}{8m} \frac{(Q^2 + 2mE(\boldsymbol{Q},\boldsymbol{P}))^2}{Q^2}\right), \quad (2)$$

with

$$E(Q,P) = \frac{(P+Q)^2}{2M} - \frac{P^2}{2M} = \frac{Q^2}{2M} + \frac{Q \cdot P}{M} \qquad (3)$$

the energy transfer in the single collision and $\beta = 1/(k_B T)$. We are not going to delve on details of the structure of such an equation. We only point out that it actually provides an example of translation-covariant master equation complying with the general mathematical result. We are however interested to show that such a structure actually describes decoherence phenomena in both momentum and position. In fact while the classical linear Boltzmann equation only describes dissipative effects, corresponding to the behavior of populations in momentum space, that is the diagonal matrix elements in the momentum representation of Eq. (1), the quantum master equation also describes coherences and therefore possibly interference phenomena and suppression thereof as a consequence of the dynamics, provided suitable quantum states given by linear superpositions states are considered.

Looking at coherence in momentum space implies considering coherent superpositions of momentum eigenstates. Such highly non classical motional states can show interference effects which are expected to be suppressed as a consequence of the interaction with the environment. As a consequence matrix elements of the form $\langle P|\rho|P'\rangle$ are quickly suppressed for $P \neq P'$, so that for long enough times the dynamics only affects the behavior of the probability density $\langle P|\rho|P\rangle$, and the master equation Eq. (1) goes effectively over to a classical rate equation for such a probability density. Due to the complexity of Eq. (1) obtaining an analytical solution is hardly feasible, so that the natural strategy is to numerically solve the master equation, relying on a so called unraveling of the master equation itself[15], to be solved by means of Monte Carlo methods. In this case setting

$$V(Q) = e^{iQ\cdot X/\hbar} \sqrt{\frac{n_{\text{gas}}}{m_*^2} \sigma(Q) S(Q,\mathsf{P})}, \qquad (4)$$

one can consider the following stochastic differential equation for the stochastic wave vector $\psi(t)$

$$d|\psi(t)\rangle = \left[-\frac{1}{2}\int dQ V^\dagger(Q) V(Q) + \frac{1}{2}\int dQ \|V(Q)|\psi(t)\rangle\|^2\right]|\psi(t)\rangle dt$$
$$+ \int dQ \left[\frac{V(Q)|\psi(t)\rangle}{\|V(Q)|\psi(t)\rangle\|} - |\psi(t)\rangle\right] dN_Q(t), \qquad (5)$$

where the field of increments satisfies

$$dN_Q(t) dN_{Q'}(t) = \delta^3(Q - Q') dN_Q(t)$$
$$\mathsf{E}[dN_Q(t)] = \|V(Q)|\psi(t)\rangle\|^2 dt,$$

so that indeed the solutions of the stochastic differential equation (5) provide unravelings of the master equation Eq. (1), in the sense that

$$\rho(t) = \mathsf{E}[|\psi(t)\rangle\langle\psi(t)|].$$

Despite the formal complexity of Eq. (5), for initial states given by momentum eigenvectors one can develop a simple algorithm to study the dynamics of such states, essentially corresponding to the Gillespie algorithm[30], leading to a pure jump process in momentum space. On similar grounds one can also study the dynamics of coherent superpositions of the form

$$|\psi(0)\rangle = \alpha_1(0)|P_1\rangle + \alpha_2(0)|P_2\rangle,$$

with $\sum_{i=1}^2 |\alpha_i(0)|^2 = 1$, which evolve in time according to

$$|\psi(t)\rangle = \alpha_1(t)|P_1(t)\rangle + \alpha_2(t)|P_2(t)\rangle,$$

where again $\sum_{i=1}^2 |\alpha_i(t)|^2 = 1$. An estimate of loss of coherence can be obtained studying the quantity

$$C(t) = \mathsf{E}\left[\frac{|\alpha_1(t)\alpha_2^\star(t)|}{|\alpha_1(0)\alpha_2^\star(0)|}\right].$$

As it turns out this measure for the coherence of the state in the momentum basis behaves for a constant scattering cross section approximately as[31]

$$C(t) = \exp[-\gamma(|P_1 - P_2|)t], \tag{6}$$

where the argument of the exponential is given by

$$\gamma(P) = \Lambda(P) - \Lambda_0 \frac{\text{erf}(P)}{P},$$

with

$$\Lambda(P) = \frac{n_{\text{gas}}}{m_*^2} \int dQ \sigma S(Q, \mathsf{P}), \tag{7}$$

$\text{erf}(x) = 2\pi^{-\frac{1}{2}} \int_0^x \exp(-t^2) dt$ denotes the error function, and Λ_0 is a reference scattering rate given by $\Lambda_0 = n_{\text{gas}} v_{\text{mp}} 4\pi\sigma$, with v_{mp} the most probable velocity for the gas particles. Eq. (6) clearly predicts an exponential loss of coherence in the momentum basis, depending on the relative distance in momentum space of the states making up the coherent superposition.

For the study of decoherence in position space we can follow a different strategy. Neglecting in Eq. (1) the dynamics of the momentum, we can replace the corresponding operator by a classical label P_0 giving the mean value of the momentum of the incoming particle. The master equation then reads

$$\mathcal{L}[\rho] = \frac{n_{\text{gas}}}{m_*^2} \int d\boldsymbol{Q}\, \sigma(\boldsymbol{Q})\, S(\boldsymbol{Q}, \mathsf{P}_0) \left[e^{i\boldsymbol{Q}\cdot\mathsf{X}/\hbar} \rho e^{-i\boldsymbol{Q}\cdot\mathsf{X}/\hbar} - \rho \right], \qquad (8)$$

corresponding to a particular realization of Eq. (4). Considering a constant scattering cross section and defining the rate $\Lambda(P_0)$ according to (7) one can introduce the following characteristic function

$$\Phi_S(\boldsymbol{X}) = \frac{n_{\text{gas}}\sigma}{m_*^2 \Lambda(P_0)} \int d\boldsymbol{Q}\, S(\boldsymbol{Q}, \mathsf{P}_0)\, e^{i\boldsymbol{Q}\cdot\boldsymbol{X}/\hbar},$$

so that the master equation (8) can be solved in the position representation as in (1), leading to

$$\langle \boldsymbol{X}|\rho_t|\boldsymbol{Y}\rangle = \exp\left(-\Lambda_0 \frac{2}{\sqrt{\pi}}\left[1 - \Phi_S(\boldsymbol{X} - \boldsymbol{Y})\right]t\right) \langle \boldsymbol{X}|\rho_0|\boldsymbol{Y}\rangle, \qquad (9)$$

where according to the general framework presented in Sect. 3 the characteristic function of a compound Poisson process appears. A suitable measure of decoherence is given in this case by

$$D(t) = \frac{\langle \boldsymbol{X}|\rho_t|\boldsymbol{Y}\rangle}{\langle \boldsymbol{X}|\rho_0|\boldsymbol{Y}\rangle}.$$

For a test particle slower than the gas particles, so that $P_0 \ll M v_{\text{mp}}$, one has

$$\Phi_S(\boldsymbol{X}) \approx {}_1F_1\left(1, \frac{3}{2}; -4\pi \frac{X^2}{\lambda_{\text{th}}^2}\right),$$

with λ_{th} the thermal de Broglie wavelength of the gas particles given by $\lambda_{\text{th}} = \sqrt{2\pi\beta\hbar^2/m}$, and ${}_1F_1$ the confluent hypergeometric function, so that

$$D(t) = \exp\left(-\Lambda_0 \frac{2}{\sqrt{\pi}}\left[1 - {}_1F_1\left(1, \frac{3}{2}; -4\pi \frac{X^2}{\lambda_{\text{th}}^2}\right)\right] t\right),$$

which for spatial distances above the thermal de Broglie wavelength $X \gamma \lambda_{\text{th}}$ is well approximated by a fixed decoherence rate $D(t) = \exp(-2\Lambda_0 t/\sqrt{\pi})$, expressing the fact that for large enough distances off-diagonal matrix elements in the position representation are uniformly suppressed.

5. Conclusions and outlook

We have given a brief presentation of how quantum dynamical semigroups can be useful for the description of decoherence in quantum mechanics, as also pursued in[32,33], coping in a quantitative way with experimentally realizable situations. This has been obtained relying on a characterization of translation-covariant quantum dynamical semigroups, leading to a quantum non-commutative generalization of the Lévy-Khintchine formula. When applied to the study of decoherence, neglecting dissipative phenomena, such a structure leads to a description of loss of coherence with a wide variety of possible behaviors, each corresponding to the characteristic function of a classical Lévy process. Despite pursued within the framework of the Markov assumption, thus supposing that the dynamics does not entail memory effects, the approach to the description of decoherence building on covariance properties, recently also followed in[34], can be of more general validity, as it appears from recent results pointing to a generalization of the Lindblad structure for the description of a class of non-Markovian evolutions[35].

Acknowledgments

The author is grateful to the organizers for the kind invitation and hospitality at CIMAT. The work was partially supported by the Italian MUR under PRIN2005.

References

1. R. F. Streater, *J. Math. Phys.* **41**, 3556 (2000).
2. A. S. Holevo, *Statistical Structure of Quantum Theory* (Springer, Berlin, 2001).
3. R. Alicki and M. Fannes, *Quantum dynamical systems* (Oxford University Press, Oxford, 2001).
4. A. S. Holevo, *Rep. Math. Phys.* **32**, 211 (1993).
5. A. S. Holevo, *Rep. Math. Phys.* **33**, 95 (1993).
6. A. S. Holevo, *Izv. Math.* **59**, 427 (1995).
7. A. S. Holevo, *J. Math. Phys.* **37**, 1812 (1996).
8. B. Vacchini, *J. Math. Phys.* **42**, 4291 (2001).
9. B. Vacchini, *J. Math. Phys.* **43**, 5446 (2002).
10. B. Vacchini, *Phys. Rev. Lett.* **95**, p. 230402 (2005).
11. E. Joos, H. D. Zeh, C. Kiefer, D. Giulini, J. Kupsch and I.-O. Stamatescu, *Decoherence and the Appearance of a Classical World in Quantum Theory*, 2nd edn. (Springer, Berlin, 2003).
12. M. Schlosshauer, *Decoherence and the Quantum-To-Classical Transition* (Springer-Verlag, Berlin, 2007).

13. R. Alicki and K. Lendi, *Quantum Dynamical Semigroups and Applications* (Springer, Berlin, 1987).
14. R. Alicki, Invitation to quantum dynamical semigroups, in *Dynamical semigroups: Dissipation, chaos, quanta*, eds. P. Garbaczewski and R. Olkiewicz, Lecture Notes in Physics, Vol. 597 (Springer-Verlag, Berlin, 2002) pp. 239–264.
15. H.-P. Breuer and F. Petruccione, *The Theory of Open Quantum Systems* (Oxford University Press, Oxford, 2007).
16. V. Gorini, A. Kossakowski and E. C. G. Sudarshan, *J. Math. Phys.* **17**, 821 (1976).
17. G. Lindblad, *Comm. Math. Phys.* **48**, 119 (1976).
18. B. Vacchini, to appear in Lecture Notes in Physics [arXiv:quant-ph/0707.0603].
19. A. S. Holevo, *J. Funct. Anal.* **131**, 255 (1995).
20. W. Feller, *An introduction to probability theory and its applications. Vol. II* (John Wiley & Sons Inc., New York, 1971).
21. L. Lukacs, *Characteristic Functions* (Griffin, London, 1966).
22. K. Hornberger, S. Uttenthaler, B. Brezger, L. Hackermüller, M. Arndt and A. Zeilinger, *Phys. Rev. Lett.* **90**, p. 160401 (2003).
23. L. Hackermüller, K. Hornberger, B. Brezger, A. Zeilinger and M. Arndt, *Nature* **427**, 711 (2004).
24. D. A. Kokorowski, A. D. Cronin, T. D. Roberts and D. E. Pritchard, *Phys. Rev. Lett.* **86**, p. 2191 (2001).
25. H. Uys, J. D. Perreault and A. D. Cronin, *Phys. Rev. Lett.* **95**, p. 150403 (2005).
26. B. Vacchini, *Phys. Rev. Lett.* **84**, 1374 (2000).
27. B. Vacchini, *Phys. Rev. E* **66**, p. 027107 (2002).
28. K. Hornberger, *Phys. Rev. Lett.* **97**, p. 060601 (2006).
29. K. Hornberger and B. Vacchini, *Phys. Rev. A* **77**, p. 022112 (2008).
30. D. T. Gillespie, *Markov Processes* (Academic Press, Boston, 1992).
31. H.-P. Breuer and B. Vacchini, *Phys. Rev. E* **76**, p. 036706 (2007).
32. R. Rebolledo, *Ann. Inst. H. Poincaré Probab. Statist.* **41**, 349 (2005).
33. R. Rebolledo, *Open Syst. Inf. Dyn.* **12**, 37 (2005).
34. J. Clark, [arXiv:math-ph/0710.1344].
35. H.-P. Breuer, *Phys. Rev. A* **75**, p. 022103 (2007).

MINIMUM BACK-ACTION MEASUREMENT VIA FEEDBACK

M. YANAGISAWA

Department of Engineering, The Australian National University,
Canberra, ACT 0200, Australia
E-mail: y.m@anu.edu.au

Measurement is based on physical interactions between a system to be observed and a measurement device. Information about the system is transferred to the measurement device through the interactions and replaced to a visible information. Then, we read the visible data. As a consequence of measurement, the system is changed. Usually the change of the system is not critical for classical systems. For quantum systems, however, it is. In this paper, we introduce feedback design to reduce the change caused by measurement in the quantum setting. Moreover, this formulation is applied to coherence control for Bose-Einstein Condensates.

Keywords: Feedback control, Measurement back-action, Bose-Einstein Condensates.

1. introduction

A quantitative analysis of measurement back-action was initiated by Heisenberg. An inequality known as Heisenberg's uncertainty relation is now given in a different form from his original analysis and it is common knowledge that the inequality has nothing to do with measurement. The uncertainty relation is the inherent property of quantum systems which is derived from the noncommutatity of quantum variables (observables) and holds true without measurement. However, Heisenberg's observation on measurement back-action still has an interesting interpretation from control theoretical point of view.

Suppose that we want to measure the position of a particle such as an atom. We usually shine photons to the particle to make a measurement. If photons hit the particle, they are scattered and go to different directions from other photons. Then, we can measure the position of the particle by detecting the scattered photons or detecting other non-scattered photons.

As a result of scattering, the particle is kicked by the photons and changes its momentum. The precision of this position measurement is determined by the wavelength of the photons so that photons with short wavelength provide more precise information about the position of the particle. In this case, the particle is strongly kicked by scattering because such photons have high energy. Thus, there is a tradeoff between the amount of information we can obtain from measurement and measurement back-action.

Usually, this kind of influence from measurement is not seriously taken into account in a classical setting. However, if the scale of a system is small enough and detection methods are limited, the measurement back-action cannot be ignored. For example, let us consider a case where we want to measure the temperature of water in a small cup with a "big" thermometer. Suppose that the water temperature is $T > 0$ and the thermometer is initially set to indicate zero. To measure the temperature, we stick the thermometer to the water. Then, the water and thermometer start to interact with each other and the heat of the water is transfered to the thermometer. After a long time, the total system, the water plus thermometer, is in an equilibrium state, and then, we read the thermometer. If the size of the thermometer was small, the effect of heat transfer would have been ignored. But it is not, now. Obviously, measurement outcomes are less than T and the true value can never be obtained unless there are many sampled prepared in the same condition and we change the initial temperature of the thermometer adaptively.

This change of temperature is caused by a physical interaction between the water and thermometer, as in the case of the particle and photons, and thought of as measurement back-action. This seems to be unavoidable as long as measurement relies on physical interactions. However, it is actually possible to reduce the measurement back-action using feedback.

To see this, let us consider the case of the position measurement. We have measurement outcomes from the photon detector now. Thus, the estimates of the position and momentum of the particle after scattering can be obtained by calculating conditional expectations from the measurement outcomes. Then, we push the particle back to the original position and momentum by designing a control input from the estimates. These estimates are still subject to the uncertainty relation so that no control can break the uncertainty relation. However, the measurement back-action is reduced to some extent by this procedure.

In the case of the temperature measurement, we can obtain the estimate of temperature in the same way if physical properties of the thermometer such as specific heat are known. If the indicator of the thermometer goes up, the water must be cooled down. Then, we can reduce this measurement back-action by heating up the water according to the estimate.

In this paper, we show measurement back-action reduction based on a simple idea given above. For simplicity, we consider quantum linear systems interacting with quantum stochastic noise such as an optical field in the free space. As an example, we introduce Bose-Einstein Condensates (BEC) systems. For a practical use of BEC, it is important to keep the coherence of atoms in the condensates.[5] However, it is destroyed by nonlinear dynamics due to atom-atom collisions. This nonlinear effect can be eliminated by measurement and feedback control.[6] Unfortunately, the measurement introduces another nonlinearity into the system in the same way as the position measurement described above. Then, it is required to reduce measurement back-action using another feedback control.

2. Intuitive model

2.1. *A Simple Model*

We first consider a linear system with a single variable to analyze the idea introduced in Introduction. Let us assume that the system is described as

$$dx = axdt + bdw, \quad x(0) = 0, \qquad (1)$$

where w is a Wiener process and the second term represents the interaction with an external system which is a part of measurement. If the system is initially in a Gaussian state, it is characterized by the first and second moments. In particular, we are interested in the second moment because it is related to a quadratic cost functional. Note that in the present case, the second moment is equivalent to the variance of x. The second moment is given by a Lyapunov equation

$$\frac{d}{dt}\langle x^2 \rangle = 2a\langle x^2 \rangle + b^2. \qquad (2)$$

The second term of the right hand side is positive so that the interaction with the external system forces the second moment to increase. The solution is given by

$$\langle x^2(t) \rangle = e^{2at}\langle x^2(0) \rangle + \frac{e^{2at} - 1}{2a}b^2. \qquad (3)$$

If the system is stable $a < 0$, the second moment converges to $-b^2/2a$. In the example of the particle described in Introduction, the high precision measurement corresponds to a large b.

To reduce the effect of the interaction with the external field, we read measurement outcomes and estimate the state of the system. Let us assume that the measurement process is given as

$$dm = cxdt + dv, \qquad (4)$$

where v is another Wiener process independent of w. The expectation of x conditioned on the measurement outcome m, $\hat{x} := \mathrm{E}[x|m]$, is fed back to the system with a gain k. Then, the system under measurement is described by a Kalman filter as[1]

$$d\hat{x} = a\hat{x}dt + k\hat{x}dt + pcd\hat{v}, \qquad (5)$$

where $\hat{v}$ is an innovation process defined as $d\hat{v} = dm - c\langle x \rangle dt$ and $p := \langle (x - \hat{x})^2 \rangle$ is the mean square error satisfying a Riccati equation

$$\dot{p} = 2ap + b^2 - (pc)^2. \qquad (6)$$

From the definition of the conditional expectation, the second moment is decomposed to

$$\langle x^2 \rangle = \langle \hat{x}^2 \rangle + p. \qquad (7)$$

The effect of the interaction with the external field is reduced by minimizing the second moment $\langle x^2 \rangle$. Note that p is independent of the feedback control. Thus, the feedback is designed to reduce the first term. The solution is given by

$$\langle \hat{x}^2(t) \rangle = e^{2(a+k)t} \langle \hat{x}^2(0) \rangle + \int_0^t e^{2(a+k)(t-s)} (p(s)c)^2 ds. \qquad (8)$$

If the feedback gain $-k$ is very large, this quantity quickly converges to zero so that the second moment after the ideal control is approximated to the solution of the Riccati equation (6):

$$\langle x^2 \rangle = p. \qquad (9)$$

The difference between the Lyapunov equation (2) and Riccati equation (6) is the last term of (6). Note that it is always negative and corresponds to a reduction in the entropy of the system due to measurement. Thus, we can reduce the influence of the interaction with the external field by the amount of information we extract from the system by measurement. This measurement back-action reduction is optimal because no control can reduce the second moment less than (9).

2.2. Even Simpler Case

Here, let us assume $a = 0$. In this case, corresponding to (2), the second moment of the state before applying feedback is expressed as

$$\frac{d}{dt}\langle x^2 \rangle = b^2. \tag{10}$$

The solution is given by

$$\langle x^2(t) \rangle = \langle x^2(0) \rangle + b^2 t. \tag{11}$$

Under the same measurement process and feedback as the previous subsection, the conditional expectation is given by[1]

$$d\widehat{x} = k\widehat{x}dt + pcd\widehat{v}, \tag{12a}$$
$$\dot{p} = b^2 - (pc)^2. \tag{12b}$$

By taking a large feedback gain $-k\gamma 1$ as in the previous case, we can approximate the second moment after control to

$$\langle x^2 \rangle = \frac{\langle x^2(0)\rangle(1 + e^{-2bct}) + \frac{b}{c}(1 - e^{-2bct})}{\langle x^2(0)\rangle(1 - e^{-2bct}) + \frac{b}{c}(1 + e^{-2bct})} \cdot \frac{b}{c}$$
$$\to |b/c| \quad \text{as } t \to \infty. \tag{13}$$

Thus, the feedback control reduces the effect of measurement back-action for $t\gamma 1$ by the amount of

$$\langle x^2(0)\rangle + b^2 t - |b/c|, \tag{14}$$

which is a great improvement. Note that this inversely depends on c. If we extract a lot of information about the system (which corresponds to a large c), then the effect of measurement back-action is more reduced, as stated before.

3. BEC example

3.1. Model

Bose-Einstein Condensate (BEC) systems consist of many cold atoms. Basically, a BEC should be described as a multi-mode system. However, if all atoms are in a condensate state, their phases are synchronized so that they behave as a single particle. Then, it is relatively a good approximation to represent a BEC system with a single mode operator.[3] Here, we denote the single mode by a which satisfies a standard bosonic commutation relation $[a, a^\dagger] = 1$.

The purpose of feedback is now to protect the system from decoherence. Suppose that the BEC system is initially prepared in a coherent state. Unlike photons, atoms interact with each other in the condensate. As a result, the initial coherence slowly disappears, and the BEC state is eventually destroyed. To keep the BEC state, we introduce feedback and eliminate the atom-atom interactions.

We use measurement for feedback design. Thus, we have to let the system interact with an optical field and obtain information about the system by detecting the outgoing optical field. Unfortunately, this destroys the coherence again for the same reason stated in Introduction. Then, we design another feedback to eliminate the decoherence from the optical field in the same way as the previous section.

Here is the detailed description of the system: The system is coupled to one quantum and two classical inputs. The quantum input is a quantum optical field represented by a mode operator b and the corresponding output is measured by a homodyne detection system. Note that b satisfies quantum Ito's formula,[2,4] e.g., $dbdb^\dagger = dt$. The resulting photocurrent is used in two ways. One is to produce a proportional feedback signal u_1 which is designed to eliminate the atom-atom interaction,[6] while the other is used for a feedback signal u_2 which is designed to reduce the influence of the interaction with the optical field.

Before the feedback is applied, the system is described by a Hamiltonian

$$H_0 = k_0 a^\dagger a^\dagger aa + (u_1 + u_2)a^\dagger a, \qquad (1)$$

where the first term of the Hamiltonian H_0 represents the atom-atom interaction and the second one is the classical control term. The interaction between the system and optical field is given by

$$H_{int} = \sqrt{\gamma} a^\dagger a (b^\dagger + b), \qquad (2)$$

where γ is a coupling constant. Due to this interaction, the information of the atom number $a^\dagger a$ is transferred to the quadrature of the optical field after interaction. Thus, we can measure the atom number by detecting the quadrature with a standard homodyne detector. The resulting output process is given by

$$dm = 2\sqrt{\gamma} a^\dagger a\, dt + d(b + b^\dagger), \qquad (3)$$

where m represents measurement outcomes. The second term represents measurement noise which comes from the optical field.

The idea of feedback for eliminating the atom-atom interaction is based on this output equation. From (3), the measurement outcomes contain the information of $a^\dagger a$. The first term of the Hamiltonian (1) can be canceled out by designing u_1 proportional to m with an appropriate gain because $u_1 a^\dagger a \sim k a^\dagger a^\dagger a a$.[6] After this feedback, the resulting dynamics of the system is given by[7]

$$da = \left(-\frac{|\lambda|^2}{2} + iu_2\right)a\,dt - (\lambda db^\dagger - \lambda^* db)a,$$
$$dm = 2\lambda_r a^\dagger a\,dt + (db + db^\dagger), \tag{4a}$$

where we have defined a constant λ as

$$\lambda = \sqrt{\gamma} + i\frac{k_0}{\sqrt{\gamma}} := \lambda_r + i\lambda_i. \tag{5}$$

Suppose that the center of the initial coherent BEC state is located at $(x_0, 0)$ in the phase space, i.e., $\langle a + a^\dagger \rangle = x_0$ and $-i\langle a - a^\dagger \rangle = 0$. Let us define new quadrature operators around the center as $a = (x_0 + \xi) + i\eta$ and assume that $x_0\gamma 1$, which implies that the number of the trapped atoms is sufficiently large. In this approximation, the system dynamics and output process can be expressed as

$$d\boldsymbol{\xi} = Bu_2 dt + G d\boldsymbol{w}, \tag{6a}$$
$$dm = C\boldsymbol{\xi}\,dt + D d\boldsymbol{w}. \tag{6b}$$

Here we have introduced

$$B = \begin{bmatrix} 0 \\ x_0 \end{bmatrix}, \quad G = 2x_0 \begin{bmatrix} 0 & 0 \\ -\lambda_i & \lambda_r \end{bmatrix}, \tag{7a}$$
$$C = \begin{bmatrix} 2\lambda_r x_0 & 0 \end{bmatrix}, \quad D = \begin{bmatrix} 1 & 0 \end{bmatrix}, \tag{7b}$$

and

$$\boldsymbol{\xi} = \begin{bmatrix} \xi \\ \eta \end{bmatrix}, \quad \boldsymbol{w} = \begin{bmatrix} w \\ v \end{bmatrix}, \tag{8}$$

where w, v are noncommutative independent Wiener processes.[2,4]

The conditional expectation of $\boldsymbol{\xi}(t)$, $\widehat{\boldsymbol{\xi}}(t)$, is given by

$$d\widehat{\boldsymbol{\xi}} = Bu_2 dt + (PC^\dagger + GD^\dagger)d\widehat{w}, \tag{9a}$$
$$\dot{P} = GG^\dagger - (PC^\dagger + GD^\dagger)(PC^\dagger + GD^\dagger)^\dagger, \tag{9b}$$

where $P := \langle (\boldsymbol{\xi} - \widehat{\boldsymbol{\xi}})(\boldsymbol{\xi} - \widehat{\boldsymbol{\xi}})^T \rangle$ is the covariance matrix of the error and $\widehat{w}$ is the innovation process.

3.2. Performance of feedback

To see the effect of the feedback input u_2, let us consider the mean square costs of ξ and η, respectively. By definition, the variance of the quadratures is decomposed into the variance of the conditional expectations and the mean square error as

$$\langle \|\boldsymbol{\xi}\|^2 \rangle = \langle \|\widehat{\boldsymbol{\xi}}\|^2 \rangle + \langle \|\boldsymbol{\xi} - \widehat{\boldsymbol{\xi}}\|^2 \rangle, \tag{10}$$

The second term of the right hand side is given by the covariance matrix P. From (9), each element of P is given by

$$\dot{P}_{11} = -(2x_0)^2 \lambda_r^2 P_{11}^2, \tag{11a}$$
$$\dot{P}_{12} = (2x_0)^2 (\lambda_r \lambda_i P_{11} - \lambda_r^2 P_{11} P_{12}), \tag{11b}$$
$$\dot{P}_{22} = (2x_0)^2 (\lambda_r^2 + 2\lambda_r \lambda_i P_{12} - \lambda_r^2 P_{12}^2). \tag{11c}$$

From the first equation, the mean square error of ξ is given by

$$P_{11}(t) = \frac{P_{11}(0)}{4x_0^2 \lambda_r^2 P_{11}(0) t + 1}. \tag{12}$$

On the other hand, the conditional expectation of ξ is given by

$$d\widehat{\xi} = 2x_0 \lambda_r P_{11} d\widehat{w}, \tag{13}$$

and therefore

$$\langle \widehat{\xi}(t)^2 \rangle = P_{11}(0) - P_{11}(t). \tag{14}$$

From these relations and (10), the variance of ξ is given by

$$\langle \xi(t)^2 \rangle = \langle \widehat{\xi}(t)^2 \rangle + P_{11}(t) \tag{15}$$
$$= P_{11}(0), \tag{16}$$

which is invariant. This implies that the coherence of the BEC in ξ-direction is preserved in time. Thus, we do not need to control ξ.

The other quadrature η fluctuates due to the interaction with the optical field and needs to be controlled. Since the covariance matrix P is independent of the input u_2, the second term of (10) cannot be changed and the feedback control is designed to reduce the first term. This is the same situation as the simple classical examples in Section 2. Thus, we consider a feedback input of the form

$$u_2 = -kx_0 \widehat{\eta}, \tag{17}$$

where $\widehat{\eta}$ is the conditional expectation of η and k is a feedback gain.

The conditional expectation $\widehat{\eta}$ obeys

$$d\widehat{\eta} = -kx_0\widehat{\eta}dt + 2x_0\lambda_r\left(P_{12} - \frac{\lambda_i}{\lambda_r}\right)d\widehat{w}. \tag{18}$$

Note that the initial condition of $\widehat{\eta}$ is the center of the initial state, i.e., $\widehat{\eta}(0) = 0$. It is easy to see

$$\langle\widehat{\eta}(t)^2\rangle = \int_0^t ds\, e^{-2kx_0(t-s)}\left[2x_0\lambda_r\left(P_{12}(s) - \frac{\lambda_i}{\lambda_r}\right)\right]^2, \tag{19}$$

where

$$P_{12}(t) = \frac{\lambda_i}{\lambda_r}\left(1 - \frac{1}{1 + (2\lambda_r x_0)^2 P_{11}(0)t}\right). \tag{20}$$

Note that $P_{12}(t) > 0$ for $t > 0$. On the other hand, P_{22} can be represented as

$$\dot{P}_{22}(t) = (2x_0)^2\left[|\lambda|^2 - \lambda_r^2\left(P_{12} - \frac{\lambda_i}{\lambda_r}\right)^2\right] \tag{21}$$

From these relations, the variance of η from the center of the initial state is given by

$$\langle\eta(t)^2\rangle = P_{22}(0) + (2x_0)^2|\lambda|^2 t \tag{22}$$
$$+ \int_0^t ds\left[e^{-2kx_0(t-s)} - 1\right]\left[2x_0\lambda_r\left(P_{12}(s) - \frac{\lambda_i}{\lambda_r}\right)\right]^2.$$

If the gain of the feedback is sufficiently large, the fluctuation in the conditional expectation of η can be reduced and the variance is approximately represented as

$$\langle\eta(t)^2\rangle \sim P_{22}(0) + (2x_0)^2|\lambda|^2 t - \frac{\lambda_i P_{12}(t)}{\lambda_r P_{11}(0)}. \tag{23}$$

On the other hand, if $u_2 = 0$, the variance is given by

$$\langle\eta(t)^2\rangle = P_{22}(0) + (2x_0)^2|\lambda|^2 t. \tag{24}$$

Thus, the variance is reduced by the amount of

$$\frac{\lambda_i P_{12}(t)}{\lambda_r P_{11}(0)} > 0 \tag{25}$$

subject to high gain feedback control. Note that this quantity is sensitive to the coupling constant with the optical field γ. If γ is small, so is the fluctuation in the BEC state and it is easy to reduce the variance by feedback. If γ is large, the fluctuation from the optical field is very strong and the control input u_2 can hardly reduce the noise. This is consistent with the example of position measurement described in Introduction. The strong coupling constant $\gamma\gamma 1$ corresponds to photons with high energy. In this case, the BEC system is strongly kicked by the photons so that it is difficult to put the atoms back to the original position and momentum.

4. Conclusion

We have shown the possibility of feedback control for reducing the effect of measurement back-action. Measurement and feedback are useful to increase the potential of quantum systems in information technologies. Meanwhile, measurement destroys the systems, so we cannot obtain satisfactory performances of control sometimes. The BEC system is a good example of this case. A feedback loop is designed to cancel the nonlinear terms in the Hamiltonian which represent the atom-atom interactions in the BEC system and improve the coherence of the system. However, this measurement introduces another nonlinearity which destroys the coherence again. Then, we design another feedback loop which reduces fluctuation noise from the optical field. Combining these two loops, the performance of the BEC system is improved and the system holds its coherent properties longer than the single loop feedback control.

References

1. M.H.A. Davis, *Linear estimation and stochastic control*, (Chapman And Hall, London, 1977).
2. C.W. Gardiner and P. Zoller, *Quantum Noise*, (Springer, Berlin, 2000).
3. E.W. Hagley, L. Deng, M. Kozuma, J. Wen, K. Helmerson, S.L. Rolston, W.D. Phillips. A Well-Collimated Quasi-Continuous Atom Laser. *Science*, 283:1706–1709, 1999.
4. R.L. Hudson and K.R. Parthasarathy. Quantum Ito's formula and stochastic evolutions. *Commun. Math. Phys.*, 93:301–323, 1984.
5. H.J. Metcalf and P. van der Straten, *Laser Cooling and Trapping*, (Springer, New York, 1999).
6. L.K. Thomsen, H.M. Wiseman. Atom-laser coherence and its control via feedback. *Phys. Rev. A* **65** 063607 (2002).
7. M. Yanagisawa, M. R. James. Multi-loop Feedback Control for Atom Laser Coherence. *Proceedings of the 17th World Congress, International Federation of Automatic Control*, to appear.

AUTHOR INDEX

Accardi, L., 1
Ando T., 11
Arnold, A., 23
Asai, N., 49

Barchielli, A., 63
Barhoumi, A., 77
Bhat, B. V. R., 93
Boukas, A., 1

Demni, N., 107

Echavarría-Cepeda, L. A., 120

Fagnola, F., 23
Fichtner, K.-H., 135
Fichtner, L., 135
Floricel, R., 145

Gibilisco, P., 157
Gregoratti, M., 63

Hangos, K. M., 212

Imparato, D., 157
Isola, T., 157

Kuo, H.-H., 165

Labuschagne, L. E., 176
Liebscher, V., 93

Magyar, A., 212
Majewski, W. A., 176

Neumann, L., 23
Novak, J. I., 190

Ouerdiane, H., 77

Pechen, A., 197
Pita-Ruiz-Velasco, C. J., 120

Rabitz, H., 197
Riahi, A., 77
Ruppert, L., 212

Schürmann, M., 225
Skeide, M., 93
Sontz, S. B., 120
Spring, W. J., 236
Stan, A. I., 242

Vacchini, B., 254
Voß, S., 225

Whitaker, J. J., 242

Yanagisawa, M., 266